Helen McGuinness

Anatomy & Physiology

Orders: Hachette UK Distribution, Hely Hutchinson Centre, Milton Road, Didcot, Oxfordshire, OX11 7HH. Telephone: +44 (0)1235 827827. Email education@hachette.co.uk Lines are open from 9 a.m. to 5 p.m., Monday to Friday. You can also order through our website: www.hoddereducation.co.uk

ISBN: 978 1510 435179

© Helen McGuinness 2018

First published in 2018 by
Hodder Education,
An Hachette UK Company
Carmelite House
50 Victoria Embankment
London EC4Y 0DZ

www.hoddereducation.co.uk

Impression number   10 9 8 7 6 5

Year   2023

Cover photo © Sebastian Kaulitzki/123RF.com

Illustrations by Barking Dog Art

Typeset in India by Integra Software Services Pvt. Ltd, Pondicherry, India

Printed and bound by CPI Group (UK) Ltd, Croydon, CR0 4YY

A catalogue record for this title is available from the British Library.

Contents

Acknowledgements

Whilst preparing the original text of this book back in the early 1990s, I never dreamed it would be in circulation for over 20 years and now be in its fifth edition, with an accompanying workbook.

Firstly, I would like to extend my most significant thanks to my husband Mark for his constant love, help, support and understanding, along with his constructive comments made in the updating of this book.

To my late mum, Valerie, whose eternal love, words of encouragement and belief in my abilities continue to motivate and inspire me to this day.

To my dear friend Dee Chase (aka Mum Dee), for her constant love, belief, support and encouragement of my work and writing.

To Dr Nathan Moss, for help in checking the accuracy of the text on pathologies.

I will always be greatly indebted to Deirdre Moynihan for her professional help and contributions throughout the preparation of the original text back in 1995 when the book was in its infancy.

Special thanks go to Linda Biles, Head of Beauty Therapy at Chichester College, who has encouraged and supported the update of this latest edition and offered her invaluable suggestions and constructive comments.

I would like to thank all the students, colleges and lecturers who have used this book over the past 20 years and who have been most encouraging and supportive of my work.

This book is devoted to our beautiful daughter, Grace.

Helen J. McGuinness

Picture credits

How to use this book

Dear Colleague,

This book, now in its fifth edition, has been designed for those studying beauty therapy, complementary therapies or any subject that requires a sound foundation knowledge of anatomy and physiology.

This edition has been completely revised to bring it in line with the latest anatomy and physiology specifications of the Technical/Advanced Level qualifications.

As well as new and updated content, additional new features include an expanded introductory chapter on how the body is organised, key word glossaries at the end of the chapter, new and improved illustrations and photographs, and expanded end-of-chapter revision summaries and test your knowledge questions.

Each chapter gives an overview of a system and why it is of significance to a therapist, has a list of learning objectives, and is full of interesting facts and information to help stimulate your learning.

At the end of each chapter there is a link to other body systems to help to put the subject into context, showing how the body systems work as a whole to keep us in balance, along with a comprehensive revision summary and test your knowledge questions in multiple choice and new exam-style formats.

Once you have studied the contents of this textbook, there is a new accompanying workbook available to help test your knowledge and prepare you for assessments and examinations. Contents of the workbook include a range of activities including additional multiple choice and exam-style questions, labelling, matching the key words, sorting, and filling in the blanks.

Anatomy and physiology is a fascinating subject and I sincerely hope that this new edition provides you with an improved learning experience.

Helen J. McGuinness

Answers are available online at www.hoddereducation.co.uk/Anatomy-and-Physiology-Extras

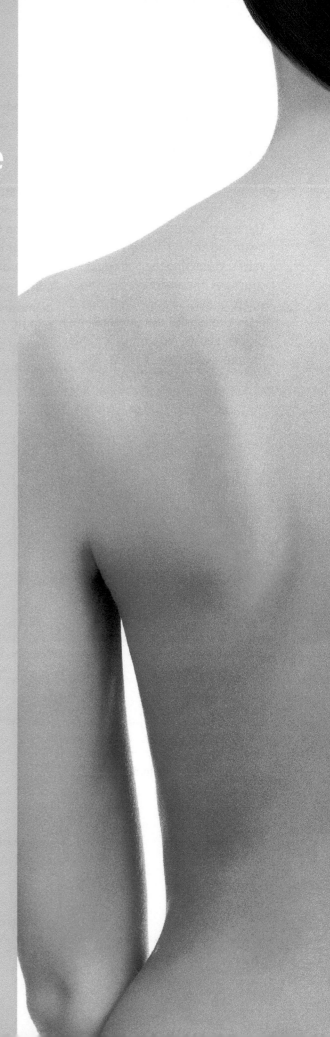

1 An introduction to anatomy and physiology: how the body is organised

Introduction

Before we begin the fascinating journey of learning about how the body works, let's look at the terminology associated with the study of the human body.

- **Anatomy** is the study of the **structure** and **location** of body parts.
- **Physiology** is the study of the **function** of body parts.

It is very important to co-ordinate these two subjects because knowledge of structure is incomplete without the knowledge of function, and the knowledge of function is incomplete without the knowledge of structure.

- **Pathology** is the study of **disease** in the body. At the end of each chapter in this book there is a section on the common pathologies associated with a system of the body.

OBJECTIVES

By the end of this chapter you will understand:

- anatomical directional terminology used to give a precise description of a body part
- anatomical planes, which divide the body into sections
- anatomical regional terms, which refer to specific areas of the body
- the main body cavities that divide the body and its internal organs into sections.

KEY FACT

Knowing where parts of the body are located will help your understanding as you build up a picture of how they function.

In practice

Think of the structure of the heart and all its chambers and valves. Visualising the individual structures (the anatomy), can help you to understand how the blood flows through the heart and how the heart beats. You are relating anatomy to function or physiology.

The body as a map

The body may be likened to a map and the key to locating and understanding the parts of the body is directional terminology.

Anatomical terminology

When studying anatomy and physiology, you should use directional terminology to give precise descriptions when referring to the exact location of a body part or structure. In anatomical terminology, all parts of the body are described in relation to other body parts using a standardised body position called the **anatomical position**.

In this position, the body is erect and facing forwards, arms to the side, palms are facing forwards with the thumbs to the side, and the feet slightly apart with toes pointing forwards. There is an imaginary line running down the centre or midline of the body.

Learning anatomical terminology is like learning a new language!

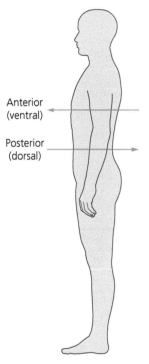

Anterior (ventral)

Posterior (dorsal)

▲ Anatomical terms

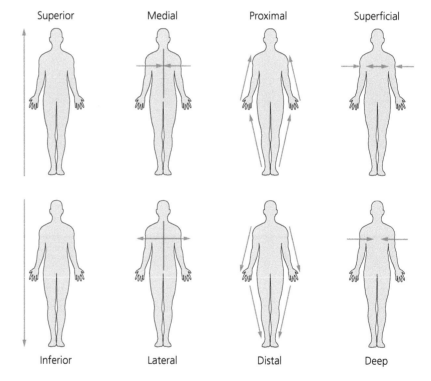

Superior   Medial   Proximal   Superficial

Inferior   Lateral   Distal   Deep

Anatomical directional terms

The anatomical terms in Table 1.1 will help you to be specific when describing the position of a body part.

Study tip

The directional terms have been organised into pairs for ease of learning; once you know one term, it is easier to learn the opposite term.

Table 1.1 Anatomical directional terms

Anatomical directional term	Definition	In practice
Anterior (or ventral)	Front surface of the body, or structure	The biceps muscle is on the anterior surface of the upper arm
Posterior (or dorsal)	Back surface of the body, or structure	The triceps muscle is on the posterior surface of the upper arm
Superficial	Near the body surface	The most superficial layer of the skin is the epidermis
Deep	Further from the body surface	The dermis is deep to the epidermis
Lateral	Away from the midline	The radius is lateral to the ulna
Medial	Towards the midline	The ulna is medial to the radius
Superior	Situated towards the head, or above a point of reference	The shoulder joint is superior to the elbow joint
Inferior	Situated away from the head or below a point of reference	The intestines are inferior to the stomach
Central	At or near the centre	The brain and spinal cord are part of the central nervous system
Peripheral	Away from the centre; outer part of the body	Peripheral vision allows us to see things out of main focus
Proximal	Nearest to the point of reference	The wrist joint is proximal to the elbow joint
Distal	Furthest away from the point of reference	The shoulder joint is distal to the wrist joint
Prone	Lying face down in a horizontal position	When receiving a back massage a client lies prone
Supine	Lying face up in a horizontal position	When receiving a facial a client lies supine
Caudal	Away from the head, or below a point of reference	The coccyx (tail bone) is an example of a caudal position
Cranial (or cephalic)	Relating to the head end or skull (cranium)	The brain is located in the cranial cavity
Palmar	Relating to the palm side of the hand	The thenar muscle is on the palmar surface of the hand
Plantar	Relating to the sole of the foot	There is a central tendon on the plantar surface of the foot
Ipsilateral	On the same side as another structure	The right radius and right humerus are ipsilateral
Contralateral	On the opposite side to another structure	The right and left kidneys are contralateral

Other directional terms

- **Longitudinal**: running in the direction of the length of the body or any of its parts.
- **Visceral**: used when referring to any internal organs, specifically those in the main body cavity (intestines, liver, stomach, for example).
- **Parietal**: used to refer to things within the body that are attached to the inside of the body cavity or a hollow structure.
- **Internal**: near the inside.
- **External**: near the outside.

Anatomical terms applied to movement

There are several anatomical terms relating to movement in the body, such as **adduction** and **abduction**. These are defined in Chapter 5, The muscular system.

The anatomical planes of the body

In the study of anatomy, there are three planes that separate the body into sections:

1 The **median or sagittal plane**: a vertical plane that divides the body lengthwise into right and left sections.

2 The **frontal or coronal plane**: divides the body into a front (anterior) portion and a rear (posterior) section.

3 The **transverse plane**: a horizontal plane that divides the body into top (superior) and bottom (inferior) sections.

> **Activity**
>
> Make up a blank template of the face and body. Design some small labels, each with an anatomical region on it (for example buccal, cervical). Attach the labels onto the facial/body template to indicate where each region is located.

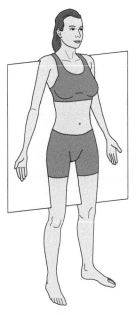
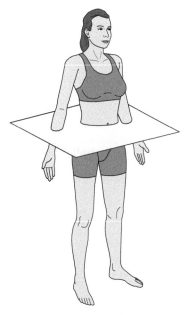

Sagittal plane   Frontal/coronal plane   Transverse plane

▲ Anatomical planes of the body

The anatomical regions of the body

Just like regions on a map, the anatomical regions of the body refer to certain body areas.

The body is divided into:

- head and neck
- trunk
- upper limbs (arms)
- lower limbs (legs).

Tables 1.2, 1.3, 1.4 and 1.5 will help you to learn the correct terminology for each region.

The head and neck

Table 1.2 Anatomical regional terms of the head and neck

Anatomical regional term	Area of body
Buccal	Cheek
Cephalic	Head
Cranial	Skull
Cervical	Neck
Facial	Face
Frontal	Forehead
Mental	Chin
Nasal	Nose
Occipital	Back of head/skull
Ophthalmic/orbital	Eyes
Oral	Mouth
Otic	Ear

The upper and lower limbs

Table 1.3 Anatomical regions of the upper limbs

Anatomical term	Area of the body
Brachial	Upper arm
Carpal	Wrist
Cubital	Elbow
Digital/phalangeal	Fingers (and toes, see Table 1.4)
Forearm	Lower arm
Palmar	Palm of hand

Table 1.4 Anatomical regions of the lower limbs

Anatomical term	Area of the body
Calcaneal	Heel
Crural	Leg or thigh
Digital/phalangeal	Toes (and fingers, see Table 1.3)
Femoral	Thigh
Patellar	Knee cap
Pedal	Foot
Plantar	Sole of foot
Popliteal	Hollow behind knee
Sural	Calf
Tarsal	Ankle

The trunk

Table 1.5 Anatomical regional terms of the trunk (thorax and abdomen)

Anatomical regional term	Area of body
Abdominal	Abdomen
Axillary	Armpit
Coeliac	Abdomen
Costal	Ribs
Gluteal	Buttocks
Inguinal	Groin
Lumbar	Lower back
Mammary	Breast
Pectoral	Chest
Pelvic	Pelvis/lower abdomen
Pericardial	Heart
Perineal	Between anus and external genitalia
Pubic	Pubis (front of pelvis)
Sacral	Sacrum
Thoracic	Thorax (chest cavity)
Umbilical	Navel
Vertebral	Spine/backbone

Other general anatomical regional terms

Cutaneous: skin

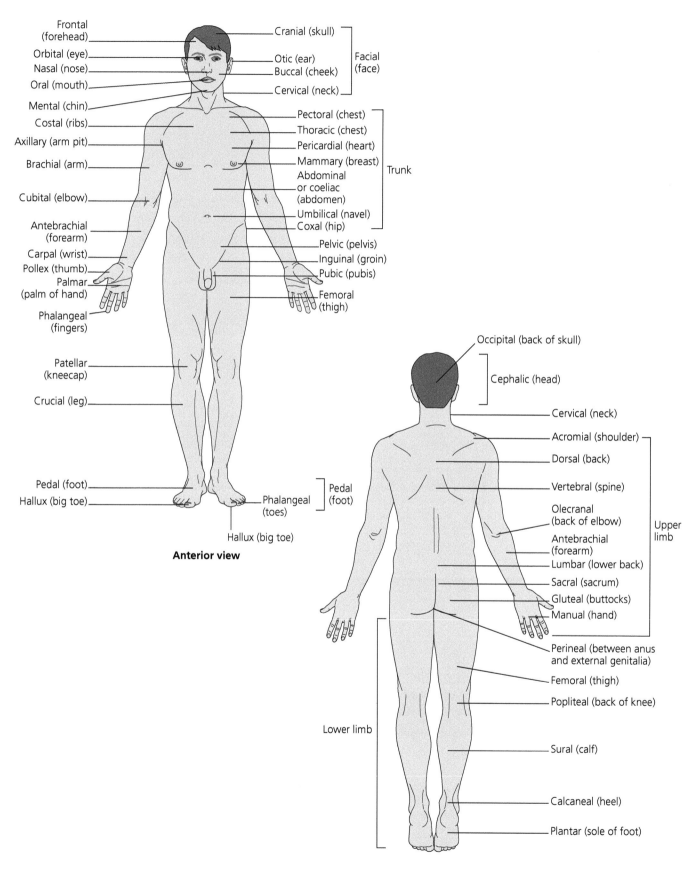

Anterior view

Frontal (forehead)
Orbital (eye)
Nasal (nose)
Oral (mouth)
Mental (chin)
Costal (ribs)
Axillary (arm pit)
Brachial (arm)
Cubital (elbow)
Antebrachial (forearm)
Carpal (wrist)
Pollex (thumb)
Palmar (palm of hand)
Phalangeal (fingers)
Patellar (kneecap)
Crucial (leg)
Pedal (foot)
Hallux (big toe)

Cranial (skull)
Otic (ear)
Buccal (cheek)
Cervical (neck)
Facial (face)
Pectoral (chest)
Thoracic (chest)
Pericardial (heart)
Mammary (breast)
Abdominal or coeliac (abdomen)
Umbilical (navel)
Coxal (hip)
Trunk
Pelvic (pelvis)
Inguinal (groin)
Pubic (pubis)
Femoral (thigh)
Phalangeal (toes)
Pedal (foot)
Hallux (big toe)

Posterior view

Occipital (back of skull)
Cephalic (head)
Cervical (neck)
Acromial (shoulder)
Dorsal (back)
Vertebral (spine)
Olecranal (back of elbow)
Antebrachial (forearm)
Lumbar (lower back)
Sacral (sacrum)
Gluteal (buttocks)
Manual (hand)
Upper limb
Perineal (between anus and external genitalia)
Femoral (thigh)
Popliteal (back of knee)
Sural (calf)
Calcaneal (heel)
Plantar (sole of foot)
Lower limb

▲ Anatomical regional terms

6

Body cavities

Body cavities are spaces within the body that contain the internal organs.

There are two main cavities in the body:

1   the **dorsal** cavity, located in the posterior (back) region of the body
2   the **ventral** body cavity, occupying the anterior (front) region of the trunk.

The dorsal cavity is subdivided into two cavities:

*   the **cranial cavity** – encases the brain and is protected by the cranium (skull)
*   the **vertebral/spinal cavity** – contains the spinal cord and is protected by the vertebrae.

The ventral cavity is subdivided into:

*   the **thoracic cavity** – surrounded by the ribs and chest muscles, the thoracic cavity contains the lungs, heart, trachea, oesophagus and thymus.

It is separated from the abdominal cavity by the diaphragm muscle

*   the **abdominopelvic cavity** – consists of both the abdominal and pelvic cavities; contains the liver, stomach, pancreas, spleen, gall bladder, kidneys, and most of the small and large intestines, as well as the bladder and the internal reproductive organs
*   the **abdominal cavity** – contains the stomach, spleen, liver, gall bladder, pancreas, small intestine and most of the large intestine. The abdominal cavity is protected by the muscles of the abdominal wall and partly by the diaphragm and rib cage
*   the **pelvic cavity** – contains the bladder, some of the reproductive organs and the rectum. The pelvic cavity is protected by the pelvic bones.

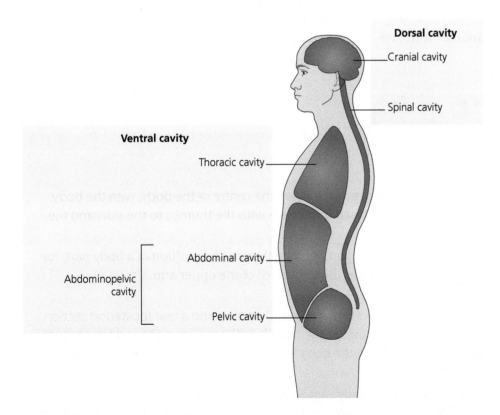

▲ Body cavities

Key words

Anatomical position: point of reference in anatomy; the body is erect and facing forwards, arms to the side, palms facing forwards with the thumbs to the side, and the feet slightly apart with toes pointing forwards

Anatomy: study of the structure and location of body parts

Anterior (ventral): front surface of the body, or structure

Caudal: away from the head, or below

Contralateral: on the opposite side to another structure

Cranial (cephalic): relating to the head end or skull

Deep: further from the body surface

Distal: furthest away from the point of reference

Frontal or coronal plane: divides the body into a front (anterior) portion and a rear (posterior) section

Inferior: situated away from the head or below

Ipsilateral: on the same side as another structure

Lateral: away from the midline

Median or **sagittal plane**: vertical plane that divides the body lengthwise into right and left sections

Medial: towards the midline

Palmar: relating to the palm side of the hand

Parietal: relating to the inner walls of a body cavity

Pathology: the study of disease in the body

Peripheral: away from the centre; outer part of the body

Physiology: the study of the function of body parts

Plantar: relating to the sole of the foot

Posterior (dorsal): back surface of the body, or structure

Prone: lying face down in a horizontal position

Proximal: nearest to the point of reference

Superficial: near to the body surface

Superior: situated towards the head, or above

Supine: lying face up in a horizontal position

Transverse plane: horizontal plane that divides the body into top (superior) and bottom (inferior) sections

Visceral: relating to the internal organs of the body

Revision summary

- **Anatomy** is the study of the structure and location of body parts.
- **Physiology** is the study of the function of those body parts.
- **Pathology** is the study of disease in the body.
- An **anatomical position** follows an imaginary line running down the centre of the body, with the body erect and facing forwards, arms to the side, palms facing forwards with the thumbs to the side, and the feet slightly apart with toes pointing forwards.
- By learning and using **anatomical terms** it will help you describe the specific position of a body part; for instance the biceps muscle is located on the **anterior** (front) surface of the upper arm.
- There are three planes that separate the body into sections:
  - **Frontal or coronal plane**: divides the body into a front (anterior) portion and a rear (posterior) section.
  - **Median or sagittal plane**: a vertical plane that divides the body lengthwise into right and left sections.
  - **Transverse plane**: a horizontal plane that divides the body into top (superior) and bottom (inferior) sections.
- There are two main cavities, or spaces within the body that contain the internal organs:
  - The **dorsal body cavity**, which is located in the posterior (back) region of the body and is subdivided into two cavities: the cranial and spinal cavity.
  - The **ventral body cavity**, which occupies the anterior (front) region of the trunk and is divided into the thoracic cavity and the abdominal and pelvic cavities.

Test your knowledge questions

Multiple choice questions

1 What does the study of physiology entail?
  a study of disease in the body
  b study of the function of body parts
  c study of the structure of body parts
  d study of anatomical terms

2 Which of the following defines the term *proximal*?
  a at or near the centre
  b away from the centre
  c nearest to the point of reference
  d furthest away from the point of reference

3 Which of these anatomical terms describes lying face down in a horizontal position?
  a caudal
  b palmar
  c prone
  d supine

4 Which of these options describes the median or sagittal plane?
  a a vertical plane that divides the body lengthwise into right and left sections
  b a horizontal plane that divides the body into top (superior) and bottom (inferior) sections
  c a vertical plane that divides the body into top (superior) and bottom (inferior) sections
  d a horizontal plane that divides the body lengthwise into right and left sections

5 To which area of the body does the term *cervical* refer?
  a the skull
  b the head
  c the neck
  d the cheek

6 *Cubital* is an anatomical term relating to which area of the body?
  a the wrist
  b the elbow
  c the forearm
  d the upper arm

7 Where is the pericardial region of the body located?
  a the lungs
  b the chest
  c the heart
  d the navel

8 Which of these anatomical terms describes the area of the body relating to the foot?
  a plantar
  b pedal
  c pelvic
  d popliteal

9 When describing a structure or body part that is above, or near to the head, which is the correct term to use?
  a inferior
  b ipsilateral
  c contralateral
  d superior

10 Which of these options is used to describe the back surface of the body or of a structure?
  a ventral
  b dorsal
  c plantar
  d distal

Exam-style questions

11 Describe the following directional anatomical terms:
  a lateral   1 mark
  b distal   1 mark
  c anterior   1 mark
  d prone.   1 mark

12 Describe the following anatomical regional terms:
  a axillary   1 mark
  b brachial   1 mark
  c cephalic   1 mark
  d inguinal.   1 mark

13 Define the following terms in relation to anatomical position:
  a median or sagittal plane   1 mark
  b coronal or frontal plane.   1 mark

14 List the two main body cavities.   2 marks

15 State the anatomical regional area to which each of the following applies:
  a calcaneal   1 mark
  b cubital   1 mark
  c gluteal.   1 mark

2 Cells and tissues

Introduction

The human body is like a universe; it is made up of very small structures that are organised to function as a whole.

It is incredible to think that the human body, a complicated and sophisticated machine, starts its journey of life as a single cell. In order to understand how the body functions as a whole, we need to consider how the structure of the body is organised on five basic levels:

1 chemical
2 cellular
3 tissue
4 organ
5 system.

All the body systems, and the tiny cells that are the basic component parts of all organs and tissues, are involved in maintaining health and keeping the body in a state of balance.

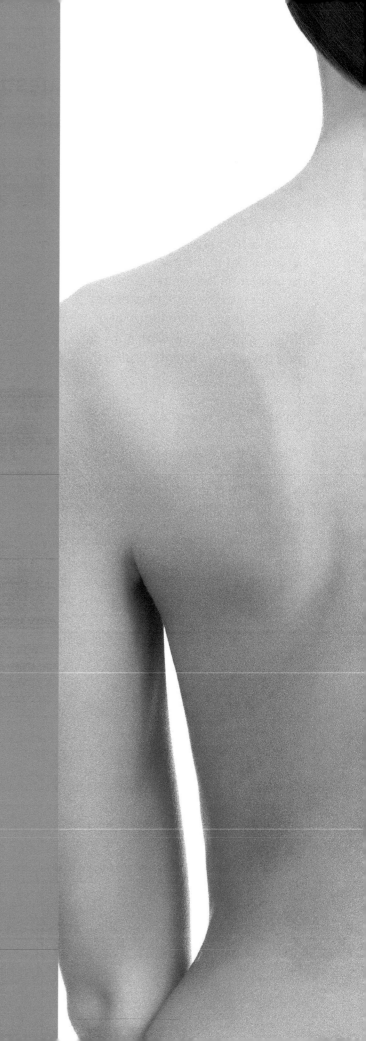

OBJECTIVES

By the end of this chapter you will understand:

- the different levels of structural organisation in the body
- the importance of metabolism and homeostasis for body function
- the structure and function of the cell and its parts
- the structure and function of the main tissue types in the body
- how to identify the major systems of the body
- the interrelationships between the cells, tissues and body systems
- common pathologies associated with cells and tissues.

In practice

In order for a therapist to carry out the most successful treatment possible for their client's needs, they should have an understanding of cells and tissues as the building blocks of the human body.

Examining cells and tissues is like looking at the body from the inside out. Understanding how the body functions at a cellular level will help you to understand how the body functions in times of health and illness, and will enable you to link the structure and function of the body's organs.

The different levels of structural organisation in the body

Humans are organisms that contain many parts making up the whole. In order to appreciate the structure of the human body, we need to study the five principal levels of structural organisation: chemical, cellular, tissue, organ and system.

Study tip

When learning the levels of structural organisation, you may find it helpful to memorise this mnemonic:

Clever **C**arol **t**akes **o**rganisation **s**eriously!

chemical	**c**lever
cellular	**C**arol
tissue	**t**akes
organ	**o**rganisation
system	**s**eriously

1 Chemical level

Every substance in the world is made up of basic particles called atoms and molecules.

Atoms and **molecules** represent the lowest level of organisational complexity in the body and are essential for maintaining life. At the chemical level, the smallest unit of matter is the atom.

- An atom is the smallest particle of an element; an example is a hydrogen or oxygen atom.

- A molecule is a particle composed of two or more atoms joined together; a common example is a water molecule (H_2O), made of one oxygen atom and two hydrogen atoms.

KEY FACT

Molecules combine to form cells.

Study tip

To understand the relationship between atoms and molecules, it can be helpful to think of the molecule as the wall, and the atoms as the bricks from which the wall is built.

2 Cellular level

Cells are the basic structural and functional unit of all living organisms, including the human body. They are, therefore, the smallest units that show characteristics of life.

There are many different types of cells in the body. These vary in structure, size and shape according to their function. An example is a white blood cell (leucocyte) which helps fight infection in the body.

KEY FACT

Cells combine to form tissues.

3 Tissue level

A **tissue** is a group of similar cells that perform a particular function. You will learn about these tissue types: epithelial, connective, muscular and nervous tissue.

KEY FACT

Two or more types of tissue combine to form organs.

4 Organ level

An **organ** is a specialised structure made up of different types of tissues that are grouped into structurally and functionally integrated units. The heart and the lungs are examples of organs.

KEY FACT

Organs combine to form systems.

5 System level

A **system** is a group of organs that work together to perform specific functions. The systems of the body include the circulatory, skeletal, skin, respiratory, reproductive, muscular, endocrine, nervous, renal and digestive systems.

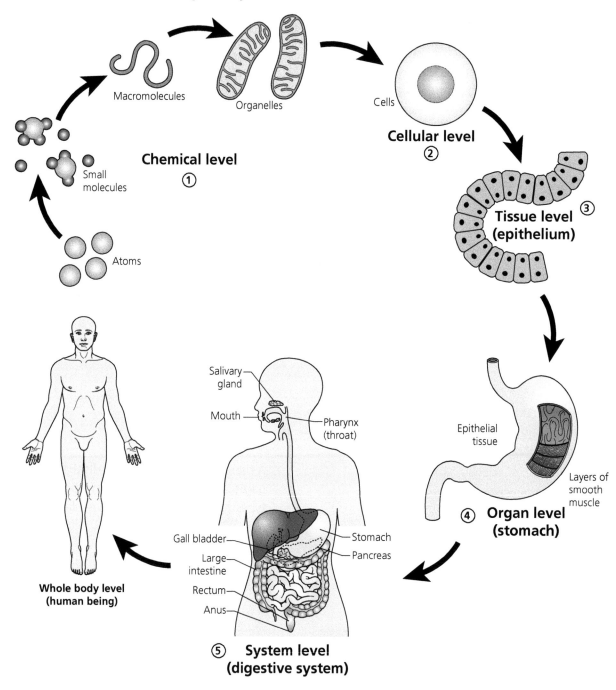

▲ Levels of structural organisation in the body

Homeostasis

The body is divided into different systems according to their specific functions. However, the purpose of all the systems working together is to maintain a constant internal environment so that each cell survives.

The body functions efficiently only when the concentrations of water, food substances, oxygen and wastes, and the conditions of heat and pressure, remain within certain limits. The process by which the body maintains a stable internal environment for its cells and tissues is called **homeostasis**; if one of the variables such as temperature, blood pressure, and levels of oxygen or carbon dioxide in the blood or tissues is not kept within normal limits, imbalance and illness may occur.

The human body is exposed to a constantly changing external environment. These external changes are counteracted by the internal environment of blood, lymph and tissue fluids that bathe and protect the cells, so that the body functions correctly. Examples of homeostasis are regulation of blood sugar level via insulin and control of body temperature via the hypothalamus in the brain.

The process of homeostasis is like an automatic fine-tuning mechanism that restores balance in the body's systems.

When the body's systems are not balanced, whether through stress, pain, infection or depleted oxygen level, the body's cells do not work optimally, leading to signs of disorder and disease. The body systems are constructed to work synergistically (together) to maintain homeostasis.

Examples of homeostatic mechanisms in the body include those that regulate:

- body temperature
- blood pressure
- blood sugar level
- pH level.

Regulating the pH balance

The pH scale is a chemical rating used to measure the acid or alkaline (base) content of a substance.

- Acids have a pH from 0 to 6.
- Alkalis (bases) have a pH of 8 to 14.

KEY FACT

The pH level of blood in the human body should be around 7.4.

If the pH level drops below 7.0 to an acidic level, a condition known as acidosis results. If the pH goes above 7.8, the condition is called **alkalosis**. Both acidosis and alkalosis can be life threatening.

In order to maintain the blood at a pH of 7.4, the body's systems work together by producing buffer substances (carbonate and bicarbonate), which function to regulate the pH level by absorbing excess hydrogen or hydrogen ions. The kidneys are significant in homeostasis as they can detect if the pH of your body's fluids is too low (too acidic).

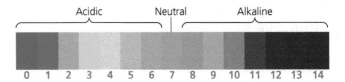

▲ The pH scale

KEY FACT

Part of the brain, the hypothalamus, is vital in homeostasis. Throughout the body, receptors in blood vessels detect the blood's pressure, temperature, glucose level and pH. These receptors send signals through the nervous systems to the hypothalamus, where changes initiate (start) to bring the body back to balance. An example is the stimulation by the hypothalamus of the pancreas to secrete insulin when the blood sugar level gets too high.

KEY FACT

Homeostasis is maintained by adjusting the metabolism of the body.

Metabolism

Metabolism is the term used to describe the physiological processes that take place in our bodies to convert the food we eat and the air we breathe into the energy we need to function. Metabolism is essentially the basic chemical workings of the body cells; through metabolism, food substances are transformed into energy or materials that the body can use or store.

Metabolism involves two processes:

1 **Catabolism** – the chemical breakdown of complex substances by the body to form simpler ones, accompanied by the release of energy. The substances that are broken down include nutrients in food (carbohydrates and proteins), as well as the body's storage products (glycogen).

2 **Anabolism** – the building up of complex molecules, such as proteins and fats, from simpler ones by living things.

Metabolic rate

The rate at which a person consumes energy for activity and body processes is known as the **metabolic** rate. The minimum energy required to keep the body alive is known as the **basal metabolic** rate.

How elements contribute to the body's chemical make-up

It is important to consider how the body's chemical make-up (its most common major elements and compounds) relates to its physiological processes.

Elements and compounds occur in one of three states: gas, liquid or solid. It takes energy to change the state of an element or a compound.

This can be illustrated by water, which is usually a liquid (although it can become solid, as ice, or turn into steam, a gas). Salt and glucose are examples of solids, and carbon dioxide is a gas.

The body stores energy in chemical bonds between the elements that make up its compounds. This energy, when released, may be used in a variety of ways: for example to initiate chemical processes, to aid movement or for the body's growth, maintenance and repair.

The chemical make-up of a cell

Chemically, a cell is composed of the major elements carbon, oxygen, hydrogen and nitrogen, with trace elements of several other elements such as sodium, calcium, chlorine, magnesium, iron, iodine, potassium, sulfur and phosphorus (Table 2.1).

> **KEY FACT**
> Cells are made up of approximately 80% water, 15% protein, 3% lipids or fats, 1% carbohydrates and 1% nucleic acids.

Table 2.1 Overview of chemical components of the cell

Major compound	Elements present	Main significance in body
Water	Hydrogen and oxygen	The body's reservoir, water provides a universal solvent for the facilitation of chemical reactions in the tissues Helps transport substances around the body
Carbohydrates	Carbon, hydrogen and oxygen	Main fuel for the body
Proteins	Carbon, hydrogen, oxygen, nitrogen (may also contain sulfur)	Main building blocks of the body's tissues
Fats/lipids	Carbon, hydrogen and oxygen	Energy source for the body's activities Energy store
Nucleic acids	Carbon, hydrogen, oxygen, nitrogen and phosphorus	Important molecules found inside cells Deoxyribonucleic acid (DNA) is the genetic material inside the nucleus

Cells

KEY FACT

Cytology is the scientific study of cells.

The cell is the fundamental unit of all living organisms and is the simplest form of life that can exist as a self-sustaining unit. Cells are, therefore, the building blocks of the human body.

Cells in the body take many forms, the size and shape being largely dependent on their specialised function. For example, some cells help fight disease, others transport oxygen or produce movement, some manufacture proteins or chemicals, and others function to store nutrients.

KEY FACT

Each type of cell has a structure that is suited to its specific function. A muscle cell is long and thin with structures that enable it to contract and shorten, while skin cells are flat and tough, providing a waterproof covering.

Cell structure

Although cells are the smallest units that show characteristics of life, they are made up of different parts.

Cell organelles

Cell organelles ('little organs') are the basic component parts of cells and are formed from molecules that combine in very specific ways. Each organelle has particular functional significance within the cell.

Study tip

When examining the function of each organelle, it is helpful to think of the cell as the 'factory' and the organelles as 'departments' within the factory. Each cell organelle is responsible for the production of a certain product or substance that is used elsewhere in the cell or body.

Despite the great variety of cells in the body, they all have the same basic structure.

Study tip

When studying cell structure it is helpful to think of three parts:

1 the outer part – the **cell membrane**
2 the inner part – containing the **nucleus**
3 the middle layer – a semi-fluid substance called **cytoplasm** which contains all of the cell's organelles.

The outer part of the cell

Cell membrane

The cell membrane, or plasma membrane, is a fine membrane that encloses the cell and protects its contents. This membrane is semipermeable, in that it selectively controls the movement of molecules into and out of the cell. Oxygen, nutrients and hormones are taken into the cell as needed and cellular waste, such as carbon dioxide, passes out through the membrane. As well as governing the exchange of nutrients and waste materials, its function is also to maintain the shape of the cell.

The inner part of the cell

Nucleus

The nucleus is the largest organelle in the cytoplasm. It acts as the control centre of the cell, regulating the cell's functions and directing most metabolic activities. The nucleus governs the specialised work performed by the cell and the cell's own growth, repair and reproduction. All cells have at least one nucleus at some time in their existence. The nucleus is significant in that it contains all the information required for the cell to function and it controls all cellular operations.

The information required by the cell is stored in DNA, the genetic material. The DNA is found in a molecule called chromatin that condenses to form thread-like structures known as chromosomes.

Chromosomes carry the genetic information in the form of genes. The nucleus of a human cell contains 46 chromosomes, 23 of which are from the mother and 23 of which are from the father. Each chromosome can duplicate an exact copy of itself at each cell division, so that every new cell formed receives a full set of chromosomes.

KEY FACT

DNA is often called the body's blueprint, as it is a record of a person's inherited characteristics – their height, bone structure, hair colour and body chemistry, for example. When cells divide and multiply, DNA passes on its hereditary information, ensuring new cells are direct copies.

If the spiral of DNA in the nucleus of just one human cell were stretched out in a single line, it would extend more than 6 feet.

Chromatin

Chromatin is the substance inside the nucleus that contains the DNA and some proteins.

Nucleolus

The nucleolus is a dense spherical structure inside the nucleus, which contains ribonucleic acid (RNA) structures that form ribosomes.

KEY FACT

RNA is the molecule that transports the genetic information out of the nucleus and allows translation of the genetic code into proteins.

Nuclear membrane

Surrounding the cell nucleus is a double-layered membrane called the nuclear membrane, or nuclear envelope. This membrane separates the nucleoplasm, or fluid inside the nucleus, from the cytoplasm, or fluid outside the nucleus.

The function of the nuclear membrane is to regulate the materials that enter or exit the nucleus.

Nuclear pores are tiny passageways through the nuclear membrane. They have a sophisticated biological entry and exit control system, only permitting selected chemicals to move in and out of the nucleus.

Study tip

A nuclear pore is a bit like a ticket gate, acting as a security control system that guards the barrier between the nucleus and the cytoplasm.

The middle part of the cell

Cytoplasm

Cytoplasm is the gel-like substance which is enclosed by the cell membrane. The cytoplasm contains the nucleus and the organelles.

Cell metabolism predominantly takes place inside the cytoplasm. The cytoplasm, as part of its function, contains elements that aid metabolic operation and break down waste.

The **centrosome** is an area of clear cytoplasm found next to the nucleus. It contains the centrioles.

Centrioles

Contained within the centrosome are the small spherical structures called centrioles. These are associated with cell division, or mitosis. During cell division, the centrioles divide and migrate to opposite sides of the nucleus to form the spindle poles.

Chromatids are pairs of identical replicated strands of a chromosome. They are joined at the **centromere** and separate during cell division.

Ribosomes

Ribosomes are tiny organelles made up of RNA and protein. They may be fixed to the walls of the endoplasmic reticulum (ER) or may float freely in the cytoplasm. Their function is to manufacture proteins for use within the cell and also to produce other proteins that are exported outside the cell.

Endoplasmic reticulum (smooth ER and rough ER)

The ER is a series of membrane tubes that are continuous with the cell membrane. It functions like an intracellular transport system, allowing movement of materials from one part of the cell to another. It also links the cell membrane with the nuclear membrane and assists the movement of materials in and out of the cell.

The ER contains enzymes and helps in the synthesis of proteins, carbohydrates and lipids. It serves to store material and to transport substances inside the cell, as well as to detoxify harmful agents. Some of the ER appears smooth, while some appears rough due to the presence of ribosomes.

Mitochondria

Mitochondria (oval-shaped organelles) lie in varying numbers within the cytoplasm and are the site of the cell's energy production. Mitochondria supply the majority of a cell's adenosine triphosphate (ATP), a compound that stores the cell's energy.

Enzymes are protein catalysts that speed up chemical change. Mitochondria contain large amounts of enzymes which power the cell's activities through cellular respiration.

Lysosome

Lysosomes resemble round sacs and contain powerful enzymes capable of digesting proteins. They are present in the cytoplasm and work to destroy parts of cells that are no longer functioning. This process is known as lysis.

Vacuole

Vacuoles are membrane-bound spaces within the cytoplasm that contain waste materials or secretions produced by the cytoplasm. Vacuoles function as temporary storage, for transportation from one part of a cell to another, and for digestive purposes in some cells.

Golgi body/apparatus

The Golgi body resembles a collection of flattened sacs and is located within the cytoplasm, typically near the nucleus and attached to the ER. The Golgi apparatus stores the protein manufactured in the ER and later transports it out of the cell.

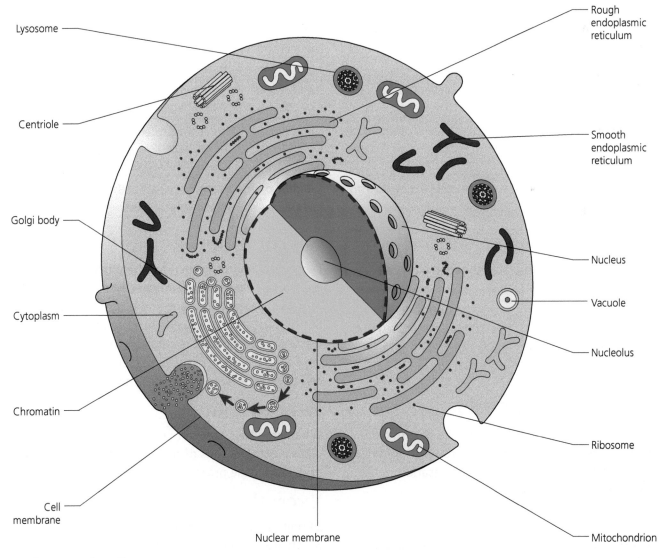

▲ Structure of a cell

Summary of function of main cell organelles

Table 2.2 Summary of main cell organelles

Cell organelle	Description	Location	Function
Cell membrane	Fine membrane that encloses the cell	Outer part of cell	Selectively controls the movement of molecules into and out of the cell
Nucleus	Largest organelle in the cytoplasm, containing DNA in chromosomes	Inner part of cell	The control centre of the cell, regulating its functions
Nuclear membrane	Double-layered membrane surrounding the cell nucleus	Inner part of cell	Regulates the materials that enter or exit the nucleus
Cytoplasm	Gel-like substance that is enclosed by the cell membrane	Middle part of cell	Substance that fills the cell Most cellular metabolism occurs within the cytoplasm
Ribosomes	Tiny organelles made up of RNA and protein	Middle part of cell	Manufacture proteins for use within the cell and also to produce other proteins that are exported outside the cell
Endoplasmic reticulum	Series of membranes continuous with the cell membrane	Middle part of cell	Intracellular transport system, allowing movement of materials from one part of the cell to another
Mitochondria	Oval-shaped organelles that lie within the cytoplasm	Middle part of cell	Provide the energy which powers the cell's activities
Golgi body	Collection of flattened sacs within the cytoplasm	Middle part of cell	Stores the protein manufactured in the endoplasmic reticulum and later transports it out of the cell
Lysosome	Round sacs located in the cytoplasm	Middle part of cell	Destroy any part of the cell that is worn out

 Activity

Draw a simple cell and label the following cell organelles:

- cell membrane
- nucleus
- nuclear membrane
- ribosome
- mitochondria
- endoplasmic reticulum
- Golgi body
- lysosome.

Functions of cells

In order for a cell to survive it must be able to carry out a variety of functions such as growth, respiration, reproduction, excretion, sensitivity/sensation, movement and nutrition.

Growth

Cells have the ability to grow until they are mature and ready to reproduce. A cell can grow and repair itself by manufacturing protein.

Respiration

Every cell requires oxygen for the process of respiration. Oxygen is absorbed through the cell's semipermeable membrane and is used to oxidise nutrient material to provide heat and energy. The waste products of cell respiration include carbon dioxide and water. These are passed out from the cell through its semipermeable membrane.

Reproduction

When growth is complete in a cell, reproduction takes place. The cells of the human body reproduce or divide by the process of mitosis.

Excretion

During metabolism, various substances are produced which are of no further use to the cell and can be damaging. These waste products are removed through the cell's semipermeable membrane.

Sensitivity/sensation

A cell has the ability to respond to a stimulus (a cause or trigger of a reaction), which may be physical, chemical or thermal. For example, a muscle fibre contracts when stimulated by a nerve cell.

Movement

Movement may occur in the whole or in part of a cell. White blood cells, for instance, are able to move freely.

Nutrition

The endoplasmic reticulum and Golgi apparatus manufacture different substances such as protein and fats, either as needed internally by the cell or according to its specific function. The cell utilises basic nutrient molecules that are either dissolved in the cytoplasm or specific substances contained within vesicles.

Carbohydrates are transported to the mitochondria, where they are broken down to yield energy. In the process, high-energy ATP molecules are manufactured and provide energy for other organelles.

> **In practice**
>
> Cell growth and reproduction requires favourable conditions such as an adequate supply of food, oxygen, water, suitable temperatures and the ability to eliminate waste.
>
> Some factors, such as smoking, sun damage and air pollution, create unfavourable conditions for the skin. These can impair cell function and cells may be destroyed, resulting in loss of skin elasticity, lines, wrinkles and dehydration.
>
> In these cases, massage treatments can be beneficial. Massage procedures encourage cell nutrition, as well as increasing elimination of waste from the cells and tissues. Practitioners can advise clients to have treatments regularly to aid cell regeneration.

The cell life cycle

For body growth and repair it is vital for living cells to reproduce. Consequently, human body cells undergo many divisions from the time of fertilisation/conception to the end of life. When a single cell undergoes division for growth or repair, it forms two daughter cells that are identical to the original cell. A cell may live from a few days to many years, depending on its type.

Depending on the purpose of division, cells divide in one of two ways:

1 **mitosis** – division of one cell into two genetically identical daughter cells

2 **meiosis** – division of one cell into four genetically different daughter cells.

1 Mitosis

Mitosis is when a single cell produces two genetically identical daughter cells. It is the way in which new body cells are produced for both growth and repair. Division of the nucleus takes place in four main phases (prophase, metaphase, anaphase and telophase) and is followed by the division of the cytoplasm to form the daughter cells.

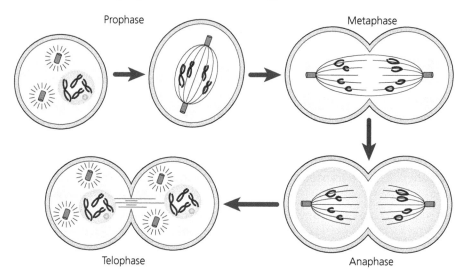

▲ Stages of mitosis

Study tip

The acronym PMAT (prophase, metaphase, anaphase and telophase) can be helpful when learning the stages of mitosis.

- **Prophase** = chromosomes <u>condense</u>
- **Metaphase** = chromosomes <u>line up</u> in the centre
- **Anaphase** = chromosomes <u>separate</u>
- **Telophase** = chromosomes <u>de-condense</u>

Interphase (preparation for mitosis)

A cell must prepare before it can divide into two identical daughter cells with equal shares of DNA and organelles. A state known as interphase (between phases) precedes mitosis. During interphase, the cell makes a copy of all its DNA. Once the cell has duplicated its DNA and organelles, it can proceed into mitosis.

Prophase

Chromatin condenses to give distinct chromosomes consisting of pairs of chromatids joined at the centromere. Centrioles duplicate and separate to form spindles.

Metaphase

Chromosomes align themselves in the centre of the cell, midway between the centrioles. The protective nuclear membrane breaks down. The centromere of each chromosome then replicates.

Anaphase

Centromeres divide and identical sets of chromosomes move to opposite poles of the cell

Telophase

This is the final stage of mitosis. A nuclear membrane forms around each set of chromosomes, giving two new nuclei. The spindle fibres disappear. The cytoplasm compresses and then divides in half in cytokinesis.

Cytokinesis

Usually, after telophase, the cytoplasm divides and separates into two identical daughter cells. Each daughter cell is an exact copy of the parent cell before the DNA was duplicated during interphase.

2 Meiosis

Meiosis is a type of cell division that produces four daughter cells, each having **half** the number of chromosomes of the original cell. Meiosis forms eggs in females and sperm in males and is

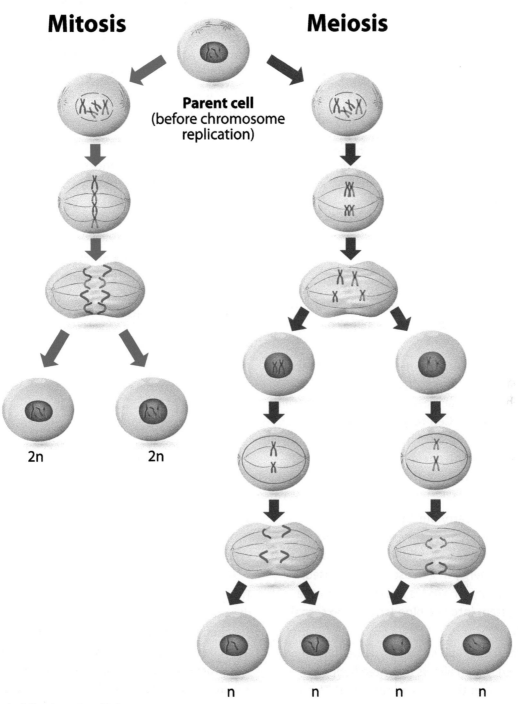

Mitosis   **Meiosis**

Parent cell
(before chromosome replication)

2n   2n

n   n   n   n

▲ Mitosis and meiosis

the preparation for formation of a new organism through the fusion of one egg and one sperm.

Before fertilisation there are 23 chromosomes present in the sperm and 23 in the egg. After fertilisation has taken place, the egg and the sperm fuse together to form a single cell called a zygote with 46 chromosomes (23 from each parent). The zygote undergoes mitosis to make more and more cells, forming first an embryo, then a foetus and eventually a baby.

KEY FACT

There is virtually no limit to the ways in which a reproductive cell's 23 chromosomes can be combined during meiosis, meaning that every sperm and every egg contains different hereditary information. Consequently, the genetic characteristics of brothers and sisters are never the same (except for identical twins, who share the same genetic code).

Cellular respiration

All cellular functions depend on energy generation and transportation of substances within and between cells.

In order to function properly, a cell must maintain a stable internal environment; therefore, the transport of materials has to be achieved without an excessive build-up of chemicals. **Cell respiration** refers to the cell's use of nutrients (such as oxygen and glucose) to activate the energy needed for the cell to function and the output of waste (such as carbon dioxide).

In order for cells to carry out their work, they need to produce enough energy. Fuel for energy is provided by glucose from carbohydrate metabolism. In order for the energy in glucose to be released, it is oxidised by oxygen that is absorbed from the respiratory system into the bloodstream (**external respiration**).

Cells are bathed in a fluid known as tissue fluid or interstitial fluid, which allows the interchange of substances between the cells and the blood. This is known as **internal respiration**.

Cell transport

As explained, cells export some materials out of the cell cytoplasm and also receive substances from the outside.

The body's internal transport system, the blood, carries oxygen from the respiratory system and nutrients such as glucose from the digestive system to the cells. These are absorbed through the cell membrane in several different ways: **diffusion**, **osmosis**, **active transport** and **filtration**. When certain molecules are needed, such as glucose, the cell will actively take these in and may discard other materials in order to preserve the equilibrium.

Diffusion

If chemicals become concentrated outside the cell, a flow of small molecules takes place through the cell membrane until there is a balance between the internal and external concentration. This process, in which small molecules move from areas of high concentration to those of lower concentration, is called diffusion. Diffusion is the basis by which the cells lining the small intestines take in digestive products to be utilised by the body.

Diffusion

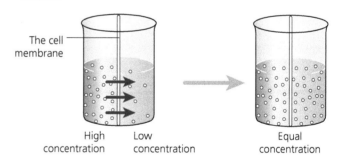

The cell membrane

High concentration | Low concentration | Equal concentration

▲ Diffusion: the process in which small molecules move from an area of high concentration to lower concentration

Osmosis

This process refers to the movement of water through the cell membrane from areas of low chemical concentration (many water molecules) to areas of high chemical concentration (fewer water molecules). This process allows for the dilution of chemicals, which are unable to cross the cell membrane by diffusion, in order to maintain equilibrium within the cell.

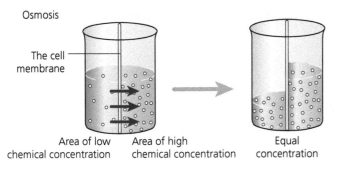

Osmosis

The cell membrane

Area of low chemical concentration

Area of high chemical concentration

Equal concentration

▲ Osmosis: the movement of water through the cell membrane from low to high chemical concentration

Active transport

This is an energy-dependent process in which certain substances (including glucose, ions, some drugs and amino acids) are able to cross cell membranes against a concentration gradient (moving from an area with a lower number of particles to an area with a higher number of particles). This is the process, using chemical energy, by which the cell takes in larger molecules that would otherwise be unable to enter in sufficient quantities. Carrier molecules within the cell membrane bind themselves to the molecules, rotate around them and release them into the cell.

Filtration

This is the movement of water and dissolved substances across the cell membrane due to differences in pressure. The force of the weight of the tissue fluid pushes against the cell membrane, thereby moving molecules into the cell. One site of filtration in the body is in the kidneys. Blood pressure forces water and small molecules through plasma membranes of cells and the filtered liquid then enters the kidneys for further processing and disposal through the renal system.

Pinocytosis and phagocytosis

Like us, cells can 'drink' (take in liquid) and 'eat' (take in particles). These processes are:

1 pinocytosis
2 phagocytosis.

> **Study tip**
> Remember:
> - pinocytosis = cell drinking
> - phagocytosis = cell eating.

1 Pinocytosis

Pinocytosis is derived from *pino*, Greek for 'to drink' and *cytosis*, 'a transport mechanism'. It is a process whereby small droplets are taken into the cell. The cell encases the droplets inside vesicles that are formed from parts of the cell membrane that have been split off. This cell process is used for taking extracellular fluid into the cell.

2 Phagocytosis

The word phagocytosis comes from the Greek *phago*, meaning 'to eat'. It describes the process by which a cell engulfs (takes in) particles such as bacteria and other micro-organisms, old red blood cells and foreign matter.

The principal human phagocytes (cells that carry out phagocytosis) include the neutrophils and monocytes, which are types of white blood cells.

Pinocytosis and phagocytosis are similar processes, but have some key differences:

- Larger materials such as bacteria, which are too big to be absorbed by pinocytosis, are absorbed by phagocytosis.
- Solid materials are absorbed through phagocytosis, while liquids and dissolved solutes are ingested through pinocytosis.
- The liquid contents of small vesicles from pinocytosis are deposited directly into the cell. However, the vesicles from phagocytosis are bigger and the contents are not deposited directly into the cell.
- In order for the vesicle contents to be broken down during phagocytosis, the vesicles combine with lysosomes.

Tissues

Tissues are defined as a group of similar cells that act together to perform a specific function.

> **KEY FACT**
> The study of tissues is known as **histology**.

Due to the complexity of the human body, it is not possible for every cell to carry out all the functions required by the body. Some cells, therefore, become specialised to form a group of cells or tissues.

There are four major types of tissues in the human body:

1 **epithelial** tissue

2 **connective** tissue

3 **muscle** tissue

4 **nervous** tissue.

Table 2.3 gives an overview of the four major types of tissue in the body.

Table 2.3 Overview of tissue types

Type of tissue	Main function
Epithelial tissue	Provides a protective covering for surfaces inside and outside the body
Connective tissue	Protects, binds and supports the body and its organs
Muscle tissue	Provides movement
Nervous tissue	Initiates and transmits nerve impulses

KEY FACT

The four types of tissue have different rates of cellular regeneration, related to their specific functions.

- Epithelial tissue is renewed constantly by the process of mitosis (cell division).

- Bone tissue and adipose connective tissue have a very good blood supply (described as being highly vascular tissue). A supply of nutrients allows tissue to repair and heal quickly through fast cell division.

- Muscle tissue takes longer to regenerate.

- Nervous tissue regenerates very slowly.

- Muscle and nerve cells divide more slowly as they are more specialised in their function. Once cells in the body become specialised, they take on structures unique to their specific functions; these structures are not always compatible with cell division.

- The less vascular forms of connective tissue, such as ligaments and tendons, are even slower to heal than muscle tissue. Cartilage is among the slowest to heal.

1 Epithelial tissue

Epithelial tissue consists of sheets of cells which cover and protect the external and internal surfaces of the body, lining the insides of hollow structures. They are specialised to move substances in and out of the blood during secretion, absorption and excretion. As these tissues are subject to considerable wear and tear, epithelial cells reproduce actively to repair surfaces very quickly.

Epithelial tissue usually has a very thin **matrix** (ground substance or base). The matrix is a continuous sheet of cells, which are held very tightly together. A thin, permeable basement membrane attaches epithelial tissues to the underlying connective tissue. Epithelial tissue cells, which are closely packed together, come in various shapes. There are two categories of epithelial tissue:

- **simple** (single-layer) epithelial tissue
- **compound** (multi-layer) epithelial tissue

Simple epithelium

Simple epithelial tissue has only one layer of cells over a basement membrane. Being thin, these tissues are fragile and are found only in areas inside the body which are relatively protected. Examples are the lining of the heart, blood vessels and the linings of body cavities.

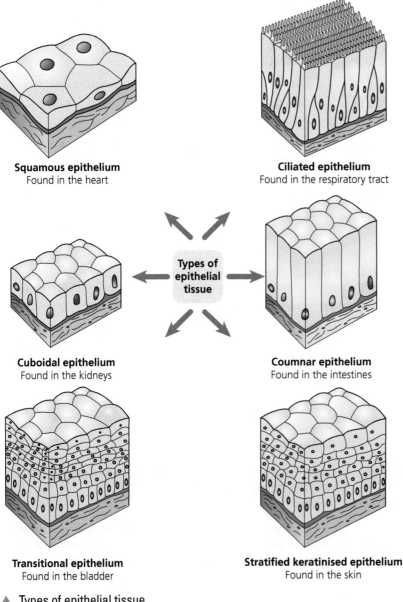

Squamous epithelium
Found in the heart

Ciliated epithelium
Found in the respiratory tract

Types of epithelial tissue

Cuboidal epithelium
Found in the kidneys

Coumnar epithelium
Found in the intestines

Transitional epithelium
Found in the bladder

Stratified keratinised epithelium
Found in the skin

▲ Types of epithelial tissue

They are also found lining the digestive tract and in the exchange surfaces of the lungs, where their thinness is an advantage for speedy cross-membrane transport. There are four different types of simple epithelium, named according to their shape and the functions they perform: simple squamous, simple cuboidal, simple columnar, simple ciliated (columnar). See Table 2.4.

Compound epithelium

The main function of compound epithelium is to protect underlying structures. Compound epithelium contains two or more layers of cells. There are two main types:

- **stratified** epithelium
- **transitional** epithelium.

Stratified epithelium

Stratified epithelium is composed of a number of layers of cells of different shapes. In the deeper layers the cells are mainly columnar in shape and as they grow towards the surface they become flattened.

There are two types of stratified epithelium:

- **Non-keratinised stratified epithelium** – this is found on wet surfaces that may be subject to wear and tear, such as the conjunctiva of the eyes, the lining of the mouth, the pharynx and the oesophagus.

- **Keratinised stratified epithelium** – this is found on dry surfaces, such as the skin, hair and nails. The surface layers of keratinised cells are dead cells. They give protection and prevent drying out of the cells in the deeper layers from which they develop. The surface layer of cells is continually being rubbed off and is replaced from below.

Transitional epithelium

Transitional epithelium is composed of several layers of pear-shaped cells which change shape when they are stretched. This type of tissue is found lining the uterus, bladder and pelvis of the kidney.

2 Connective tissue

Connective tissue is the most abundant type of tissue in the body. It connects tissues and organs by binding the various parts of the body together, giving protection and support.

Connective tissue has three components: a matrix, cells called fibroblasts and fibres made of protein.

Connective tissue cells are often more widely separated from each other than those forming epithelial tissue; the space between cells is larger and is filled with non-living matrix. There may or may not be fibres in the matrix, which can be either a semi-solid, jelly-like consistency or dense and rigid, depending on the position and function of the tissue.

Table 2.4 The structure, location and function of the different types of simple epithelial tissue

Type	Structure	Location	Function
Simple squamous	A single layer of flat, scale-like cells, each with a central nucleus. The cells fit closely together, rather like a pavement, producing a very smooth surface	Lining the alveoli of the lungs Lining blood and lymphatic vessels and the heart	Allows for exchange of nutrients, wastes and gases
Simple cuboidal	Single layer of cube-like cells	Ovaries, kidney tubules, thyroid gland, pancreas and salivary glands	Secretion and absorption
Simple columnar	Single layer of tall, cylindrical column-shape cells, each with a nucleus situated towards the base of the cell	Lining the small and large intestine, stomach and gall bladder	Secretion and absorption
Simple ciliated (columnar)	A form of columnar epithelium Single layer of rectangular cells that have hair-like projections (cilia) on their surface	Lining the upper part of the respiratory system Lining the uterine tubes	Beating cilia carry unwanted particles, along with mucus, out of the system Helps propel the ova towards the uterus

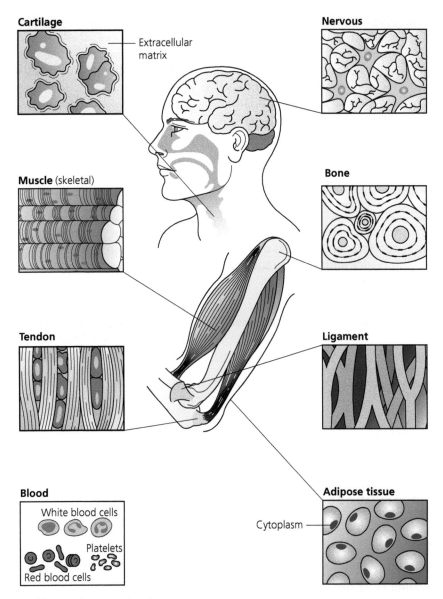

▲ Types of connective tissue

There are several different types of connective tissue, which are summarised in Table 2.5.

Table 2.5 Types of connective tissue

Type	Structure	Location	Function
Areolar	Most widely distributed type of connective tissue in the body A loose, soft and pliable tissue Contains collagen, elastin and reticular fibres	Under the skin, between muscles, supporting blood vessels and nerves, and in the alimentary canal	Provides strength and elasticity Connects and supports organs
Adipose	A type of areolar tissue containing fat cells (adipocytes)	Surrounds organs such as kidneys and the heart Under the skin (subcutaneous layer) Between bundles of muscle fibres, in the yellow bone marrow of long bones and as a padding around joints	Provides insulation, support and protection Emergency energy reserve

Type	Structure	Location	Function
White fibrous	Strong connecting tissue made up of mainly closely packed bundles of white collagenous fibres, with very little matrix Contains cells called fibrocytes between the bundles	Forms tendons which attach muscle to bone, ligaments which tie bones together and as an outer protective covering for some organs, such as kidney and bladder	Provides strong attachment between different structures
Yellow elastic	Consists of branching yellow elastic fibres with fibrocytes in the spaces between the fibres	Arteries, trachea, bronchi and lungs	To allow the stretching of various organs, followed by a return to their original shape and size
Lymphoid	This tissue has a semi-solid matrix with fine branching fibres The cells contained within this tissue are specialised and are called lymphocytes	In the lymph nodes, spleen, tonsils, adenoids, walls of the large intestine and glands of the small intestine	Forms part of the lymphatic system whose function is to protect the body from infection
Blood	Also known as liquid connective tissue, it contains the blood cells erythrocytes, leucocytes and thrombocytes which float within a fluid called plasma	Contained within blood vessels	Helps maintain homeostasis by transporting substances throughout the body, by resisting infection and by dispersing heat
Bone	Hardest and most solid of all connective tissues Consists of tough, dense compact bones and slightly less dense cancellous bone	Bones	Protects and supports other organs and soft tissues
Cartilage	Very firm tissue; matrix is quite solid	See Table 2.6	See Table 2.6

Cartilage

For descriptive purposes, cartilage is divided into three types (see Table 2.6):

1 **hyaline** cartilage

2 **white fibrous** cartilage

3 **yellow elastic** fibrocartilage.

Table 2.6 Types of cartilage

Type of cartilage	Description	Location	Function
Hyaline cartilage	Most abundant type of cartilage found in the body A smooth, bluish-white, glossy tissue Contains numerous cells called chondrocytes which produce cartilage	Found on the surfaces of the parts of bones which form joints Forms the costal cartilage which attaches the ribs to the sternum Forms part of the larynx, trachea and bronchi	Provides a hard-wearing, low friction surface within joints Provides flexibility in the nose and trachea
White fibrous cartilage	This type of cartilage is tough but slightly flexible It is composed of bundles of collagenous white fibres in a solid matrix with cells scattered among them	It is found as pads between the bodies of the vertebrae, called the intervertebral discs, and in the symphysis pubis which joins the pubis bones together	Its function is one of support and to join together or fuse certain bones

Type of cartilage	Description	Location	Function
Yellow elastic fibrocartilage	Consists of yellow elastic fibres running through a solid matrix, between which chondrocytes are situated	It forms the pinna (lobe of the ear) and the epiglottis	To provide support and to maintain shape

Muscle tissue

Muscle tissue is very elastic and has the unique ability to provide movement by shortening as a result of contraction. This tissue is made up of contractile fibres, usually arranged in bundles and surrounded by connective tissue. There are three types of muscle tissue:

1 **voluntary** (skeletal) muscle

2 **involuntary** (smooth) muscle

3 **cardiac muscle**.

The different types of muscle tissue are discussed in more detail in Chapter 5, The muscular system.

3 Nervous tissue

Nervous tissue consists of cells called neurones which can convert stimuli into nerve impulses that are transmitted as electrical signals. Nervous tissue has the characteristics of excitability and conductivity. Its functions are to co-ordinate and regulate body activity.

Nervous tissue and neurones are discussed in more detail in Chapter 9, The nervous system.

4 Membranes

A membrane is a thin, soft, sheet-like layer of tissue that covers a cell, organ or structure, that lines tubes or cavities or that divides and separates one part of a cavity from another. There are three basic types of membranes in the body (see Table 2.7):

● **mucous** membrane

● **serous** membrane

● **synovial** membrane.

Table 2.7 Types of membrane

Type of Membrane	Description	Location	Function
Mucous membrane	Type of membrane that lines body cavities and outer layer of organs	Lining the respiratory, digestive, urinary and reproductive tracts	As well as lining the openings to the external environment, secretes a viscous slippery fluid called mucus that coats and protects underlying cells
Serous membrane	Type of membrane that lines body cavities which are not open to the external environment and that covers many of the organs Serous membranes consist of two layers: ● a parietal layer – lines the wall of body cavities ● a visceral layer – provides an external covering to organs in body cavities	The pericardium of the heart Pleural membranes in the lungs The peritoneum lining the abdominal organs	As well as lining body cavities that are not open to the external environment, they secrete a thin, watery (serous) fluid that lubricates organs to reduce friction as they rub against one another and against the wall of the cavities
Synovial membrane	Type of membrane that lines the cavities of freely movable joints	Lining the spaces around certain joint cavities (shoulder, hip and knee)	Secretes synovial fluid which provides nutrition and lubrication to joints so they can move freely without undue friction

Create a mind map that identifies the different type(s) of tissue found in the following body parts:

- skin
- hair
- nail
- lungs
- heart
- stomach
- pancreas
- gall bladder
- small intestine
- large intestine
- spleen
- kidneys
- bladder
- uterus
- ovaries.

System level of organisation of the body

A system is a group of organs that work together to perform specific functions. Cells and tissues are the building blocks that make up the body's systems.

Although the subject matter on each body system is organised into a separate chapter in this textbook, it is important to realise no one system works independently of another. All the systems require the support of and co-ordination with other systems to form a healthy living human body.

Towards the end of each chapter you will find a section called 'Interrelationships with other systems'. The information in these sections will help you to understand how the body systems work together to maintain balance.

Table 2.8 provides an overview of each body system's function.

Table 2.8 Overview of body systems

Body system	Function
Integumentary system (skin, hair and nails)	The body's largest organ Protects us from the outside world by providing the first line of defence against bacteria, viruses and other pathogens
Skeletal system	Provides shape, support and protection for the body
Muscular system	Enables movement of the body
Cardiovascular system	Provides a transport system of blood around the body
Lymphatic system	Provides a filtering system for the blood and tissues
Respiratory system	Enables gaseous exchange through the lungs
Nervous system	Responsible for receiving and interpreting information from inside and outside the body
Endocrine system	Secretes hormones to help regulate body processes
Reproductive system	Responsible for producing human offspring to continue the species
Digestive system	Responsible for the breakdown of food and absorption of nutrients
Urinary system	Provides elimination of metabolic waste through the kidneys via the production of urine

Common pathologies of cells and tissues

Cancer/abnormal cell division

Cancerous diseases are characterised by the growth of abnormal cells that invade surrounding tissues, followed by the process of **metastasis** (the spread of cancerous cells to other parts of the body).

When cells in an area of the body divide without control, the excess tissue develops into a tumour, growth or neoplasm.

A cancerous growth is called a malignant tumour and a non-cancerous growth is called a benign tumour. Cancerous growths may eventually be fatal. Benign tumours do not spread to other parts of the body, but they may be removed if they interfere with a normal body function or are disfiguring.

The name of the cancer is derived from the type of tissue in which it develops. Cancers are further classified according to their microscopic appearance and the body site from which they arise.

Most human cancers are **carcinomas**. These are malignant tumours that arise from epithelial cells. Examples are:

- **melanoma** – a cancerous growth of melanocytes, the skin cells that produce the pigment melanin
- **sarcoma** – a general term for any cancer arising from muscle cells or connective tissues (for example, osteogenic sarcoma (bone cancer) is the most frequent type of childhood cancer; this destroys normal bone tissue and eventually spreads to other areas of the body)
- **leukaemia** – a cancer of blood-forming organs characterised by rapid growth and distorted development of leucocytes
- **lymphoma** – a malignant disease of lymphatic tissue such as the lymph nodes, e.g. Hodgkin's disease.

In practice

The study of tumours is called **oncology** and a physician who specialises in this field is called an **oncologist**.

Interrelationships with other systems

Cells and tissues

Cells and tissues are found in the following body systems.

Skin

Keratinised stratified epithelium (a type of tissue containing layers of cells) is found on dry surfaces such as the skin, hair and nails.

Skeletal

Bone is the hardest and most solid type of connective tissue in the body. It is needed for building the structures of the skeletal framework.

Muscular

There are three types of muscle tissue:
1 skeletal muscle, which controls voluntary movements
2 smooth muscle, which controls involuntary movements
3 cardiac muscle, which makes up the heart.

Circulatory

Blood is a form of liquid connective tissue that transports substances to and from the cells in different parts of the body.

Respiratory

A type of tissue called ciliated columnar epithelium lines the respiratory tract. The cilia carry unwanted particles, such as bacteria, pollen and dust, out of the system.

Nervous

Neurones and neuroglia are the specialised cells that form the nervous tissue. They enable the body to receive and transmit nerve impulses in order to regulate and co-ordinate body activities.

Endocrine

The endocrine glands are made from epithelial tissue. They secrete hormones directly into the bloodstream to influence the activity of other organs or glands.

Digestive

The digestive system is lined with epithelial tissue. This contains goblet cells which secrete mucus to aid the flow of material through the digestive system.

Urinary

The bladder is lined with transitional epithelium which allows the bladder to expand when full and deflate when empty.

Key words

Cells

Active transport: the process by which molecules move across a cell membrane from a lower to a higher concentration using energy

Atom: the smallest particle of an element

Cell: the basic unit of all living organisms

Cell membrane: a fine membrane that encloses the cell and protects its contents

Cell respiration: the chemical processes that generate most of the energy in the cell

Centrioles: small spherical structures which are associated with cell division

Centromere: the portion of a chromosome where the two chromatids are joined

Centrosome: an area of clear cytoplasm around the nucleus which contains the centrioles

Chromatid: a pair of identical strands of a replicated chromosome that are joined at the centromere and which separate during cell division

Chromatin: the substance inside the nucleus that contains the DNA

Chromosome: the thread-like DNA structure in the cell nucleus that carries the genetic information in the form of genes

Cytokinesis: the cytoplasmic division of a cell at the end of mitosis or meiosis, bringing about the separation into two daughter cells

Cytoplasm: a gel-like substance enclosed by the cell membrane that contains organelles

Diffusion: the process in which small molecules move from areas of high concentration to those of lower concentration

Endoplasmic reticulum: a series of membranes that are continuous with the cell membrane and which allow movement of materials from one part of the cell to another

Filtration: the movement of water and dissolved substances across the cell membrane due to differences in pressure

Golgi body: a collection of flattened sacs within the cytoplasm that is involved in protein production and transport

Homeostasis: the process by which the body maintains a stable internal environment for its cells and tissues

Lysosome: round sacs in the cytoplasm containing powerful enzymes that are capable of digesting proteins

Meiosis: the division of one cell into four genetically different daughter cells

Metabolism: a physiological process in the body that converts energy from food into fuel

Mitochondria: oval-shaped organelles for energy production, located in the cell's cytoplasm

Mitosis: the division of one cell into two genetically identical daughter cells

Molecule: a particle composed of two or more atoms joined together

Nuclear membrane: a double-layered membrane surrounding the nucleus which regulates the materials that enter or exit the nucleus

Nuclear pore: a tiny passage through the nuclear membrane forming the system that controls entry and exit into the nucleus

Nucleolus: a dense spherical structure inside the nucleus, containing RNA

Nucleus: the control centre of the cell that regulates the cell's functions

Organ: a specialised structure made up of different types of tissue

Osmosis: the movement of water through the cell membrane from areas of low chemical concentration to areas of high chemical concentration

pH scale: a chemical rating scale used to measure the acid or alkaline (base) content of a substance

Phagocytosis: the process by which a cell engulfs particles such as bacteria

Pinocytosis: the method by which a cell absorbs small droplets outside the cell and brings them inside

Ribosome: a tiny organelle that may be attached to the endoplasmic reticulum or may be within the cytoplasm; ribosomes manufacture proteins for use within the cell

System: a group of organs that work together to perform specific functions

Tissue: a group of similar cells that perform a particular function

Vacuole: a membrane-bound space for storage within the cytoplasm of the cell

Tissues

Adipose tissue: a type of connective tissue containing fat cells

Areolar tissue: the most widely distributed type of connective tissue in the body, providing strength and elasticity

Blood: a type of liquid connective tissue, containing the blood cells erythrocytes, leucocytes and thrombocytes in a fluid called plasma

Bone: the hardest and most solid of all connective tissues

Cartilage: a tough, elastic, firmer type of fibrous connective tissue contained within joints

Ciliated epithelium: a form of columnar epithelium with a single layer of rectangular cells that contain hair-like projections (cilia) from its surface

Columnar epithelium: a type of simple epithelium with a single layer of tall, cylindrical column cells, each with a nucleus situated towards its base

Compound epithelium: a type of epithelium containing two or more layers of cells

Connective tissue: a type of tissue that connects tissues and organs by binding the various parts of the body together

Cuboidal epithelium: a type of simple epithelium with a single layer of cube-like cells, adapted for secretion and absorption

Epithelial tissue: sheets of cells which cover and protect the external and internal surfaces of the body

Histology: the study of tissues

Homeostasis: the process by which the body maintains a stable internal environment for its cells and tissues

Lymphoid tissue: a type of connective tissue with a semi-solid matrix and fine, branching fibres; contains specialised lymphocytes

Muscular tissue: a type of tissue made up of contractile fibres, usually arranged in bundles and surrounded by connective tissue

Mucous membrane: a type of membrane that lines body cavities and makes up the outer layer of organs

Nervous tissue: a type of tissue that initiates and transmits nerve impulses

Serous membrane: a type of membrane lining body cavities that are not open to the external environment (for example the heart has a serous membrane called the pericardium)

Simple epithelium: a type of tissue with only one layer of cells over a basement membrane

Squamous epithelium: a type of simple epithelium; a single layer of flat, scale-like cells with a central nucleus

Stratified epithelium: a type of compound epithelium composed of a number of layers of cells of different shapes

Synovial membrane: a type of membrane that lines the joint cavities of freely movable joints

Tissue: a group of similar cells that perform a particular function

Transitional epithelium: a type of compound epithelium composed of several layers of pear-shaped cells which change shape when they are stretched

White fibrous tissue: strong, connecting tissue made up of white, collagenous fibres to provide strong attachment between different structures

Yellow elastic tissue: type of connective tissue consisting of branching yellow elastic fibres to allow the stretching of various organs, followed by a return to the original size and shape

Revision summary

Cells

- The human body has five levels of structural organisation – atoms and molecules, cells, tissues, organs and systems.
- **Atoms and molecules** are the lowest level of organisational complexity in the body.
- **Cells** are the smallest units that show characteristics of life.
- A **tissue** is a group of similar cells that perform a certain function.
- An **organ** consists of tissues grouped into a structurally and functionally integrated unit.
- A **system** is a group of organs that work together to perform a specific function.
- **Homeostasis** is the process by which the body maintains a stable internal environment for its cells and tissues.
- Homeostatic mechanisms in the body include the regulation of body temperature, blood pressure, blood sugar level and pH.
- The term **metabolism** is used to describe the physiological processes that take place in our bodies to convert the food we eat and the air we breathe into the energy our cells need to function.
- The minimum energy required to keep the body alive is known as the **basal metabolic rate**.
- Major elements and compounds are involved in the body's make-up (e.g. **oxygen**, **carbon**, **hydrogen**, **nitrogen**, **calcium** and **phosphorus**).
- Chemically, a cell is composed of the major elements carbon, oxygen, hydrogen and nitrogen.
- Cells are made up of approximately: 80% water, 15% protein, 3% lipids or fats, 1% carbohydrates, 1% nucleic acids.
- A **cell** is the basic, living, structural and functional unit of the body.
- The principal parts of the cell are the **cell membrane** and its **organelles**, which play specific roles in cellular growth, maintenance, repair and control.
  - The **cell membrane** encloses the cell and protects its contents. It is semipermeable and governs the exchange of nutrients and waste materials.
  - The **nucleus** controls the cell's activities and contains the genetic information.

- The **cytoplasm** is the substance inside the cell between the cell membrane and the nucleus.
- The **ribosomes** are sites of protein synthesis.
- The **endoplasmic reticulum** links the cell membrane with the nuclear membrane and assists movement of materials out of the cell.
- The **Golgi body** processes, sorts and delivers proteins and lipids (fats) to the cell membrane, via lysosomes and secretory vesicles.
- The **lysosome** is a round sac in the cytoplasm that contains powerful enzymes to help destroy waste and worn-out cell materials.
- The **mitochondria** are the 'powerhouses' of the cell where respiration takes place to produce energy.
- The **centrosome** is a dense area of cytoplasm, containing the centrioles.
- The **centrioles** are paired small spherical structures associated with mitosis (cell division).
- The **chromatids** are a pair of identical strands of a chromosome that are joined at the centromere and separate during cell division.
- The **centromere** is the portion of a chromosome where the two chromatids are joined.

- Functions of cells include **respiration**, **growth**, **excretion**, **movement**, **sensitivity** and **reproduction**.
- Cell division is the process by which cells reproduce themselves.
- **Mitosis** is cell division that results in an increase in body cell numbers and involves division of nuclei.
- **Meiosis** is reproductive cell division and results in the formation of eggs and sperm.
- Cells function through the exchange of fluids, nutrients, chemicals and ions which are carried out by passive processes such as **diffusion**, **osmosis** and **filtration**, and active processes such as **active transport**.
- **Cell respiration** is the use of nutrients such as oxygen and glucose, and output of waste such as carbon dioxide, by the cell to produce the energy needed for the cell to function.
- The fuel required by cells is provided by **glucose** from carbohydrate metabolism and **oxygen** absorbed from the respiratory system into the bloodstream.
- Cells are bathed in a fluid known as **tissue fluid**, or interstitial fluid, which allows the interchange of substances between the cells and the blood, known as **internal respiration**.

Tissues

- A **tissue** is a group of similar cells that are specialised for a particular function.
- The tissues of the body are classified into four main types: **epithelial**, **connective**, **muscular** and **nervous**.
- **Epithelial tissue** covers and lines many organs and vessels.
  - There are two categories of epithelial tissue: **simple** (single layer) and **compound** (multi-layer).
  - There are four different types of simple epithelium: **squamous**, **cuboidal**, **columnar** and **ciliated**.
  - There are two different types of compound epithelium: **stratified** and **transitional**.
- **Connective tissue** is the most abundant type of body tissue. It connects tissues and organs to give protection and support.
- Connective tissue consists of the following different types: **areolar**, **adipose**, **white fibrous**, **yellow elastic**, **lymphoid**, **blood**, **bone** and **cartilage**.
- **Muscle tissue** is elastic and is modified for contraction. It is found attached to bones (skeletal muscle), in the wall of the heart (cardiac muscle) and in the walls of the stomach, intestines, bladder, uterus and blood vessels (smooth muscle).
- **Nervous tissue** is composed of nerve cells called neurones, which pick up and transmit nerve signals.
- **Membranes** are thin, soft, sheet-like layers of tissue.
- **Mucous membranes** line cavities that open to the exterior, such as the digestive tract.
- **Serous membranes** line body cavities that are not open to the external environment (such as the lungs and the heart).
- **Synovial membranes** line joint cavities of freely movable joints such as the shoulder, hip and knee.

Test your knowledge questions

Multiple choice questions

1 What is the name of the process by which the body maintains a stable internal environment of its cells and tissues?
   a physiology
   b metabolism
   c homeostasis
   d anatomy

2 A system is a
   a group of similar cells that perform a particular function
   b group of organs that work together to perform specific functions
   c a specialised structure made up of different types of tissues
   d a specialised structure made up of different types of cells.

3 What is the simplest form of life that can exist as independent self-sustaining units?
   a tissues
   b cells
   c atoms
   d organs

4 Which is the correct order of the five levels of organisation in the human body, from the lowest to the highest?
   a cells, atoms and molecules, organs, tissues, systems
   b tissues, cells, atoms and molecules, organs, systems
   c atoms and molecules, cells, tissues, organs, systems
   d atoms and molecules, tissues, cells, organs, systems.

5 Where is non-keratinised stratified epithelium found?
   a bladder
   b intestines
   c lining of mouth
   d hair

6 Where would you find squamous epithelium?
   a in the lungs
   b in the brain
   c in the kidneys
   d in the ovaries

7 Where would you find ciliated epithelium?
   a in the kidney tubules
   b in the eyes
   c in the respiratory system
   d in the pancreas

8 Where would you find columnar epithelium?
   a in the small and large intestine
   b in the stomach
   c in the gall bladder
   d all of the above

9 Which is the most widely distributed type of connective tissue in the body?
   a areolar tissue
   b adipose tissue
   c epithelial tissue
   d white fibrous tissue

10 The control centre of the cell that directs nearly all metabolic activities is the:
   a mitochondria
   b Golgi body
   c nucleus
   d cell membrane.

Exam-style questions

11 List the five levels of organisation of the body.
                                                    5 marks

12 List three functions of cells.   3 marks

13 State three functions of the cell membrane.
                                                    2 marks

14 Which part of the cell contains the genetic materials for replication?   1 mark

15 List the four major types of tissues in the human body.   4 marks

16 List the four different types of simple epithelium.   4 marks

3 The skin, hair and nails

Introduction

The skin is the largest of the body's organs, making up approximately 16% of a person's body weight and with a surface area of 1.8 m². It is a dynamic organ in a constant state of change, as cells from the outer layers are continuously shed and replaced by inner cells moving up to the surface. Like a cell membrane, the skin provides a barrier between the external environment and our internal organs. Located within its layers are several types of tissues that carry out specific functions, such as protection, temperature regulation and excretion.

A change in the surface appearance and texture of the skin may be the first sign of a lack of balance between skin and its related systems.

The nail is an appendage (an attachment or a projecting part) of the skin and is a modification of the stratum corneum (horny) and stratum lucidum (clear) layers of the epidermis. Nails are non-living tissue. Their two main functions are protection for the fingers and toes and as tools for the manipulation of objects.

Hair is also an appendage of the skin and grows from a sac-like depression in the epidermis called a hair follicle. The primary function of hair is protection.

OBJECTIVES

By the end of this chapter you will understand:

- the structure and functions of the skin
- how the skin repairs itself
- the characteristics of the different skin types and skin conditions
- the structure and functions of hair
- the structure and functions of nails
- common pathologies associated with the skin, hair and nails
- the interrelationships between the skin and other body systems.

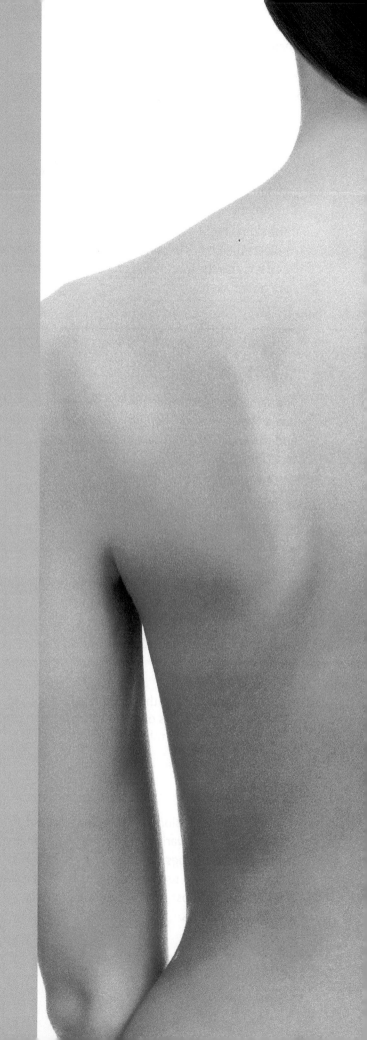

The functions of the skin

The skin is so much more than an external covering. It is a highly sensitive boundary between our bodies and the environment. The skin has several important functions:

- sensitivity
- heat regulation
- absorption
- protection
- excretion
- storage
- vitamin D production
- melanin formation.

Study tip

It is helpful to think of the mnemonic *SHAPES VM* (from the initial letters of each of the words in the list) to help you remember the functions of the skin.

Sensitivity

The skin is an extension of the nervous system. It is very sensitive to various stimuli due to its many sensory nerve endings, which can detect changes in temperature and pressure and which can register pain. The skin is most sensitive on the fingertips, the tongue and the lips.

Heat regulation

The skin helps to regulate body temperature.

- When the body is losing too much heat:
  - The blood capillaries near the skin surface contract, to keep warm blood away from the surface of the skin and closer to major organs.
  - The erector/arrector pili muscles raise the hairs and trap air next to the skin so that heat is retained.
  - The adipose tissue in the dermis and the subcutaneous layer helps to insulate the body against heat loss.
- When the body is too warm:
  - The blood capillaries dilate to allow warm blood to flow near to the surface of the skin, so heat can be lost in order to cool the body.
  - The evaporation of sweat from the surface of the skin also assists in cooling the body.

Absorption

The skin has a limited ability to absorb materials. Substances that can be absorbed by the epidermis include fat-soluble molecules such as oxygen, carbon dioxide, fat-soluble vitamin A and steroids, along with small amounts of water.

KEY FACT

The skin is capable of absorbing small particles of substances such as essential oils due to the fact that they contain fat-soluble and water-soluble components.

Protection

The skin acts as a protective organ:

- The film of sebum and sweat on the surface of the skin, known as the **acid mantle**, is an antibacterial agent that acts as a barrier, preventing the multiplication of micro-organisms on the skin.
- The fat cells in the subcutaneous layer of the skin provide padding to help protect bones and major organs from injury.

- Melanin, which is produced in the stratum germinativum (basal cell layer) of the skin, helps to protect the body from the harmful effects of ultraviolet radiation.
- The cells in the horny layer of the skin overlap like scales to stop micro-organisms from penetrating the skin and to prevent excessive water loss from the body.

> ### In practice
> The mantle of the skin is acidic and varies in pH between 4.5 and 6.2 on the pH scale. It is important for therapists to consider variations in skin pH, as treatment products with either a high or low pH may be harmful to the skin and may disrupt the barrier function, resulting in irritation.

Excretion

The skin functions as part of the excretory system, eliminating waste through perspiration. The eccrine glands of the skin produce sweat, which helps to remove some waste materials from the skin such as urea, uric acid, ammonia and lactic acid.

Storage

The skin also acts as a storage depot for fat and water. About 15% of the body's fluids are stored in the subcutaneous layer.

Vitamin D production

The skin synthesises vitamin D when exposed to ultraviolet light. Modified cholesterol molecules in the skin are converted by the ultraviolet rays in sunlight to vitamin D. This is then absorbed by the body for the maintenance of bones and to facilitate the absorption of calcium and phosphorus from the diet.

Melanin formation

In the human skin, melanin formation is initiated by exposure to ultraviolet light and helps to prevent damage caused by this form of radiation. Melanin forms in specialised cells called melanocytes. Melanocytes are especially abundant in the stratum germinativum (basal layer) of the epidermis and underlying dermis and they are responsible for the pigmentation of the skin.

The structure of the skin

Before looking at the structure of the skin, let's consider a few facts:

- The skin is a very large organ covering the whole body (making up approximately 16% of your total body weight).
- It varies in thickness on different parts of the body. It is thinnest on the lips and eyelids, where it must be light and flexible, and thickest and roughest on the soles of the feet and palms of the hands, where friction is needed for walking and gripping.
- The epidermis varies in thickness from 0.1 mm on the eyelids to more than 1 mm on the palms and soles of the feet.
- As the skin is the external covering of the body, it is easily irritated and damaged, and certain symptoms of disease and disorders may be observed.
- Each client's skin varies in unique combinations of colour, texture and sensitivity.

> ### In practice
> The appearance of the skin reflects a client's physiology. Observation of a client's skin can give an indication of their nutritional and immune statuses, circulation, age, and genetics, as well as their exposure to environmental factors. These all play a significant role in the skin's colour, condition and tone.

There are two main layers of skin:

1 the **epidermis**, which is the **outer**, thinner layer
2 the **dermis**, which is the **inner**, thicker layer.

Below the dermis is **the subcutaneous layer,** which attaches to underlying organs and tissues.

Although the skin is technically a single organ, these two main layers, the epidermis and the dermis, do have different structures and functions.

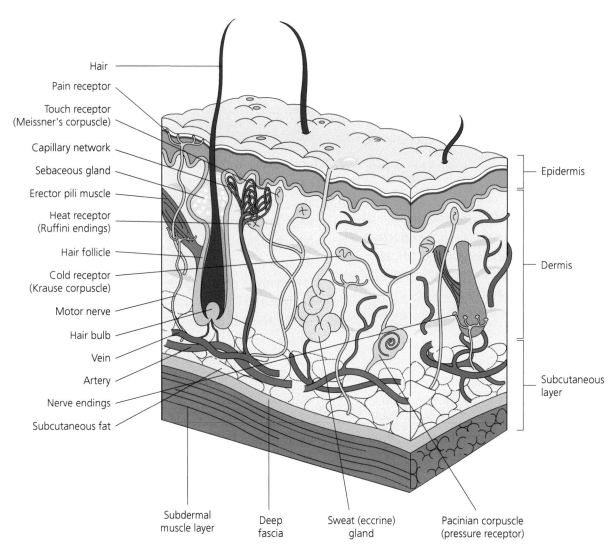

Hair
Pain receptor
Touch receptor (Meissner's corpuscle)
Capillary network
Sebaceous gland
Erector pili muscle
Heat receptor (Ruffini endings)
Hair follicle
Cold receptor (Krause corpuscle)
Motor nerve
Hair bulb
Vein
Artery
Nerve endings
Subcutaneous fat

Epidermis
Dermis
Subcutaneous layer

Subdermal muscle layer
Deep fascia
Sweat (eccrine) gland
Pacinian corpuscle (pressure receptor)

▲ The structure of the skin

1 The epidermis

The epidermis is the most superficial layer of the skin and consists of five layers of cells:

- the stratum germinativum (basal cell layer) – the **innermost layer**
- the stratum spinosum (prickle cell layer)
- the stratum granulosum (granular layer)
- the stratum lucidum (clear layer)
- the stratum corneum (horny layer) – the **outermost layer**.

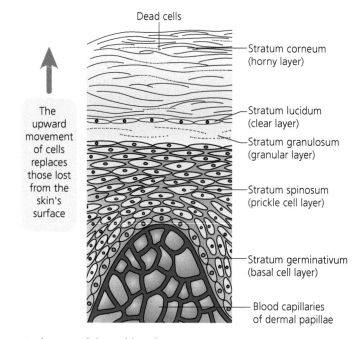

Dead cells
Stratum corneum (horny layer)
Stratum lucidum (clear layer)
Stratum granulosum (granular layer)
Stratum spinosum (prickle cell layer)
Stratum germinativum (basal cell layer)
Blood capillaries of dermal papillae

The upward movement of cells replaces those lost from the skin's surface

▲ Layers of the epidermis

The thickness of the epidermis varies in different types of skin and on different parts of the body. The average thickness of the epidermis is 0.1 mm, which is about the same thickness as one sheet of paper. The epidermis holds a large amount of water, with the basal cell layer being approximately 80% water and the horny layer being approximately 15% water. Water is held in the spaces between the cells, and the inner layers contain more water than the outer layers. The skin's capacity to retain water decreases with age, making the skin more vulnerable to dehydration and wrinkles.

Functions of the epidermis

The epidermis has three primary functions:

1 protecting the body from the external environment, particularly from sunlight
2 preventing excessive water loss from the body
3 protecting the body from infection.

Stratum germinativum

The stratum germinativum or **basal cell layer** is the deepest of the five layers, at the base of the epidermis. Also known as the Malpighian layer, it consists of a single layer of column cells on a basement membrane which separates the epidermis from the dermis (the dermal–epidermal junction). This layer of the epidermis is concerned with cellular regeneration. The basal cells within this layer are constantly producing new cells by division. As new cells are formed, they push adjacent cells towards the skin's surface.

Approximately 95% of the cells within the epidermis are keratinocytes, which deposit a protein called keratin. Keratinocyte stem cells reside at the bottom of the rete ridges at the base of the basal cell layer. Keratin makes the epidermal cells more resilient and protective as they are pushed towards the skin's surface.

At intervals between the column cells, which divide to reproduce, are large star-shaped cells called melanocytes. These form the pigment melanin, the skin's main colouring agent. This layer also contains tactile (Merkel) discs that are sensitive to touch.

Stratum spinosum

The stratum spinosum or **prickle cell layer** is the thickest layer of the epidermis. It is known as the prickle cell layer because each of the rounded cells contained within it has short projections that make contact with the neighbouring cells, giving them a prickly appearance. The tiny, hair-like structures on the prickle cells eventually become desmosomes, which are small disc-shape attachments that provide strength and integrity by holding the upper level of epidermal cells together.

The cells of this layer are living and are therefore capable of dividing by the process of mitosis.

The cells of this layer make special fats called sphingolipids. When these cells reach the top layer (stratum corneum) these lipids play an important role in the retention of moisture by the skin.

It is in the stratum spinosum that the keratinocyte has a major role to play in the skin's barrier defence. Keratinisation refers to the process that skin cells undergo when they change from living cells with a nucleus to dead cells without a nucleus. The keratinisation process begins in the upper cells of this layer as the cells migrate upwards.

The stratum spinosum is also responsible for cellular transport of melanosomes (organelles containing melanin). In this layer and in response to sunlight, melanocytes synthesise melanosomes full of melanin and transfer them to neighbouring keratinocytes, providing protection against harmful ultraviolet rays. The melanin becomes part of the keratinocyte, which then continues its journey upwards to the horny layer and onwards to desquamation (shedding of dead skin cells).

This layer also includes Langerhans cells, which set up an immune response to foreign bodies. These special defence cells are spread out among the keratinocytes and their role is to attack invading foreign bodies that have found their way into the body through the skin, then transporting them to be neutralised by specialised white blood cells.

Stratum granulosum

The stratum granulosum or **granular layer** consists of distinctly shaped cells that resemble granules. They are filled with keratin and produce intercellular lipids (the substances that fill the spaces between the upper epidermal cells) from structures called lamellar bodies.

These lipids form a strong cement-like substance that prevents the absorption of harmful substances by the skin and helps maintain hydration of the lower layers. This layer has an appearance of a wall of bricks (the cells) and mortar (lipids).

As the cells move into the stratum granulosum, a number of changes occur to the keratinocyte. The keratinocyte becomes less flexible and more granular in appearance and the keratin within the cell hardens completely, thereby completing the keratinisation process. The granular hardened structure of the keratinocyte helps prevent absorption of many substances by the skin and assists maintaining hydration of lower layers.

As the cells move further up into the granular layer, further changes occur in the keratinocytes. The desmosomes begin to dissolve, preparing the corneocyte (dead skin cell) for desquamation.

In practice

The intercellular lipids between the epidermal cells are responsible for hydration, epidermal firmness and smoothness. They protect against transepidermal water loss (TEWL), which can result in dehydration. They also provide protection against substances that may damage the skin.

Ceramides belong to this group of intercellular lipids, and they are also contained within the ingredients of some skin care products.

It is important to recognise that the cell renewal process is responsible for the production of these essential lipids. If the cell renewal process slows down, the production of lipids also reduces, resulting in dryness and dehydration.

Stratum lucidum

The stratum lucidum or **clear layer** consists of transparent cells that let light through. The cells in this clear layer are filled with eleidin, a substance that is produced from keratohyalin and which is involved in the keratinisation process.

This layer is an important transitional stage in the development of the top layer of the epidermis (the stratum corneum, see below). Keratinisation is complete by the time the keratinocyte reaches this layer, and the cell is almost ready to become a flexible, mature corneocyte.

Cells in this clear layer release lipids from the bilayers (glue that holds the corneocytes together). These lipids play an important role in skin barrier defence by slowing down transdermal water flow and maintaining hydration of epidermis.

The stratum lucidum is very shallow in facial skin, but thick on the soles of the feet and the palms of the hands, and is generally absent in hairy skin.

Stratum corneum

The stratum corneum or **horny layer** is the outermost skin layer, consisting of dead, flattened keratinocytes, which are now known as corneocytes. The cells of this layer form a waterproof covering for the skin and help to prevent the penetration of bacteria.

The acid mantle rests on this outer layer and is the skin's first line of defence. The acid mantle is a natural hydrolipidic film which covers the entire surface of the skin and is formed by substances secreted by the sweat and sebaceous glands, epidermal lipids and substances known as the natural moisturising factor.

The stratum corneum is the result of the changes that occur when new living cells are produced in the basal layer and are pushed upwards by newer cells until they reach the surface, where they die, dry out and are sloughed off.

The outer layer of dead cells is continually being shed; this process is known as desquamation.

In practice

Knowledge of the stratum corneum layer of the epidermis is crucial to understanding skin problems, as it is the part of the skin that is directly affected by the external environment.

It also plays a key role in helping to retain moisture in the rest of the skin, and in regulating the natural moisture flow from the deeper layers to the skin surface where it is lost by evaporation. This natural moisture flow is known as TEWL.

Without adequate retained moisture, skin can become dry and unhealthy.

Under normal conditions, up to 15% of the horny layer consists of water, which is vital to enable the stratum corneum to work effectively. The natural functions of the skin are not as efficient when the horny layer contains less than 10% of water, and it becomes dry.

The process of skin renewal

The process of renewal of the epidermal cells is continuous throughout life and occurs in four stages:

- **Stage 1 – cell regeneration** and **mitosis** in the deepest layer of the epidermis (stratum germinativum)
- **Stage 2 – cell maturation** and migration from the stratum germinativum to the stratum spinosum
- **Stage 3 – keratinisation** in which the cells undergo change and die in the stratum granulosum
- **Stage 4 – desquamation** when dead cells are sloughed off.

In normal skin, it takes approximately 28–30 days for a cell produced by the stratum germinativum to move through the epidermis to the surface. The rate of regeneration is partly determined by the rate at which the outer layer is being desquamated.

With age, the speed of this process is greatly reduced. By the age of 50, it may take in the region of 37 days to complete the same process.

Cell regeneration through the epidermal layers

Keratinocytes are the predominant cell type of the epidermis. In their upward journey, they undergo a series of chemical changes, transforming from soft cells into flat scales that are constantly rubbed off.

As keratinocytes move up the layers of the skin, they go through several changes forming many other skin components, all of which play an important role in the skin's barrier defence. It is important to remember that healthy skin can only be formed by healthy keratinocytes. A balance of all the essential skin substances (water and lipids, for example) is needed in order to ensure that the health of the keratinocytes and skin is not impaired.

> **In practice**
>
> When the cells of the horny layer are lost quickly (for instance due to skin injury or sunburn), the renewal process speeds up as the cells are replaced more quickly from below.
>
> Removing the outer layers of the skin with a chemical peel also speeds up replacement.

Table 3.1 Functions of epidermal layers in cell regeneration

Epidermal layer	Function	Significance
Stratum germinativum (innermost layer)	Cellular regeneration	Keratinocytes begin their life cycle in this layer As new cells are formed by division, they push adjacent cells towards the skin's surface Formation of the pigment melanin begins here, which helps protect skin against harmful ultraviolet
Stratum spinosum	Cellular transport of the melanosome	Keratinisation begins here In this layer and in response to sunlight, melanocytes synthesise melanosomes and melanin, transferring them to neighbouring keratinocytes to provide protection against harmful ultraviolet rays The melanin becomes part of the keratinocyte, which then continues its journey upwards towards the horny layer and eventual desquamation
Stratum granulosum	Keratinisation, cellular change and lipid formation	As cells move into the granular layer a number of changes occur in the keratinocytes Keratinisation becomes complete – the granular hardened structure of the keratinocyte helps prevent absorption of many substances by the skin and assists in maintaining hydration of lower layers

Epidermal layer	Function	Significance
Stratum lucidum	Lipid release	An important transitional stage in the development of the horny layer, cells in this layer release lipids from the bilayers (glue that holds the corneocytes together), maintaining hydration of the epidermis Keratinisation is complete by the time cells have reached the clear layer
Stratum corneum (outermost layer)	Acting as skin barrier defence	The keratinocyte reaches its final destination, where it is now a corneocyte, acting as first line of defence against injury, invasion of bacteria and as a waterproof covering

Stem cells

Everyday activities cause wear and tear of skin. The epidermis and skin appendages must be renewed constantly to keep the skin in good condition. If you cut or damage your skin, it has to repair efficiently in order to protect your body from the outside environment.

Skin stem cells make all this possible; they are responsible for constant renewal (regeneration) of your skin and for healing wounds. Stem cells are found in the stratum germinativum (basal cell layer) of the epidermis. They are active during skin renewal, which occurs throughout life, and in skin repair after injury.

There are several different types of skin stem cells:

- **Epidermal stem cells** – responsible for everyday regeneration of the different layers of the epidermis. These stem cells are found in the stratum germinativum (basal layer) of the epidermis.
- **Hair follicle stem cells** – ensure constant renewal of the hair follicles. They can also regenerate the epidermis and sebaceous glands if those tissues are damaged. Hair follicle stem cells are found throughout the hair follicles.
- **Melanocyte stem cells** – responsible for regeneration of melanocytes, a type of pigment cell. Melanocytes produce the pigment melanin, and play an important role in skin and hair follicle pigmentation.

2 The dermis

The dermis lies below the epidermis and is the deeper thicker layer of the skin. It can be as much as 3 mm thick.

The dermis contains several types of tissue that provide a supporting framework to the skin, as well as blood vessels, nerves, hair roots, sweat and sebaceous glands. It is where the structural integrity and density of the skin is determined by the presence of collagen, elastin and the extracellular matrix.

The functions of the dermis

The functions of the dermis include:

- providing nourishment to the epidermis
- removing waste products from the epidermis
- giving a supporting framework to the tissues and holding all its structures together
- contributing to skin colour.

The dermis has two layers: a superficial papillary layer and a deeper reticular layer.

The papillary layer

The superficial papillary layer is made up of fatty connective tissue and is connected to the underside of the epidermis by cone-shaped projections called dermal papillae. These contain nerve endings and a network of blood and lymphatic capillaries.

The fine network of capillaries in this layer brings oxygen and nutrients to the skin, and carries waste away. The many dermal papillae of the papillary layer form indentations in the overlying epidermis, giving it an irregular or ridged appearance. It is these ridges that leave fingerprints on objects that are handled.

> **KEY FACT**
> The key function of the papillary layer of the dermis is to provide vital nourishment to the living layers of the epidermis above.

The reticular layer

The deeper reticular layer is formed of tough fibrous connective tissue, which gives the skin strength and elasticity and helps to hold all structures in place.

The protein collagen, which accounts for about 75% of the weight of the dermis, is organised in bundles running horizontally throughout the dermis and is buried in a jelly-like material called the ground substance. Collagen is responsible for giving the skin resilience and elasticity.

The collagen bundles are held together by elastic fibres running through the dermis. These are made of a protein called elastin that makes up less than 5% of the weight of the skin. Elastin contributes to the elasticity of the skin by holding the collagen fibres together. Both collagen and elastin fibres are made by cells called fibroblasts, which are located throughout the dermis.

Hyaluronic acid is an important substance which forms part of the tissue that surrounds the collagen and elastin fibres. This special substance has the ability to attract and bind hundreds of times its weight in water. In this way, it acts as a natural moisturising ingredient that plumps the skin's tissues. Glycoproteins, found in the ground substance of the dermis, are also capable of holding large amounts of water.

KEY FACT

Damage to collagen and elastin fibres as they break down is the primary cause of skin ageing and the appearance of wrinkles. Also, the amount of hyaluronic acid and glycoprotein produced in the skin reduces with age. Hence the skin becomes less resilient and loses elasticity.

In addition to fibroblasts, other cells present in the dermis include:

- **mast** cells, which secrete histamine (associated with allergies) causing dilation of blood vessels to bring blood to the area
- **phagocytic** cells (macrophages), a type of white blood cell, which are able to travel around the dermis destroying foreign matter and bacteria.

The extracellular matrix (ECM)

The dermis layer of skin has three crucial components: collagen, elastin and glycosaminoglycans (GAGs). These form the bulk of an important support system called the extracellular matrix (ECM). This matrix, consisting of structural proteins (collagen and elastin), glycosaminoglycans and proteoglycans, is often referred to as the 'ground substance'.

Every skin layer depends on the ECM; if one part fails to function correctly there will be a knock-on effect to all other parts of the skin system.

In both the papillary and reticular layers of the dermis, fibroblasts and fibrocytes (immature fibroblasts) are responsible for laying down and maintaining the ECM and, hence, determining the structure of the dermis.

The ECM gives the dermis shape, structure and support, providing the structural scaffolding and maintaining the tissue architecture. The ECM is also important in cell-to-cell signalling, wound repair, cell adhesion and tissue function.

Glycosaminoglycans

Glycosaminoglycans (GAGS) are polysaccharides of repeating amino sugars (a sugar linked with a protein). These water-binding molecules, found in the dermis, give skin its plumpness. Along with water, GAGs create a fluid that fills the space between the collagen and elastin fibres in the dermis, giving it turgidity (swelling due to high fluid content).

There are various GAGs in the dermis. The most common ones are: hyaluronic acid, chondroitin sulfate, keratin sulfate, dermatan sulfate, heparin sulfate, and heparin.

KEY FACT

GAGs are water-binding substances. The GAGs in the ground substance of the dermis attract water (brought to the dermis by blood vessels), which diffuses to the lower layers of the epidermis and eventually migrates upward through the epidermal layers.

In practice

In skin care products, water-binding ingredients are known as humectants. They help keep skin moist by attracting water from the atmosphere and the lower layers of the epidermis. This counteracts the continuous loss of moisture from the epidermis as water evaporates into the air. TEWL amounts to several ounces a day.

Skin also loses water through sweating (from the sweat pores). Without constant rehydration, skin dehydrates and wrinkles more easily.

Hyaluronic acid is a glycosaminoglycan that exists naturally in the dermis layer of skin. It can hold 1000 times its molecular weight in water, which is why it is such a popular hydrating and saturating ingredient in skin care products.

Blood supply

Unlike the epidermis, the dermis has an abundant supply of blood vessels which run through the dermis and the subcutaneous layer.

Arteries and arterioles carry oxygenated blood to the skin and these enter the dermis from below, branching into a network of capillaries around active or growing structures. These capillary networks are found in the dermal papillae, where they provide the basal cell layer of the epidermis with nutrients and oxygen. They also surround the sweat glands and erector pili muscles, two appendages of the skin.

The capillary networks drain into venules, small veins which carry the deoxygenated blood away from the skin and remove waste products.

The dermis is, therefore, well supplied with capillary blood vessels to bring nutrients and oxygen to the germinating cells in the basal cell layer of the epidermis and to remove their waste products.

KEY FACT

The network of capillaries in the face and neck is much denser than in the torso or the limbs. As a result, the face and neck are the first body parts to reflect alterations in blood flow by changing colour.

Lymphatic vessels

There are numerous lymphatic vessels in the dermis. They form a network through the dermis, facilitating the removal of waste from the skin's tissues. The lymphatic vessels in the skin generally follow the course of the veins and are found around the dermal papillae, glands and hair follicles.

Nerves

Nerves are widely distributed throughout the dermis. Most nerves in the skin are **sensory**, meaning they are sensitive to heat, cold, pain, pressure and touch and send such information to the brain. Branched nerve endings, which lie in the papillary layer and hair roots, respond to touch and temperature changes. Nerve endings in the dermal papillae are sensitive to gentle pressure and those in the reticular layer are responsive to deep pressure.

Sensory nerves

There are several different types of sensory nerve endings in the skin that are responsible for sensing touch, pressure, vibration, temperature and pain.

The sensory nerve endings are also called cutaneous receptors, because they receive information about the skin.

The different types of cutaneous receptors include:

- **mechanoreceptors** – detect sensations such as pressure, vibrations and texture
- **thermoreceptors** – detect sensations of heat or cold
- **pain receptors (nociceptors)** – detect discomfort or injury.

These receptors have overlapping roles, as seen in Table 3.2.

Table 3.2 Summary of cutaneous receptors

Category of cutaneous receptor	Type	Sensation
Mechanoreceptors	Merkel's disks	Sustained touch and pressure
	Meissner's corpuscles	Changes in texture, slow vibrations
	Ruffini's corpuscles	Stretching of the skin
	Pacinian corpuscles	Deep pressure, fast vibrations
Thermoreceptors	Ruffini's corpuscles	Detect hot temperatures
	Krause corpuscles	Detect cold temperatures
Pain receptors (nociceptors)	Free nerve endings	Detect pain that is caused by mechanical stimuli (cuts), thermal stimuli (burns), or chemical stimuli (poison from insect stings)

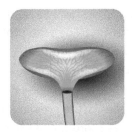
RUFFINI ENDINGS

MERKEL DISKS

PACINIAN CORPUSCLE

MEISSNER CORPUSCLE

FREE NERVE ENDINGS

KRAUSE END BULBS

ROOT HAIR PLEXUS

▲ Sensory receptors in skin

KEY FACT: MECHANORECEPTORS

Merkel's disks and Meissner's corpuscles are generally found in non-hairy skin such as on the palms, lips, tongue, soles of feet, fingertips, eyelids and the face. Merkel's disks are slowly adapting receptors, while Meissner's corpuscles are rapidly adapting receptors. This is so the skin can detect the moment you touch something and whether you continue holding it.

The ridges that make up your fingerprints are full of sensitive mechanoreceptors which provide your brain with an enormous amount of information about the texture of objects.

KEY FACT: THERMORECEPTORS

Thermoreceptors are found throughout the skin and are able to detect sensations of heat and cold. The number and location of thermoreceptors determines the degree of sensitivity of the skin to temperature change.

The highest concentration of thermoreceptors can be found in the face and ears; hence why your nose and ears get cold faster than the rest of your body.

KEY FACT: PAIN RECEPTORS

There are over three million pain receptors throughout the body, found in skin, muscles, bones, blood vessels and some organs. This type of receptor is able to detect pain caused by mechanical (injury), chemical (exposure to strong chemicals) or thermal (touching a hot item) stimuli.

Motor nerve endings

The dermis also has motor nerve endings that relay impulses from the brain and that are responsible for the dilation and constriction of blood vessels, the secretion of perspiration from the sweat glands and the contraction of the erector pili muscles attached to hair follicles.

The subcutaneous layer

The subcutaneous layer, or **hypodermis**, is a thick layer of connective tissue found below the dermis. The type of tissue (fatty tissue) found in this layer helps support delicate structures such as blood vessels and nerve endings. It also cushions the dermis from underlying tissues such as muscles and bones.

The subcutaneous layer contains the same collagen and elastin fibres as the dermis and also the major arteries and veins which supply the skin by forming a network throughout the dermis. The fat cells contained within this layer help to insulate the body by reducing heat loss.

The subdermal muscle layer lies below the subcutaneous layer.

KEY FACT

As we grow older, the amount of fat starts to decrease in the subcutaneous layer and eventually results in a bonier look to the facial contours.

How the skin repairs itself

The layers of normal skin form a protective barrier against the external environment. Once the skin is broken, the process of wound repair starts. The skin healing process involves four specific sequential, yet overlapping, phases.

- **Phase 1 – haemostasis**
  - Occurs within the first few minutes of an injury.
  - Platelets (thrombocytes) join together at the injury site to form a fibrin clot, which reduces active bleeding (haemostasis).
  - This helps stop the bleeding and creates a temporary barrier that prevents pathogens from getting into the open wound.

- **Phase 2 – inflammation**
  - From 0 to 3 days after injury.
  - The inflammatory phase is both a defence mechanism and a crucial component of the healing process in which the wound is cleaned and rebuilding begins.
  - The tissues are red, inflamed, swollen and tender. During this phase, bacteria and cell debris are removed from the wound by white blood cells, and the wound is cleansed by breakdown of the damaged tissue.

- Once the clot starts to harden and dry out, a scab is formed. The scab protects the wound, giving the skin underneath a chance to repair during the proliferation phase.
- Removing scabs too early disrupts the newly regenerated tissue growing underneath and causes more skin damage, resulting in a larger scar.

- **Phase 3 – proliferation**
  - From 1 to 24 days after injury.
  - This phase begins with the skin cells laying down a foundation for long-term wound repair.
  - Collagen is deposited by fibroblasts to strengthen the wound and new cells migrate across the foundation tissue to close the wound.
  - Connective skin tissue forms to replace the damaged skin (re-epithelialisation).
  - If a scab is present, it will loosen and fall off.
  - In this phase, the wound contracts and reduces in size. A large wound can become 40% to 80% smaller after contraction.

- **Phase 4 – maturation**
  - From 14 to 365 days after injury.
  - In this phase, the tissues are remodelled.
  - The dermis and epidermis connect and contract to close the wound.
  - Maturation is the final phase when scar tissue is formed.
  - A stronger type of collagen is laid down to replace the initial collagen that is now degrading. These new collagen fibres are aligned in one direction, as opposed to the random basket-weave formation found in normal tissue.
  - The unusual composition of scar tissue gives scars a different appearance, texture and flexibility to normal skin.

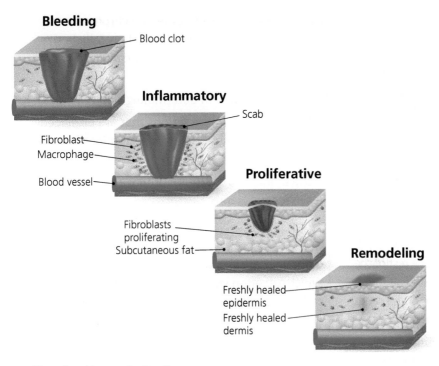

Bleeding
— Blood clot

Inflammatory
— Scab
Fibroblast
Macrophage
Blood vessel—

Proliferative
Fibroblasts
proliferating
Subcutaneous fat—

Remodeling
Freshly healed—
epidermis
Freshly healed—
dermis

▲ How the skin repairs itself

Appendages of the skin

The appendages are accessory structures that lie in the dermis of the skin and project onto the surface through the epidermis. These include the hair, erector pili muscle, sweat and sebaceous glands and the nail.

Hair

Hair is an important appendage of the skin which grows from a sac-like depression in the epidermis called a hair follicle. Hair growth, a sexual characteristic, occurs all over the body, with the exception of the palms of the hands and the soles of the feet.

One of the primary functions of hair is physical protection. The eyelashes act as a line of defence by preventing the entry of foreign particles into the eyes and helping shade the eyes from the sun's rays. Eyebrow hairs help to divert sweat away from the eyes. Hairs lining the ears and the nose trap dust and help to prevent bacteria from entering the body. Body hair acts as a protective barrier against the sun and helps to protect us against the cold with the assistance of the erector pili muscle.

Another function of hair is to prevent friction. Underarm and pubic hair cushions against friction caused by movement. The structure and functions of hair will be covered in more detail later in this chapter.

Erector (arrector) pili muscle

This is a small and weak smooth muscle. It is attached at an angle to the base of a hair follicle and contracts to make the hair stand erect in response to cold, or when experiencing emotions such as fright and anxiety.

Sweat glands

There are two types of sweat glands in the skin: **eccrine** and **apocrine**. The majority are eccrine glands which are simple coiled tubular glands that open directly onto the surface of the skin. There are several million of them distributed over the skin's surface, although they are most numerous in the palms of the hands and the soles of the feet.

The function of eccrine glands is to regulate body temperature and help eliminate waste products. Their active secretion of sweat is under the control of the sympathetic nervous system. Heat-induced sweating tends to begin on the forehead and then spreads to the rest of the body. Emotionally-induced sweating, stimulated by fright, embarrassment or anxiety, begins on the palms of the hands and in the axilla, then spreads to the rest of the body.

Apocrine glands are connected to hair follicles and are only found in the genital and underarm regions. They begin to function at puberty, producing a fatty secretion. Breakdown of the secretion by bacteria leads to body odour.

Sebaceous glands

Sebaceous glands are small sac-like pouches found all over the body, except for the soles of the feet and the palms of the hands. They are more numerous on the scalp, face, chest and back.

Sebaceous glands commonly open into a hair follicle, but some open onto the skin surface. They produce an oily substance called sebum which contains fats, cholesterol and cellular debris.

Sebum is mildly antibacterial and antifungal, and coats the surface of the skin and the hair shafts. It reduces water loss, lubricates and softens the horny layer of the epidermis, and conditions the hair. The secretion of sebum is stimulated by the release of androgen hormones.

Skin types and their characteristics

Skin type is largely determined by genetics and so is linked to ethnicity.

Genes (DNA sequences carried by chromosomes) determine skin characteristics such as follicle size, skin thickness, and distribution of circulation and nerve endings.

The primary factors in determining skin type are:

- level of lipid (fat) secretions that are produced between the skin cells (this determines how well the skin retains moisture)
- the amount of secretion produced by the sebaceous glands.

> **In practice**
>
> Skin types are classified into broad categories which are often streamlined in the skin care industry to allow clearer marketing of product lines.

Skin is generally classified into five main types:

1 normal
2 dry
3 oily
4 combination
5 sensitive.

1 Normal skin

Few clients have normal skin, as this skin type is very rare. Normal skin is balanced in that it has a good oil and water balance. Children from birth up until puberty usually have normal skin.

Distinguishing features of normal skin:

- neither too dry, nor too oily and so has perfect hydration
- soft and supple to the touch
- smooth texture, which is neither too thick nor too thin
- feels slightly warm due to a good blood supply
- has a clear, even surface that is free from blemishes
- fine, almost non-apparent pores
- firm to the touch and generally has good elasticity.

In practice

Clients with normal skin usually report very few skin problems.

The aim when treating this skin type is to maintain the skin's balance and to protect it from damage.

2 Dry skin

A dry skin is either lacking in sebum or moisture, or both. It develops as a result of under activity of the sebaceous glands.

The skin's natural oil, sebum, lubricates the stratum corneum and in the absence of this oily coating the dead cells start to curl up and flake. The sebum coating usually helps to prevent moisture loss by evaporation. If there is a lack of sebum, dry skin does not retain inner moisture.

Although dry skin is hereditary, it can also develop as a result of the ageing process.

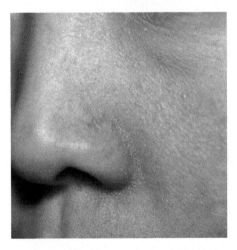

▲ Dry skin

Distinguishing features of dry skin:
- dry, often parched
- looks and feels papery
- thin and coarse in texture with patches of flaking skin
- often sensitive and prone to the formation of dilated capillaries and milia around the eye and upper cheek area
- ages prematurely due to dryness, with fine lines becoming evident around the eyes even as early as the early to mid-20s

- small and tight pores due to the lack of sebum production
- lacks elasticity.

In practice

When questioned, clients with this skin type usually report that their skin feels tight and dry. They may also complain of sensitivity and premature ageing.

The primary aim when treating dry skin is to balance the moisture and oil content of the skin, soften the texture of the skin, hydrate and moisturise. Dry skin also needs a lot of sun protection.

3 Oily skin

Oily skin is hereditary, and develops due to an overproduction of sebum from the sebaceous glands.

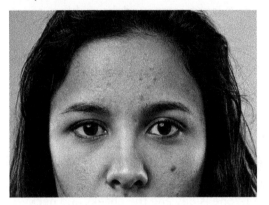

▲ Oily skin

Distinguishing features of oily skin:
- large and noticeable pores due to a build-up of sebum causing them to stretch open
- oily, with a characteristic shine, especially in the T-zone (forehead, nose and chin area)
- firm, thick and coarse to the touch
- uneven texture
- sallow in colour as a result of excess sebum production and dead corneum cells building up on the outer surface
- blemishes are often very apparent, with blocked pores, comedones, papules and pustules all being present to a greater or lesser degree
- some scarring maybe evident from previous blemish sites, leading to a very uneven surface colour
- oily skin ages least prematurely
- elasticity is generally good.

In practice

When questioned, clients with oily skin often report that their skin develops a 'shine' during the course of a day and make-up runs or 'slips,' with foundation changing to a more orange colour. They may complain that their skin often feels thick and dirty, due to the accumulation of sebum and dead cells clogging the surface. They will also suffer from blemishes.

The aim when treating an oily skin is to help balance it by bringing the oil secretions under control thorough cleansing and exfoliation. It is still important to protect oily skin by moisturising the surface with a water-based hydrating product designed for oily skin.

There is a tendency for clients to over treat their oily skin. However, this can compound the problem, as excessive stimulation strips and irritates the skin causing it to become dry and unbalanced. The skin's natural protection mechanism then responds by producing more oil.

4 Combination skin

This is actually the most common skin type.

As its name suggests, this skin is a bit of a mixture. Typically the T-zone is oily and the cheeks and neck are dry or normal. Combination skin can, therefore, be both dry and oily at the same time.

Distinguishing features of combination skin:

- dry on the cheeks and neck, and oily on the T-zone
- a mix of dry areas, feeling rough and fine, and oily areas that are thicker and coarse
- a patchy colour
- the T-zone may have blemishes, such as blocked pores, comedones, papules and pustules
- fine and small pores on the cheeks and neck, but larger pores in the T-zone
- milia may be present around the dryer skin areas, with some sensitivity and dilated capillaries evident too
- the skin's tone and elasticity will vary, being poor in the dry areas but good in the oily areas.

In practice

When questioned, clients with combination skin usually report that they have all the problems of an oily skin in the T-zone, but dryness and tightness on the cheeks, neck and around the eyes.

5 Sensitive skin

While sensitivity is a condition that may affect any skin type, sensitive skin is more commonly referred to in its own classification. Most product lines include products that are marketed specifically for this skin type.

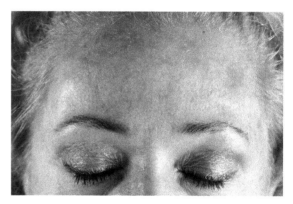

▲ Sensitive skin

Distinguishing features of sensitive skin:

- pink tone, with or without dilated capillaries
- thin and translucent appearance
- warm to the touch
- high colouring, even after a gentle cleanse
- prone to dry, flaky patches
- easily irritated by skin products and other external factors such as heat, cold, wind and sun
- reddens easily following any form of stimulation.

In practice

When questioned, clients with sensitive skin usually say their skin reacts to external stimuli by becoming red and blotchy, and may feel uncomfortable when touched.

The aim when treating sensitive skin is to soothe and calm the skin. Avoid harsh products and forms of treatment or conditions (such as heat) that may cause irritation.

Male skin

Although there is not a typical 'male' skin type, there are important general differences between male and female skins. Testosterone, the male hormone, gives men a thicker epidermis (approximately 2 mm compared to 1.5 mm in women). Male skin does have a tendency to be tougher, more elastic and less sensitive than female skin, although daily shaving can increase the risk of skin rashes, infections and in-growing hairs. It is also more acidic, has a more efficient blood supply and greater sebum production. This means it tends to age better than women's skin, remaining softer, firmer and more supple.

Factors that affect the skin

There are many factors, both **external** and **internal**, which affect a client's skin.

Internal factors

Age

The natural process of ageing affects the skin, as cell regeneration slows with age (see Table 3.3 on page 60 for information about the effects of ageing on skin).

Free radicals

Free radicals are highly reactive and damaging atoms that contribute to skin ageing. External factors, such as ultraviolet radiation, nicotine and substances in unhealthy foods, create free radicals from oxygen molecules. Because they are so reactive, free radicals bond with many other chemical substances; they attack collagen fibres, cellular membranes and the lipid layer of the skin. Free radicals also bond with DNA, changing the inherited properties that are stored in the cell nucleus, so that the quality of newly formed skin cells deteriorates.

Stress and lifestyle

Long-term stress and regular tension can cause sensitivity in the skin, as well as encouraging the formation of lines around the eyes and the mouth.

Hormones

The body's natural glandular changes have an effect on the condition of the skin throughout life.

- During puberty, the sex hormones stimulate the sebaceous glands, which may cause some imbalance in the skin.
- At the onset of menstruation, hormones may cause the skin to erupt.
- During pregnancy, pigmentation changes may occur, but these usually disappear after the birth.
- During the menopause, the activity of the sebaceous glands is reduced and the skin becomes drier.

Smoking

Smoking has been linked to premature ageing and wrinkling of the skin. Nicotine in cigarette smoke weakens the blood vessels that supply blood to the tissues; this deprives the tissues of essential oxygen, making the skin appear dull and grey in colour.

Smoking creates free radicals and destroys vitamins B and C, which are both important for a healthy skin. It dulls the skin by polluting the pores and increases the formation of lines around the eyes and the mouth.

Medication

Medication can affect the skin by causing dehydration, sensitivity and allergies.

Diet

The skin is a barometer of the body's general health. These vitamins are crucial to the maintenance of skin health:

- **vitamin A** – helps repair the body's tissues and helps prevent dryness and ageing
- **vitamin B** – improves the circulation and the skin's colour, and is essential to cellular oxidation
- **vitamin C** – essential for healing and maintaining levels of collagen in the skin
- **vitamin E** – helps to heal damaged tissues and can promote healing of structural damage to the skin.

Water consumption

The skin is approximately 70% water. Drinking an adequate amount of water (approximately 6–8 glasses per day) aids the digestive system and helps to prevent a build-up of toxins in the skin's tissues.

Alcohol consumption

Alcohol consumption has a dehydrating effect on the skin by drawing essential water from the tissues. Excessive consumption causes the blood vessels in the skin to dilate, resulting in a flushed appearance.

Exercise

Regular exercise promotes good circulation, increasing oxygen intake and blood flow to the skin.

Sleep

Sleep is essential to physical and emotional wellbeing, and is one of the most effective regenerators of the skin.

External factors

Photoageing

Photoageing is the accelerated ageing of skin after exposure to ultraviolet light. Exposure to sunlight (which contains ultraviolet radiation) is therefore a dominant factor in how the skin ages.

As we age, the collagen and elastin fibres in the dermis weaken. This natural weakening is accelerated by frequent exposure to ultraviolet light, causing wrinkling and sagging of the tissues.

It is important to note that tanning machines and sunbeds can also cause accelerated ageing of the skin due to the fact that they produce large quantities of long-wave ultraviolet light (UVA). Overexposure of skin to artificial ultraviolet tanning procedures may carry the same risks of photoageing as overexposure to natural sunlight, and also an increased risk of some skin cancers.

Environmental exposure

Exposure to adverse weather conditions, pollutants, or poor air quality can affect the condition of the skin, often resulting in dryness and dehydration.

Occupation

The client's occupation can be a factor in their skin condition. For instance, they might be working in a hot or humid environment, or in dusty and dirty conditions.

Poor skin care routine

Lack of proper skin care, or incorrect skin care for the skin type, can adversely affect the skin. The use of products that are too aggressive can strip the skin and damage its barrier function. The correct use of sunscreens provides the best protection against premature ageing.

General terms associated with the skin

Allergic reaction

This disorder occurs when the body becomes hypersensitive to a particular allergen. When irritated by an allergen, the immune system produces histamine as part of the body's defence reaction. The effects of different allergens are diverse, and they impact different tissues and organs. For example, some cosmetics and chemicals can cause skin irritation and rashes. Certain allergens such as pollen, fur, feathers, mould and dust can cause reactions in the respiratory system, leading to asthma and hayfever. If severe, allergies may be extremely serious can result in anaphylactic shock. Symptoms occur in different parts of the body at the same time and may include rashes, swelling of the lips and throat, difficulty breathing, a rapid fall in blood pressure and loss of consciousness.

In practice

Some clients may have a nut allergy. In this case, the therapist must avoid nut-based ingredients (which are common in skin care products).

Therapists should be aware that it is possible for clients to develop allergies to products after long-term use. This can be confusing for clients who may not understand that the product they have been using for years is suddenly responsible for a reaction.

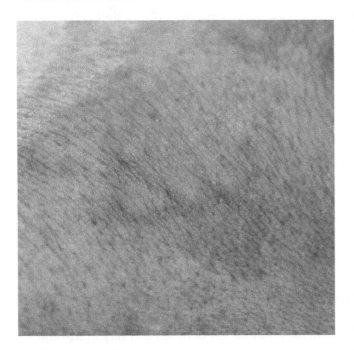

▲ Allergic reaction

If, during treatment, a client experiences a skin rash, severe redness or a burning sensation, the appropriate action is to remove all traces of the product immediately from the skin and to apply cool, wet compresses to soothe the skin.

The client should be advised to discontinue all use of products, preferably including make-up, until the reaction has stopped and all symptoms of the allergy have gone. Then, one by one, the client may be encouraged to use products on the skin, each day adding another product to see if the offending product or ingredient may be identified. In the event of a severe allergic reaction, advise the client to seek immediate medical advice.

Blocked pores

This happens when sebum begins to build up in pores. The pore will appear enlarged and the sebaceous matter inside will be evident. The excess sebum needs to be released to prevent further build-up within the pore.

Enlarged pores

Pores become larger if excess oil and debris become trapped in the follicles, or they may expand due to loss of elasticity.

Comedone

A comedone is a skin blemish caused by the accumulation of waste particles, such as sebum and keratinised cells, in the hair follicle. Comedones are commonly referred to as either 'blackheads' (an open comedone contained within the follicle) or 'whiteheads' (closed comedones trapped underneath the skin's surface).

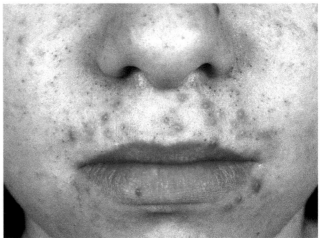

▲ Comedones

In practice

A comedone is a form of skin blockage and may be released manually, or with the use of a comedone extractor.

Crow's feet

These are fine lines around the eyes caused by habitual facial expressions and movements. They are associated with the ageing of muscle tissue; premature formation may be due to overexposure to ultraviolet light or eye strain.

Cyst

This is an abnormal sac containing liquid or a semi-solid substance. Most cysts are harmless.

Erythema

This is reddening of the skin due to the dilation of blood capillaries just below the epidermis in the dermis.

Fissure

This is a crack in the epidermis exposing the dermis.

Keloid

A keloid is the overgrowth of an existing scar to a size that is much larger than the original wound. The surface may be smooth, shiny or ridged. Formation is gradual and is due to an accumulation of collagen in the immediate area. The colour varies from red, fading to pink and white.

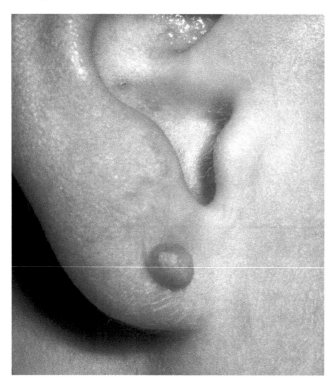

▲ Keloid scar

Lesion

A zone of tissue with impaired function, as a result of damage by disease or wounding, is called a lesion.

Macule

A macule is a small flat patch of increased pigmentation or discolouration, such as a freckle.

Milia

Milia is sebum trapped in a blind duct with no surface opening. Usually found around the eye area, they appear as pearly, white and hard nodules under the skin. Milia may be removed with a sterile microlance.

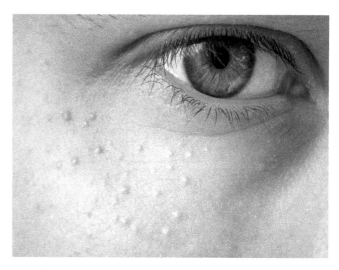

▲ Milia

Mole

Moles are also known as a pigmented naevi. They appear as round, smooth lumps on the surface of the skin. They may be flat or raised and vary in size and colour from pink to brown or black. They may have hairs growing out of them.

▲ Mole

Naevus

This is a mass of dilated capillaries and may be pigmented, as in the case of a birthmark.

▲ Naevus

Papule

A papule is a small raised elevation on the skin, less than 1 cm in diameter, which may be red in colour. It often develops into a pustule.

Pustule

This is a small raised elevation on the skin which contains pus.

Skin tag

These are small growths of fibrous tissue that stand up from the skin and which are sometimes are pigmented (black or brown).

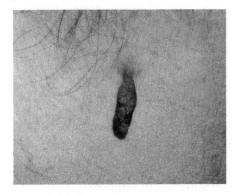

▲ Skin tag

In practice

Skin tags may be surgically removed or may be cauterised by a qualified electrologist with advanced training. They do not contraindicate skin care treatments, although care should be taken to avoid catching them and causing discomfort.

Scar

A scar is a mark that is left on the skin after a wound has healed. Scars are formed from replacement tissue that is deposited during the healing of a wound. Depending on the type and extent of damage, the scar may be raised (hypertrophic), rough and pitted (ice pick) or fibrous and lumpy (keloid). Scar tissue may appear smooth and shiny or can have a depression in the surface.

Telangiectasia

This term is used to describe persistent **vasodilation of capillaries** in the skin. Often, it is caused by extremes of temperature and overstimulation of the tissues; sensitive and fair skins are more susceptible to this condition.

▲ Telangiectasia

Tumour

A tumour is formed by an overgrowth of cells. Almost every type of cell in the epidermis and dermis is capable of benign or malignant tumour formation. Tumours are lumpy and often can be felt underneath the surface of the skin even when they cannot be seen.

Ulcer

An ulcer is a break in the skin that may extend to every layer, causing an open sore.

Urticaria

This condition is also known as 'hives'. Red weals appear rapidly and turn white or disappear again within minutes or more. The process may happen more gradually over a number of hours. The area can be itchy or may sting.

Urticaria may be caused by an allergic reaction to certain foods such as strawberries and shellfish, or to other triggers such as penicillin, house dust and pet fur. Other causes include stress and sensitivity to light, pressure, heat or cold.

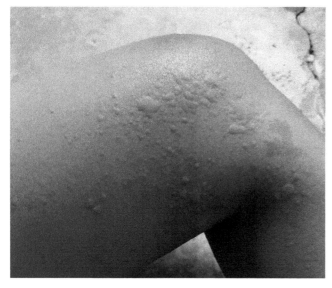

▲ Urticaria (hives)

Vesicles

These are small sac-like blisters. A bulla, commonly called a blister, is a vesicle that is larger than 50 mm across.

Wart

Warts are well-defined benign tumours that vary in size and shape. (See viral infections for more information.)

Weal

A weal is a raised area of skin that contains fluid. These may be white in the centre with a red edge and are commonly seen in the condition urticaria.

Wrinkles

When the underlying structure of the skin has become damaged and the skin starts to lose its elasticity, exaggerated lines and wrinkles start to form on the face and neck. These may form as a result of the normal ageing process, but often occur prematurely due to sun and environmental damage.

Skin conditions

All skins, regardless of their type, may be affected by one or more of the following skin conditions.

Ageing

Like all organs, the skin is affected by the ageing process. Signs of skin ageing include:

- a decrease in the thickness of the skin (thinning epidermis)
- loss of elasticity (elastosis) and muscle tone, resulting in sagging, loose skin, lines and wrinkles
- decreased tensile strength due to loss of collagen and elastin
- dehydration caused by a reduction in the skin's ability to bind water
- rough texture because of reduced cell turnover.

Congestion

Congested skin presents with enlarged pores, blocked pores and comedones due to build-up of dead skin cells, sebum and cell debris. Congested skin tends to appear thick and dull. The solution is regular cleansing and gentle exfoliation.

Dehydration

Water intake is necessary for the healthy functioning of all cells, including skin cells. Dehydration means that there is a lack of moisture in the intercellular system of the skin. Key indicators of dehydration are visible fine lines and a feeling of tightness on the skin.

Any skin type can become dehydrated, even oily skin. Oily skin can be dehydrated by use of products that are too harsh, stripping the skin of its protective coating of sebum.

Many skin types can suffer from temporary dehydration, such as that caused through illness, medication, overexposure to the elements (cold, wind, and heat), central heating and the effects of dehydrating drinks such as caffeine and alcohol.

Dehydrated skin presents a parched, dry looking rough surface and tends to soak up any applied product very quickly.

Pigmentation problems

Pigmentation may be associated with a client's ethnic skin type, or may be caused by environmental or other factors. Pigmentation problems may result from the uneven distribution of melanin over the skin's surface, either due to an accumulation of pigment or because of uneven production by the melanocytes.

The melanocyte cells are located in the basal cell layer of the epidermis, and they surround a large number of regular skin cells (keratinocytes). The melanocytes are responsible for producing melanin, the black–brown pigment that gives the skin its colour. Melanin production varies from individual to individual, and is greater in those with darker skins. Melanin has a protective function, in that it acts as a filter to help protect the skin against the harmful effects of the sun's radiation. Any surface irritation of the skin (including exposure to the sun) is capable of increasing melanin production.

Hyperpigmentation

This is an overproduction of the skin pigment melanin, resulting in brown discolouration or darkening of the skin. Overproduction of pigment occurs when melanocytes produce a greater amount of melanin in a given area of the skin, and/or when the melanin is not properly absorbed by the keratinocytes.

Hyperpigmentation is usually caused by exposure to the sun or environmental damage, although another type of hyperpigmentation may occur during pregnancy (chloasma). Hyperpigmentation may also result from injuries, rashes or chemical irritation, and is especially common in clients with darker skins.

Hypopigmentation

This condition presents as white, colourless skin areas (an absence of pigmentation) resulting from less than the normal melanin production. It may be due to long-term sun exposure or irritation, which causes a dysfunction in the melanocytes.

Sensitivity

Sensitivity is a condition that may affect any skin type, and should not be confused with genetically predisposed sensitive skin. Sensitivity usually manifests as redness, itching or burning. Sensitivity reactions are very complex and individual to each client.

Because the chemical composition of each person's skin varies, one client may react to an ingredient, while another may not. A particular ingredient may not be sensitising as a rule, but can still cause sensitivity in an individual. An ingredient can only be deemed sensitising if *most* clients react to it. Sensitivity and allergies may occur due to exposure to specific product ingredients, misuse of products, medication, diet, or other internal or external factors (hot and cold weather, and the wind). Ingredients that are known to cause sensitivity are fragrances, preservatives and some chemical sunscreens.

How ageing affects the skin throughout life

▲ Ageing skin

Table 3.3 Characteristics of skin ageing

Age	Skin ageing characteristics
Childhood	Smooth, healthy and undamaged
Adolescence	Increased sebaceous gland activity, which may result in spots, comedones and pustules and for some may be the start of acne
20s	Collagen starts to diminish (by approximately 1% per year), causing the start of fine lines and loss of elasticity in skin of upper eyelid
30s	Sagging due to stretching of the skin and continued reduction in collagen, more fine lines and wrinkles, loss of hydration, moderate decrease in dermal repair
40s	Loss of elasticity is more apparent, lines deepen in nasolabial folds, skin sags at the jaw line, forehead wrinkles deepen, noticeable drop in skin hydration level
Menopause	Loss of oestrogen accentuates wrinkles and loss of elasticity, slowing of ability to synthesise collagen, lipid production is affected causing dehydration
50s	Wrinkles and loss of elasticity in the neck more apparent, reduction in supporting fat leads to bonier appearance of face, skin tends to be drier
60s and beyond	Loss of subcutaneous fat, skin becomes thinner, increased sagging, dilated capillaries often present, uneven pigmentation, age spots, skin tags, low production of sebum and collagen, compromised dermal repair, many wrinkles and deep lines

Summary of the main skin types

Table 3.4 Summary of skin types

Skin type	Main recognition factors	Pore size	Elasticity
Normal	Soft, supple, smooth, Free from blemishes	Fine	Good; firm
Dry	Papery, thin, flaky	Small and tight	Generally not good
Oily	Shiny, thick, coarse and uneven Sallow colouring Blocked pores, comedones, papules and pustules may be present	Enlarged	Good; firm
Combination	Dry on cheeks and neck Oily/blemished in T-zone	Variable; enlarged in T-zone and fine and small on cheeks	Poor in dry areas; good in oily areas
Sensitive	Warm to touch, thin, dry and flaky High colouring, easily irritated	Variable; tend to be small and tight	May be poor in areas of sensitivity

Different skin tones

All skins vary in the degree of melanin they produce. Although different skins of different tones have the same number of melanocytes cells, black skins tend to have melanocytes capable of making larger amounts of melanin.

Table 3.5 Characteristics of skin tones

Skin tones type	Colouring	Characteristics
White skin	Generally, a fair/pale complexion	Relatively small amounts of melanin present Very susceptible to sensitivity, irritation and sunburn Greater chance of skin cancer Ages faster than black skin
Olive skin	Typically olive with light to dark brown tones, may also have reddish/pink tones.	Good degree of melanin present, which obscures the colour of the blood vessels Tends to have a generous coating of sebum and is therefore oily Tans easily and deeply with less sun damage and premature ageing Skin is usually thicker, which often means fewer wrinkles
Brown skin	Very dark skin tone which is deeply pigmented with melanin	Smooth and supple with minimal signs of ageing Sweat glands are larger and more numerous in this skin type which gives a sheen to the skin that is often mistaken for oiliness Deeply pigmented, it does not reveal the blood capillaries Signs of ageing appear very late Skin cancer is rare
Black skin	Darker skin with a higher degree of melanin	Open pores Oily with higher degree of sebaceous glands Thick and tough Desquamates easily Forms keloid scars when damaged Skin cancer is very rare Signs of ageing appear very late

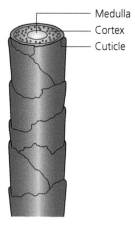

Medulla
Cortex
Cuticle

▲ Structure of hair

The structure and functions of the hair
The structure of hair

Hair is composed mainly of the protein keratin and, therefore, is a dead structure. Longitudinally, the hair is divided into three parts:

1 **hair shaft** – the part of the hair lying above the surface of the skin
2 **hair root** – the part found below the surface of the skin
3 **hair bulb** – the enlarged part at the base of the hair root.

Internally, the hair has three layers (Table 3.6) which all develop from the matrix (the actively growing part of the hair).

Table 3.6 The layers of the hair

Hair layer	Location	Description	Function
Cuticle	Outer layer	Made up of transparent protective scales which overlap one another	Protects the cortex and gives the hair its elasticity
Cortex	Middle layer	Made up of tightly packed keratinised cells containing the pigment melanin, which gives the hair its colour	Helps to give strength to the hair
Medulla	Inner layer	Made up of loosely connected keratinised cells and tiny air spaces	Determines the sheen and colour of hair due to the reflection of light through the air spaces

Hair colour is due to the presence of melanin in the cortex and medulla of the hair shaft. In addition to the standard black colour, the melanocytes in the hair bulb produce two colour variations of melanin: brown and yellow. Blond, light-coloured and red hair has a high proportion of the yellow variant. Brown and black hair possesses more of the brown and black melanin.

The structure of a hair follicle

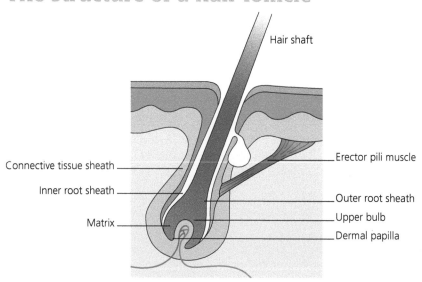

Hair shaft

Connective tissue sheath
Inner root sheath
Matrix

Erector pili muscle
Outer root sheath
Upper bulb
Dermal papilla

▲ A hair in a follicle

The individual parts of a hair's structure are described in Table 3.7.

Table 3.7 The structure of the hair

Structure	Location	Function
Connective tissue sheath	Surrounds hair follicle and sebaceous gland The connective tissue sheath is separated from the outer root sheath by a vitreous membrane	Supplies follicle with nerves and blood Main source of sustenance for the follicle
Vitreous or glassy membrane	Forms the basement membrane of the outer root sheath; separates the outer root sheath from the connective tissue sheath	Separates the two; thickens during the catagen phase of hair growth
Outer root sheath	Forms the follicle wall and is continuous with the stratum germinativum (basal cell layer) of the epidermis	Provides a permanent source of growing cells (hair germ cells) to enable the follicle to grow and renew cells during its life cycle
Dermal papilla	Elevation at the base of the hair bulb, which contains a rich blood supply	Crucial source of nourishment for hair, providing the hair cells with food and oxygen
Inner root sheath	Originates from the dermal papilla at the base of the follicle and grows upwards with the hair (ceasing to grow when level with the sebaceous gland) The inner root sheath is made of three layers: <ul><li>Henle's layer (outermost)</li><li>Huxley's layer (middle)</li><li>the cuticle (innermost).</li></ul>	Shapes and contours the hair, helping to anchor it into the follicle
Hair bulb	Enlarged part at the base of the hair root	Area where the cells grow and divide by the process of mitosis
Matrix	Lower part of the hair bulb	Area of mitotic activity of the hair cells

Types of hair

There are three main types of hair in the body: **lanugo**, **vellus** and **terminal** hair (Table 3.8).

Table 3.8 Types of hair

Type of hair	Description	Where found in body
Lanugo	Fine, soft hair; often unpigmented	Found on a foetus Grows from around the third to the fifth month of pregnancy and is eventually shed to be replaced by secondary vellus hairs, around the seventh to eight month of the pregnancy
Vellus	Soft downy hair; often unpigmented Do not have a medulla, or a well-developed bulb Lie close to the surface of the skin and, therefore have a shallow follicle	Found all over the face and body, except for the palms of the hands, soles of the feet, eyelids and lips
Terminal	Longer, coarser hairs, most are pigmented Vary greatly in shape, diameter, length, colour and texture Deeply seated in the dermis and have well-defined bulbs	Found on the scalp, under the arms, eyebrows, pubic regions, arms and legs

Hair growth

- Hair begins to form in the foetus from the third month of pregnancy.
- The growth of hair originates from the matrix, which is the active growing area where cells divide and reproduce by mitosis.
- Living cells, which are produced in the matrix, are pushed upwards away from their source of nutrition. They die and are converted to keratin to produce a hair.
- Hair growth takes approximately four to five months for an eyelash hair and approximately four to seven years for a scalp hair.

Factors that affect hair growth

Hair growth is affected by several factors, which are outlined below.

Congenital

Hair growth patterns can be passed on from parents: for example, having a heavy beard may be inherited. Hair distribution, the type of hair and rate of growth can also vary with ethnicity. For example, European people tend to have a higher number of hair follicles than Japanese and Chinese people.

Topical

The use of X-rays and the application of cortisone creams can stimulate blood supply to the hair, causing increased growth.

Climate

In hot climates hair tends to grow faster in order to protect the skin from heat, so a fine growth may be increased.

Hormonal (natural glandular changes)

If stimulated by an increase in blood circulation resulting from hormonal changes in the body (such as during puberty, pregnancy or menopause) or medication, the shallow follicle of a vellus hair can grow downwards to become a coarse, dark terminal hair.

During puberty, hormone levels are unstable. The appearance of both pubic and axillary hair is due to the level of adrenocortical androgens circulating in the blood. Androgens (male hormones) increase in men and women during puberty.

During pregnancy, levels of female hormones are raised and excess androgens are also produced, resulting in fine hair growth on the lip, chin, neck and sides of the face. After the birth, the hormone balance is usually restored and the excess hair growth disappears.

During the menopause, the ovaries slowly cease to respond to stimulation by the gonadotrophic hormones of the anterior pituitary, allowing the male hormones (androgens) to become more dominant. This results in hyperstimulation of the adrenal cortex by pituitary hormones, producing an excess of hair-promoting androgens.

Abnormal hormonal changes

Polycystic ovary syndrome is an example of abnormal ovarian function, which leads to the development of ovarian cysts and to the secretion of large quantities of androgens.

Symptoms include excessive hair growth in the male sexual pattern, irregular menstrual periods and weight gain.

Emotional stress

When the body is under stress, activity of the adrenal glands is increased. The hypothalamus triggers the anterior pituitary to produce more adrenocorticotrophic hormone. This, in turn, stimulates the adrenal glands to produce adrenaline, which increases androgen production.

When this kind of stimulation occurs over a prolonged period of time, superfluous hair growth may occur.

Medication

Certain medications are known to stimulate hair growth. Some examples include cortisone (steroids) and the birth control pill.

The growth cycle of hair

Each individual hair has its own growth cycle and undergoes three distinct stages of development: **anagen**, **catagen** and **telogen**.

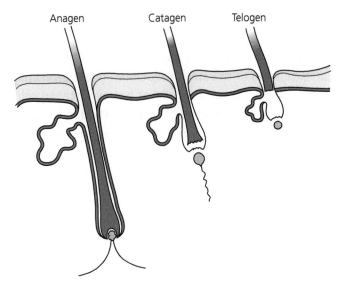

Anagen   Catagen   Telogen

▲ The hair growth cycle

Anagen growth stage

- This is an active growing stage, which lasts from a few months to several years.
- Hair germ cells reproduce at the matrix.
- A new follicle is produced which extends in depth and width.
- Hair cells pass upwards to form the hair bulb.
- Hair cells continue rising up the follicle. As they pass through the bulb, they differentiate to form the individual structures of hair.
- The inner root sheath grows up with the hair, anchoring it into the follicle.
- When cells reach the upper part of the bulb they become keratinised.
- Two-thirds of its way up the follicle, the hair leaves the inner root sheath and emerges on to the surface of the skin.

Catagen stage

- This transitional stage from active to resting lasts approximately two to four weeks.
- Hair separates from the dermal papilla and moves slowly up the follicle.
- The follicle below the retreating hair shrinks.
- The hair rises to just below the level of the sebaceous gland. Here the inner root sheath dissolves and the hair can be brushed out.

Telogen stage

- Hair is shed on to the skin's surface.
- There is a short period during which the shortened follicle rests.
- The follicle is stimulated once more and a new replacement hair begins to grow.

Different types of hair growth

As there is a continuous cycle of hair growth, the amount of hair on the body remains fairly constant. However, hair growth varies from client to client and from body area to area. A client wishing a hair removal treatment should be made aware of the fact that hair growth occurs in three stages, so there will be hair present at different lengths both above and below the skin.

In practice

While carrying out hair removal treatments, it is important to remember that the hair follicle is part of the skin's structure. Therefore, any treatment which affects the hair also affects the skin.

Once a hair has been removed, blood is directed straight to the area being treated to heal and protect the skin. This is a normal reaction of the skin and causes reddening. Extra blood that has been sent to the treated area is diverted away within a few hours of treatment.

The treated area of skin will have open follicles which offer bacteria easy entry into the body. It is vital that clients adhere strictly to aftercare advice to avoid risk of infection.

Common pathologies of the hair

Alopecia areata, alopecia totalis and alopecia universalis

- Alopecia areata – a form of hair loss that can occur at any age. It presents as patches of hair loss which may be present on the scalp, beard, eyelashes, eyebrows, and on the body or limbs.

- Alopecia totalis – affects the whole scalp, causing total hair loss on the head.
- Alopecia universalis – the whole body and scalp are affected by hair loss.

Alopecia can be genetic, but it can also be caused by stress, shock, illness or medication (as in the case of chemotherapy, for example).

Androgenetic (or pattern) alopecia

This is a genetically determined pathology – often referred to as male-pattern or female-pattern baldness – that is characterised by gradual hair loss. It is an extremely common condition which affects both men (50% over 50 years of age) and women (50% over 65 years of age). In women, the severity varies from widespread thinning to complete baldness.

The main hormone responsible for androgenetic alopecia is dihydrotestosterone, which causes the affected hairs to be reduced in diameter and length and to become lighter in colour. Eventually, the follicle shrinks and ceases to produce hair.

Hypotrichosis

This is a condition of abnormal hair patterns, such as hair loss or reduction. It occurs most frequently as the growth of vellus hair in areas of the body that normally produce terminal hair. Typically, the individual's hair growth is normal after birth, but soon the hair is shed and replaced with sparse, abnormal hair growth. The new hair is fine, short and brittle, and may lack pigmentation. Baldness may be present by the age of 25.

Hypertrichosis

Hypertrichosis is an abnormal amount of hair growth over the body. The two distinct types of hypertrichosis are generalised hypertrichosis, which occurs over the entire body, and localised hypertrichosis, which is restricted to a certain area.

Hypertrichosis can be either congenital (present at birth) or may occur later in life.

The excess of hair appears in areas of the skin that do not have androgen-dependent hair growth (that is areas not in the pubic, face and axillary regions).

Nails

The nail is an important appendage of the skin and is an extension of the stratum lucidum (clear layer) of the epidermis. It is composed of horny flattened cells which undergo a process of keratinisation, making the nail quite hard. It is the protein keratin that helps to make the nail a strong but flexible structure. The visible part of the nail is dead as it has no direct supply of blood, lymph or nerves. All nutrients are supplied to the invisible living part of the nail via the dermis.

The functions of the nail

The nail has two important functions:

- It forms a covering at the ends of the phalangeal joints of the fingers and the toes, helping to protect and support their delicate networks of blood vessels and nerves.
- It is a useful tool enabling us to touch and manipulate small objects, and scratch surfaces.

The structure of the nail

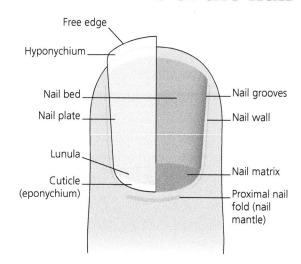

▲ Cross-section of a nail

The nail has several important anatomical regions as shown in Table 3.9.

Table 3.9 The anatomical regions of the nail

Structure	Location	Function
Nail matrix (germinal matrix)	Situated immediately below the cuticle (eponychium)	The nail's most important feature, this is the area where the living cells are produced Has a good blood supply which delivers oxygen to the nail and is vital in the production of new cells Area that determines the health of the nail
Nail mantle or proximal nail fold	Deep fold of skin above the matrix	Protects matrix or nail root from physical damage
Nail bed	Located immediately below the nail plate, this is a continuation of the matrix Part of the skin on which the nail plate rests	Provides nourishment and protection for the nail Richly supplied with blood vessels, lymph vessels and nerves from the underlying dermis
Cuticle	A fold of overlapping skin that surrounds the base of the nail There are three areas of the cuticle: ● the eponychium is the dead cuticle that adheres to the base of the nail, near the lanula ● the perionychium is the cuticle that outlines the nail plate ● the hyponychium is the cuticle skin found under the free edge of the nail	Protects the matrix and nail bed from infection by stopping dirt and bacteria from getting under the nail plate, forming a waterproof barrier
Lunula	Light-coloured semicircular area of the nail, that lies between the matrix and the nail plate Also known as the 'half-moon'	Area of the nail where cells start to harden; the cells here are in a transitional stage (between hard and soft) Forms a bridge between the living cells of the matrix and the dead cells of the nail plate
Nail plate	Main visible part of the nail which rests on the nail bed and ends at the free edge Is made up of tightly packed and hard keratinised epidermal cells Cell layers are packed closely together, and are bound together by sulfur bonds, moisture and fat	Offers protection for the nail bed
Nail wall	The folds of skin overlapping the sides of the nails	Protects the edges of the nail plate from external damage
Nail groove	Deep ridges under the sides of the nail	Guides the growth of the nail up the fingers/toes, helping the nail to grow straight
Free edge	Part of the nail plate that extends beyond the nail bed and fingertip	This is the part of the nail that is filed Protects the finger tips and the hyponychium

KEY FACT

The underside of the nail plate is grooved – it has longitudinal ridges and furrows that help keep it in place. A normal healthy plate curves in two directions:

1 transversely – side to side across the nail

2 longitudinally – from base to the free edge.

The nail plate contains three distinct layers:

1 The dorsal layer is the uppermost layer and is the hardest.

2 The intermediate layer is the thickest layer making up 70–75% of the nail plate.

3 The ventral layer is the bottom layer that is attached to the nail bed. It is only one or two cells thick and is composed of soft keratin.

The process of nail growth

KEY FACT

A good blood supply is essential for healthy nail growth; oxygen and nutrients are supplied to the living cells of the nail matrix and nail bed.

KEY FACT

Protein and calcium are good sources of nourishment for the nails.

The nail growth process is characterised by these steps:

- Nail growth occurs from the nail matrix by cell division.
- As new cells are produced in the matrix, older cells are pushed forwards and are hardened by the process of keratinisation to form the hardened nail plate.
- Translucent cells, when they first emerge from the matrix are plump and soft, but they keratinise (become harder and flatter) as they move towards the free edge.
- As the nail grows, it moves along the nail grooves at the sides of the nail. These grooves help to direct growth along the nail bed.
- The top layers of the epidermis (stratum lucidum and stratum corneum) form the nail plate.
- The remaining three layers of the epidermis (stratums germinativum, spinosum and granulosum) form the nail bed.

- It takes approximately six months for cells to travel from the lunula to the free edge of the nail for fingernails, and approximately nine to 12 months for toenails.

Factors affecting nail growth

Nail growth may be affected by the factors which are discussed below.

KEY FACTS

The rate of growth of a nail is faster in the summer due to an increase in cell division as a result of exposure to ultraviolet radiation.

Fetal nails start growing before the fourth month of pregnancy.

The growth rate of nails varies from person to person and from finger to finger, with the nail of the index finger generally being the fastest to grow.

The average growth rate of a nail is approximately 3 mm per month.

Activity

Nails grow more quickly on the dominant hand due to increased activity and the resulting faster blood flow (so if you are right handed, the nails on your right hand will grow faster than those on your left).

Age

During ageing, nail growth slows down due to the fact that the blood vessels supplying the matrix become less efficient.

Due to slower nail growth, imperfections in the nail plate, such as ridges, become more visible with age.

Diet and nutrition

A poor diet lacking in nutrients, vitamins and minerals affects both the growth and appearance of the nail.

Hormones levels

Some hormonal changes can enhance nail growth, for instance pregnancy can speed up nail growth. However, some conditions such as thyroid disorders can adversely affect the health of the nail.

Illness and medical condition

During a period of ill health, the body may reduce the blood supply to the nails as it attempts to heal other parts. This can result in changes to the nail structure (they may become ridged and paler) and growth overall is diminished.

Medication

Some medication can affect nail growth, making it faster or slower, and may also influence nail strength.

Climate

Nails grow faster in the summer when it is warmer and circulation is improved, and growth rate slows down in winter when circulation is reduced. This means that nails grow at different rates depending on the time of year.

Lifestyle and stress level

A healthy lifestyle, along with a low stress level, helps nails to grow quickly and strongly. Poor health or an unhealthy lifestyle, coupled with a high stress level, will impact the nail structure and result in poor growth.

KEY FACT
Unlike hair growth, nail growth does not follow a growth cycle – nail growth is continuous throughout life. Toe nails have a slower rate of growth than fingernails.

Occupation

Certain occupations may have an enhanced effect on nail growth (manual occupations can encourage nail growth as they promote blood flow). Some occupations may have a negative effect on nail growth (chemicals used in jobs like cleaning can have a detrimental effect).

Trauma

Any type of trauma (accidental or through poor manicuring techniques) can slow the rate of nail growth if it affects the matrix. If a heavy pressure is applied when using a cuticle knife, damage may be caused to the matrix cells resulting in ridges to the nail. This damage may be temporary if new cells produced in the matrix replace the affected ones and the ridges may eventually grow out.

Accidental damage, such as shutting a finger in a door, may result in bruising and bleeding and even complete removal of the nail. It could result in permanent malformation of the nail if the nail bed is damaged.

Skin texture

As the nail is an appendage of the skin and its blood supply is via the dermis, if the skin is in a poor condition this will affect the health and growth of the nail cells.

Current hand and nail care routine

A regular and effective nail care routine will help nails to grow stronger and longer. Correct filing of the free edge, frequent use of cuticle oil and regular manicures all help to maintain healthy nail growth.

Smoking

As smoking deprives the body of oxygen, nails suffer as a result. Slow growth, brittle, dry, cracked, hooked, and peeling nails are all common in smokers.

Alcohol

Excess alcohol in the blood can deprive the nails of vital nutrients. Brittle, pale and peeling nails are a sign that a person is consuming unhealthy amounts of alcohol.

Chemotherapy

During chemotherapy treatment, clients may notice some changes in the colour and thickness of their nails, and changes around the nail bed. The nail may look bruised, turning black, brown, blue or green.

The nail may develop blemishes, such as horizontal or vertical lines, or small indentations. These marks may reflect the timing of chemotherapy. They are not permanent and will grow out.

The nail may become dry, thin or brittle, and will tend to break more easily.

Radiotherapy

During a course of radiotherapy, nails, may become discoloured or weak, and may break or lift off. Sometimes nails develop ridges which will grow out over time. For most people, nail changes are temporary. It may take about six months after completion of treatment for nails to return to their usual condition.

Natural nail shapes

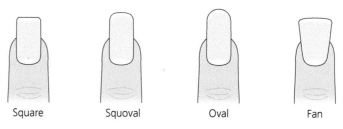

| Square | Squoval | Oval | Fan |

▲ Nail shapes

Table 3.10 Natural nail shapes

Nail shape	Description
1 Square	The nail is filed straight across at the free edge, creating a square shape Very popular shape Suits longer fingers with narrow nails and those with shorter nails or wide nail beds Can be rounded at the edges to create the 'squoval' shape, see below Low maintenance
2 Squoval	This is a square shape with rounded edges Shape helps to prevent breaks Low maintenance
3 Oval	The free edge is rounded and the nail forms an oval shape A natural and highly desirable nail shape Compliments shorter and wider nails
4 Fan	The nail grows outwards at the free edge creating a fan-like shape Fan nails are narrow at the eponychium and wide at the free edge
5 Hook (claw)	The nail grows downward in a hook-like shape See onychogryphosis (page 72)
6 Spoon	The nail curves up and grows outwards at the free edge See koilonychia (page 74)

Nail conditions

It is important to remember that nails originate in the epidermal layer of the skin. Like hair, their condition is dependent on the individual's overall health.

Table 3.11 Nail conditions

Nail condition	Recognition factors
Bitten	Nail shape appears deformed with a bitten down and ragged free edge Cuticle and skin may be chewed, and may appear red and sore Nail biting is a nervous habit which may lead to permanent damage There is a risk of transferring germs from fingers to mouth, which could cause disease
Brittle	Nails are very hard and inflexible May often break or crack easily Have a thicker nail plate May become curved as they grow longer

Nail condition	Recognition factors
Damaged	Extremely thin, weak and soft (bendy) May peel easily Free edge disintegrates when filed Growth rate is slow Sensitive to the touch
Dehydrated	Flaky, peeling and dull in appearance Often have superficial ridges Sometimes have a thickened nail plate May be discoloured Have a dragging feeling to the touch
Discoloured	Nail may appear yellow, green, black, blue or brown, depending on cause Stains may be caused by cigarette smoking, use of hair dyes or touching highly coloured vegetables Discolouration can be caused by fungal infections or other medical conditions
Dry	Dull in appearance, with a dragging feeling to the touch Often have a flaking free edge. May peel easily and have superficial ridges The cuticles are usually very dry
Misshapen	Nails can become misshapen due to: ● mechanical damage (poor treatment) ● an accident (shutting a hand in a door) ● a nail pathology such as koilonychia ● natural growth patterns (spoon-shaped nails)
Pitted	Small depressions (shallow or deep) appear on the surface of the nails Often found in clients suffering from psoriasis
Ridged	Vertical ridges or furrows run from the tip of the fingernail down to the cuticle. Often develop in older age, possibly due to a slowing of cell turnover If vertical ridges are accompanied by a change in colour and texture, the cause maybe a medical condition or iron deficiency (anaemia) Deep horizontal ridges, called Beau's lines, may be symptoms of a more serious condition such as acute kidney disease, diabetes or thyroid disease
Split	Can appear as a single horizontal split between layers of the nail plate at the growing end or as multiple splits and loosening of the growing edge of the nail plate Horizontal nail splitting may occur along with onychorrhexis, or with longitudinal ridging and Horizontal splits at the origin of the nail plate may be seen in people with psoriasis, lichen planus or in people who use oral medications made from vitamin A

Nail pathologies

Diseases of the nail occur as a direct result of bacteria, fungi, parasites or viruses attacking the nail or surrounding tissues. Nail diseases contraindicate manicure or pedicure treatments due to the risk of cross-infection.

Nail disorders may be caused by illness, physical and chemical damage, by general neglect or by poor manicuring techniques. Nail disorders do not contraindicate treatments.

In practice

A therapist must be able to recognise diseases and disorders so that the correct treatment or advice may be given.

Nail diseases

Anonychia (an-uh-NIK-ee-uh)

This is a congenital disorder in which there is a total absence of a nail. It may affect all or some of the fingernails or toenails. Although typically congenital, it may also be caused by infection, a severe allergic reaction or trauma.

Onychatrophia (on-ee-chat-tro-fi-ah)

This is a condition in which the nail plate atrophies (wastes away). Initially, the nail loses its lustre, becomes smaller and may then separate completely from the nail bed. The condition may be caused by an injury to the matrix or by a disease.

▲ Onychatrophia

Onychia (on-nik-ee-uh)

This is a generic term used to describe any disease of the nail, but more specifically refers to inflammation of the nail bed and a bacterial infection of the nail fold. In this condition the nail matrix appears red with swelling, tenderness and pus formation. The condition may lead to shedding of the nail. Causes include wearing false nails for too long, harsh manicuring, chemical applications, a variety of infections and physical damage.

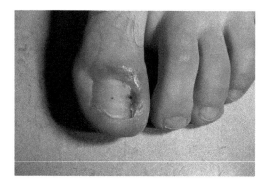

▲ Onychia

Onychogryphosis (on-e-koh-gri-foh-siss)

This is the technical term for **claw-shaped nails**, in which the nails present as crooked, curved and thickened.

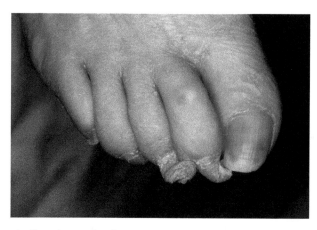

▲ Onychogryphosis

Onycholysis (on-ee-KOL-e-sis)

This condition is characterised by loosening or separation of part or all of a nail from its bed. Causes include disease and physical damage (such as insertion of a sharp instrument used under the free edge), or it may occur spontaneously without any apparent cause. Penetration of the flesh line can allow bacteria and other infective agents to enter the nail bed.

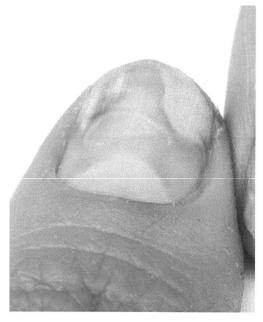

▲ Onycholysis

Onychomycosis (on-i-koh-me-KOH-sis)

This is a term given to the fungal infection of the nail commonly called **tinea unguium** or **ringworm**. It attacks the nail bed and nail plate, presenting as white or yellow scaly deposits at the free edge, which may then spread down to invade the nail walls or bed. The nails become thickened, brittle, opaque or discoloured. The nail plate appears spongy and furrowed.

In its advanced stages, the nail plate may separate from the nail bed (a condition known as onycholysis, see above). There may also be accompanying dryness and skin scaling at the base of the fingers and on the palms.

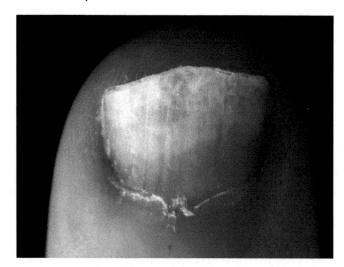

▲ Ringworm of the nail

Paronychia (parr-uh-NIK-ee-uh)

This is inflammation of the skin surrounding the nail, commonly caused by bacterial, viral or fungal infection. The tissues may be swollen and pus may be present, which can develop into an abscess.

Prolonged immersion of the hands in water, poor manicure techniques, nail biting, and picking at the cuticle or the nail wall can all increase the risk of infection.

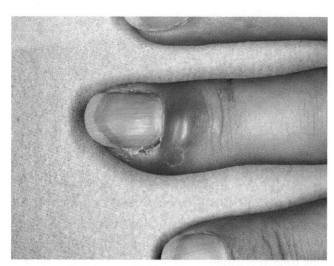

▲ Paronychia

Nail disorders and conditions

Beau's lines and other nail ridges

Beau's lines are deep, waxy, horizontal lines across the nail, often called transverse furrows, corrugations or ridges. Ridges in the nail may occur due to irregular formation of the nail or to physical or chemical injury of the nail matrix.

Beau's lines can be indicative of abnormal nail growth, a symptom of body malfunction or disease. Deep Beau's lines, especially when present across all the nails, are associated with serious illness.

Vertical or longitudinal ridges are another type of ridge – these are common in healthy nails due to uneven development of the nail tissue, poor manicuring techniques and the effects of harsh chemicals.

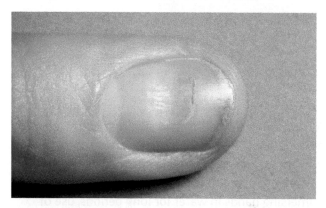

▲ Beau lines

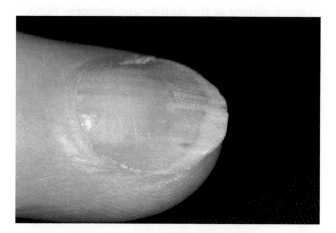

▲ Longitudinal furrows

Bruised nail

A bruised nail, resulting from a heavy blow or persistent trauma (for example in long distance runners), presents as a blackened area where dried blood is visible beneath the nail plate.

The bruised area travels with the nail growth and will eventually grow out. However, a severely bruised nail can cause the nail plate to lift from the nail plate.

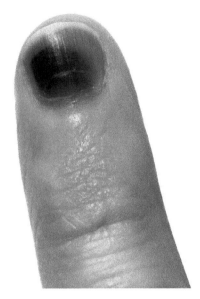

▲ Bruised nail

Hang nail

A hang nail is a small strip of skin that hangs loosely at the side of the nail, or a small portion of the nail itself that splits away. A hang nail may develop due to dry, torn or split cuticles. Common causes are immersing hands in water for long periods, use of detergents and other chemicals, cutting the nails too close, picking at the cuticles and improper filing.

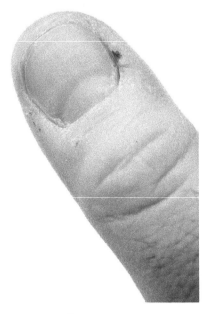

▲ Hang nail

Koilonychia (keel-oh-NIK-ee-uh)

This is the term given to concave, **spoon-shaped nails**, resulting from abnormal growth at the nail matrix. In this condition the nails are thin, soft and hollowed. Koilonychia may be congenital or it may be due to lack of iron or other minerals in the diet.

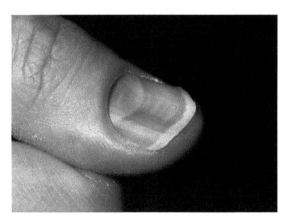

▲ Koilonychia

Leukonychia (lou-con-ik-ee-uh)

This is a term given to **white or colourless nails**, or nails with white spots, streaks or bands. There may also be evidence of ridging. Leukonychia may be caused as a result of injury to the matrix or the effects of disease. The white spots usually disappear as the nail grows.

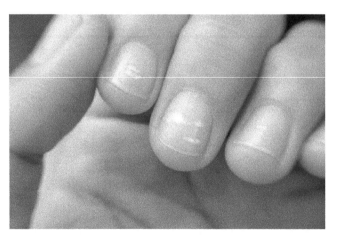

▲ Leukonychia

Onychocyanosis (on-ee-choc-an-o-sis)

In this nail condition, which is also called **blue nail**, the nail presents with a blue tinge, rather than a healthy pink tone. It is usually the result of poor circulation, a heart condition or other circulatory disorder.

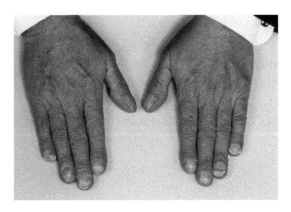

▲ Onychocyanosis

Onychophagy (on-i-kof-uh-jee)

This is the technical term for **nail biting** in which the free edge, nail plate and cuticle are bitten to leave the hyponychium exposed and the cuticle and surrounding skin ragged, inflamed and sore. Nail biting is usually a nervous or stress-induced habit.

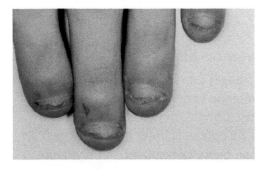

▲ Onychophagy

Onychauxis (on-chor-ex-is)

This is the medical term for extreme **thickening of the nail plate**.

Onychauxis may be a natural part of ageing, but can also be caused by trauma (injury) or from wearing tight shoes. However, it can also be caused by more serious issues affecting the nail bed and the skin, such as fungus or yeast infection.

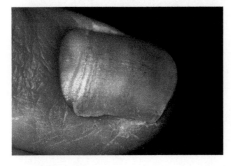

▲ Onychauxis

Onychorrhexis (on-i-ko-rek-sis)

This is the term given to **dry, brittle nails**. In this condition the nails lose their moisture, becoming dry, and the free edges may split. The nails may easily peel into layers. There may be transverse or longitudinal splitting of the nail plate and inflammation, tenderness, pain, swelling and infection may also be present. Frequent immersion in water and contact with detergents and chemicals contribute to this condition. It may also indicate an iron deficiency, anaemia, or incorrect filing which causes the nail plate to split.

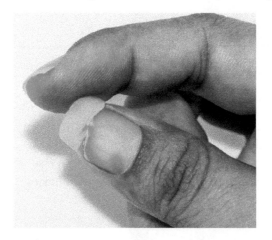

▲ Onychorrhexis

Onychocryptosis (on-i-koh-krip-toe-sis)

This is a term given to an **ingrown** fingernail or toe nail. The first signs are inflammation, followed by tenderness, swelling and pain around the side of the nail. Infection may aggravate the condition. It is caused by ill-fitting shoes, cutting or filing nails too short or too close to the skin.

It may also be due to a malformation of the nail when it was beginning to grow.

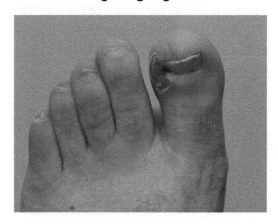

▲ Onychocryptosis

Onychomalacia (on-ee-chom-al-a-c-ah)

This is a term given to **eggshell nails** – thin, white nails that are more flexible than normal. In this condition the nail separates from the nail bed and curves at the free edges. The condition may be associated with illness, poor diet, mineral deficiency or stress.

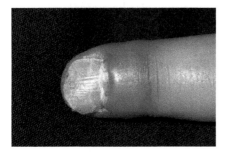

▲ Onychomalacia

Onychoschizia (ony-cho-sc-it-zee-a)

Onychoschizia or **lamellar dystrophy** is a term given to nails which present as soft, thin, split or brittle. Indicators of this condition include a dryness in the nail plate, separation of nail layers, along with flaking and peeling at the free edge.

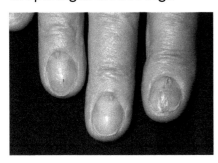

▲ Onychoschizia

Pitting of the nail plate

This nail condition presents with small depressions, which resemble pin pricks, on the surface of the nail. The depressions may be superficial or deep, and are often found in clients suffering with psoriasis or eczema.

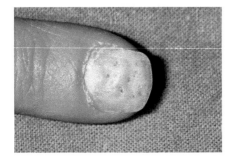

▲ Pitting of the nail plate

Pterygium (terr-e-gee-um)

This is a condition in which the cuticle becomes **overgrown** and grows forwards up the nail. The cuticle at the base of the nail becomes dry and split, and sticks to the nail plate. Pterygium may be due to faulty nail care or lack of nail care (neglect).

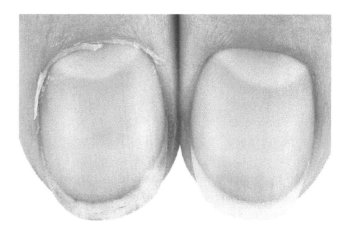

▲ Pterygium

Splinter haemorrhage of nail

These haemorrhages present as thin, red to reddish-brown lines of blood under the nails, running in the direction of nail growth. They look as if there is a splinter in the nail plate. Splinter haemorrhages can develop after an injury or trauma to a fingernail or toenail. Stubbing a toe or injuring a finger can damage blood vessels along the nail bed and trigger bleeding under the nail. If caused by injury, they are not a reason for concern as they clear up as the injury heals. However, if a splinter haemorrhage persists it may indicate an underlying disease or disorder.

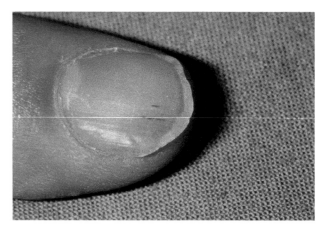

▲ Splinter haemorrhage

Common pathologies of the skin

The skin is a complex organ and when symptoms of disease and disorders occur, it can affect a client's health and wellbeing.

In practice

It is essential that therapists are able to recognise skin conditions that require medical attention, to advise the client to access treatment and to avoid possible cross-infection in the salon.

Therapists should also be knowledgeable about skin lesions and disorders so that treatment and advice is appropriate and referral to a dermatologist can be made if necessary.

Bacterial infection

Many types of bacteria can survive and reproduce on the skin. Bacterial skin infections can affect a small (localised) area or the whole of the body. Whatever the extent of an infection, they are contagious. If you treat a client with a bacterial infection you:

- may make the client's condition worse
- could contract the condition yourself, which may prevent you from working
- risk passing it on to other clients.

In practice

In the case of an infectious skin condition, no treatment can be carried out until all signs of infection have ceased. This is to prevent cross-infection and to prevent the condition spreading and/or worsening.

Blepharitis

Blepharitis is an inflammation of the eyelids, commonly caused by *Staphylococcus* bacterial infection. The condition presents as eyelids that are red, swollen, itchy and sore at the edges. Eyelids may be crusty or greasy, causing the eyelids to stick together. Eyes feel gritty and sensitive to light.

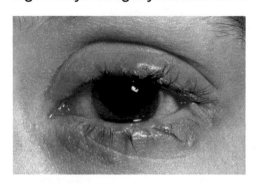

▲ Blepharitis

Boil

A boil or **furuncle** occurs when a hair follicle becomes deeply infected with *Staphylococcus* bacteria, usually appearing suddenly as a painful pink or red bump. It begins as a small inflamed nodule which then forms a large painful pustule around the base of a hair follicle or at a break in the skin.

Local injury or lowered immune resistance may encourage the development of boils, along with stress and poor hygiene. A **carbuncle** is a cluster of boils.

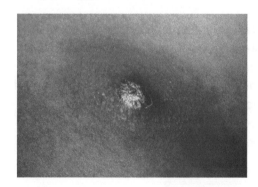

▲ Boil

Conjunctivitis

Conjunctivitis is a bacterial (staphylococcal) infection following irritation of the conjunctiva of the eye. The inner eyelid and eyeball appear red and sore and there may be a pus-like discharge from the eye. The infection spreads by contact with secretions from the eye of an infected person.

Symptoms include an itchy or gritty feeling inside the eye, along with a sensitivity to light.

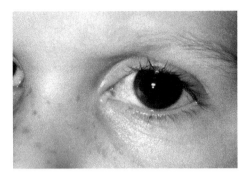

▲ Conjunctivitis

Folliculitis

This bacterial infection, usually *Staphylococcus aureus*, causes a small pustule at the base of a hair follicle. There is redness, swelling and pain around the follicle.

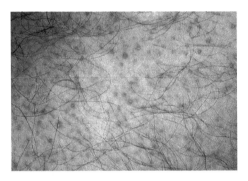

▲ Folliculitis

Impetigo

This is a contagious inflammatory disease in which weeping blisters are visible on the surface of the skin, particularly around the face, mouth and ears. Blisters dry to form honey-coloured crusts. Impetigo is caused by *Streptococcus* and *Staphylococcus* bacteria which are transmitted by dirty fingernails and contact with towels, for example.

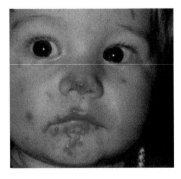

▲ Impetigo

Stye

A stye or **hordeolum** is an acute *Staphylococcus* inflammation of the gland at the base of an eyelash. The gland becomes hard and tender, and a pus-filled cyst develops at the centre.

▲ Stye

Viral infections of the skin

Virus particles need a host (such as a person) to reproduce, whereas bacteria can persist in the environment.

Viral infections are very contagious so clients should not receive treatments until the condition has cleared. Remember that some viruses remain dormant in the body even after signs of the infection have gone. Note also that viral skin conditions are more aggressive in people with compromised immune systems.

Herpes simplex

Cold sores are caused by the herpes simplex virus. They are normally found on the face and around the lips, beginning with an itching sensation, followed by erythema and a group of small blisters which weep and form crusts. This condition generally persists for approximately two or three weeks but may reappear at times of stress, ill health or exposure to sunlight.

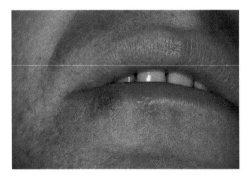

▲ Herpes simplex (cold sore)

Herpes zoster

Shingles is a very painful infection along the sensory nerves due to herpes zoster, the virus that causes chicken pox. Lesions resemble herpes simplex with erythema and blisters along the lines of the nerves. Affected areas are usually the back and upper chest. Severe pain may persist at the site of the shingle infection for months or even years after the apparent healing of the skin.

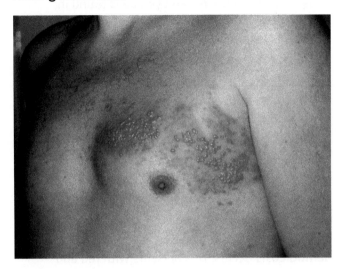

▲ Herpes zoster (shingles)

Warts

A wart is a benign (non-harmful) growth on the skin, caused by infection with the **human papilloma virus (HPV)**.

Plane warts are smooth in texture with a flat top and are usually found on the face, forehead, back of the hands and the front of the knees.

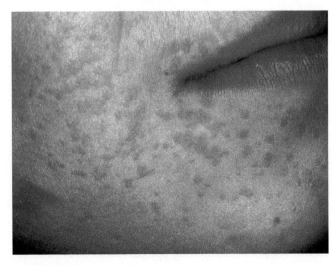

▲ Plane wart

Plantar warts or verrucae occur on the soles of the feet and are usually the size of a pea.

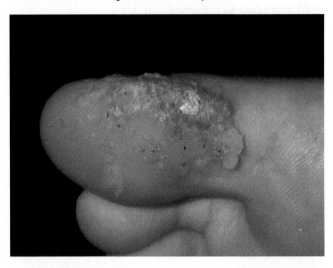

▲ Plantar wart

Facial warts present as skin-coloured, benign growths with a long thread-like appearance (similar to a skin tag). They are common on the eyelid, neck and surrounding areas.

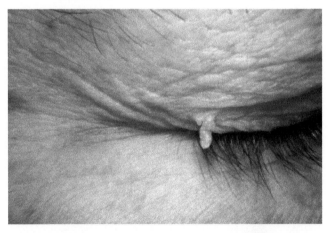

▲ Facial wart

Fungal infections of the skin

Common fungal skin infections are caused by yeasts, which multiply in damp, moist conditions. The affected area of the skin usually presents as itchy, red or scaly. It is essential to avoid contact with a fungal infection until all signs of the infection have cleared.

Tinea corporis

Tinea corporis or **ringworm of the body** is a fungal infection of the skin which begins as small red papules that gradually increase in size to form a ring. The affected areas on the body vary in severity from mild and scaly to inflamed and itchy.

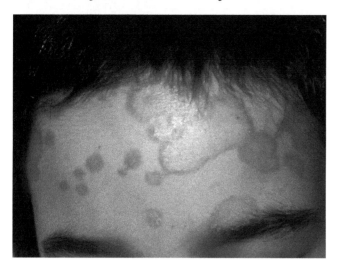

▲ Ringworm

Tinea capitis

This is **ringworm of the scalp**. It appears as painless, round, hairless patches on the scalp. Itching may be present and the lesion may appear red and scaly.

Tinea pedis

Tinea pedis or **athletes' foot** is a highly contagious fungal condition which is easily transmitted in damp, moist conditions such as swimming pools, saunas and showers. Athletes' foot appears as flaking skin between the toes, which becomes soft and soggy. The skin may also split and the soles of the feet may occasionally be affected.

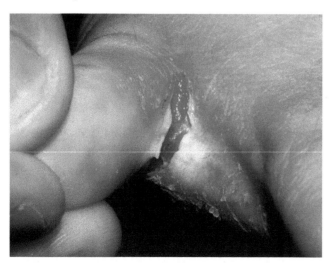

▲ Tinea pedis

Infestation disorders of the skin

Infestation disorders involve parasites that live on or in the skin. These conditions are highly contagious.

Demodex

- **Demodex brevis** – a type of demodex mite (a short-tailed, eight-legged parasite) that is invisible to the naked eye and is found in the sebaceous gland of hair follicles. It feeds on sebum produced by the sebaceous gland. *Demodex brevis* is known to make conditions such as eczema and rosacea worse.

- **Demodex folliculorum** – a longer tailed demodex mite that lives in the hair follicles.

Demodex is mostly found on the face, around the eyelids and eyelashes. It causes an increase in the number of skin cells in hair follicles, giving the appearance of scaly skin.

Demodex does not usually produce symptoms but if there is a serious infestation, a condition called **demodicosis** occurs. This presents with red and scaly skin, skin irritation and rashes, eye irritation, thickening of the eyelid and loss of eyelashes.

▲ Demodex mite

Pediculosis

This condition is commonly known as **lice**, a contagious parasitic infection. The lice live off the blood sucked from the skin. Head lice are frequently seen in young children and, if not dealt with quickly, may lead to a secondary infection of impetigo as a result of scratching. With head lice, nits (egg cases) may be found in the hair. These are pearl-grey or brown oval structures found on the hair shaft close to the scalp. The scalp may appear red and raw due to scratching.

▲ Pediculosis (lice)

Body lice are rarely seen as they burrow into the surface of the skin; however they may cause intense itching. They occur on individuals with poor personal hygiene and live and reproduce in seams and fibres of clothing, feeding off the skin. Lesions may appear as papules, scabs and, in severe cases, as pigmented, dry and scaly skin. Secondary bacterial infection may be present. A client who is affected by body lice may complain of itching, especially in the shoulder, back and buttock area.

Scabies

This is a contagious parasitic skin condition caused by the female mite burrowing into the horny layer of the skin, where she lays her eggs. The first noticeable symptom of this condition is severe itching, which worsens at night. Papules, pustules and crusted lesions may also develop. Common sites of infestation are the ulnar borders of the hand, palms of the hands and between the fingers and toes. Other sites include the axillary folds, buttocks, breasts in the female and external genitalia in the male.

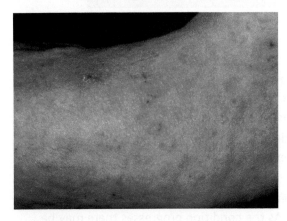

▲ Scabies

Acne

Acne is a chronic inflammatory disorder of the sebaceous glands which leads to the overproduction of sebum. It involves the face, back and chest and is characterised by the presence of greasy, oily skin with enlarged pores, inflammation in and around the sebaceous glands, papules, pustules and, in more severe cases, cysts and scars.

Acne is primarily androgen induced and appears most frequently at puberty, often persisting for a considerable period of time. Although acne is commonly associated with teenage and adolescent skin, it can actually affect people in all age groups at different stages of life.

There are two main categories of acne: **acne simplex** (non-inflammatory acne); and **acne vulgaris** (inflammatory acne).

The typical stages of acne development are as follows:

- Acne starts to develop when an increase in hormone production (commonly at puberty) stimulates the sebaceous glands.
- Excess sebum production causes additional cell build-up in the follicles, which become comedones (plugs of sebum and dead cells).
- Blockage of the follicle opening results in inflammation and irritation, and in the formation of papules.
- The blockage of sebum and dead skin cells prevents oxygen reaching the bottom of the follicle. In these conditions, particular bacteria multiply (see Key fact). The infected papules become pustules.
- The bacteria excrete an inflammatory fatty acid by-product which eventually blocks the follicle completely.
- The skin forms hardened tissue as it attempts to prevent the spread of bacteria, creating cysts.
- The damage to collagen and elastin in the dermis can lead to depressed and raised scars (the scars resulting from cysts are called ice-pick scars).

KEY FACT

The scientific name of the bacteria that causes acne vulgaris is *Propionibacterium acnes.* These bacteria are anaerobic, which means that they do not need oxygen to survive and grow. Although these bacteria are present in all follicles in small numbers, they are prevented from excessive reproduction by the oxygen that is provided by an open follicle. However, once the follicle becomes blocked and the circulation of oxygen ceases, these bacteria multiply and feed off the sebum produced by the overactive sebaceous glands.

There are four different grades of acne, the grade being dependent on the severity of the disorder (Table 3.12).

Table 3.12 Grades of acne

Grade	Characteristics
Grade I (acne simplex – non-inflammatory)	Minor breakout: presence of a few papules and pustules Mainly open comedones present, with some closed comedones Typical in a teenager just beginning puberty
Grade II (acne simplex – non-inflammatory)	Greater incidence of papules and pustules Presence of many closed comedones and more open comedones
Grade III (acne vulgaris – inflammatory)	Skin appears very red and inflamed, with many papules and pustules present
Grade IV (acne vulgaris – inflammatory)	Cysts present with comedones, papules, pustules Skin appears inflamed

In practice

Acne is a complex skin condition and can range in severity from mild breakouts to disfiguring scars and cysts. It requires specialist products and treatment.

Clients with acne which is acutely inflamed (grade III or IV) should be referred to their GP and/or to a dermatologist so that they receive the correct treatment. This might include medication for any infection that has become impacted at the base of the follicles.

Therapists should liaise with other skin care professionals (such as dermatologists) to ensure the correct aesthetic advice and treatment is given to the client.

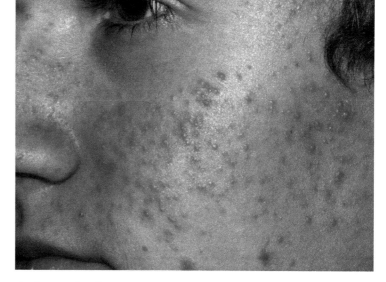

▲ Acne vulgaris

Rosacea

This is a chronic inflammatory disease of the face in which the skin appears abnormally red. The condition usually occurs in adults after the age of 40, but can begin as early as age 20.

The condition develops gradually, beginning with a seeming tendency to blush easily, a red complexion or an extreme sensitivity to cosmetic products.

The distinctive redness appears in a characteristic butterfly pattern across the nose and cheeks. As the condition progresses there may be papules and pustules present. Although the condition may resemble acne, unlike acne, the condition rosacea is rarely if ever accompanied by comedones.

The other distinguishing factors of rosacea are the dry flaky patches that may accompany dry or oily skin.

KEY FACT

As many of the symptoms of rosacea look like those of acne, the condition is often misdiagnosed.

The redness of rosacea often persists after exposure to cold or to irritants like soap. Over time, small blood vessels become more prominent, making the redness more noticeable. Many patients feel stinging or burning sensations and the skin feels tight (like mild sunburn) when smiling, frowning, or squinting. For some clients, all products sting, burn and irritate the face.

A progressive stage of rosacea is characterised by growth and swelling of the nose and central facial tissues. The ears may also be involved. This condition is known as rhinophyma and it can be very disfiguring.

Aggravating factors of rosacea include hot and spicy foods, hot drinks, alcohol, menopause, weather elements and stress.

> ## In practice
>
> Rosacea is a skin condition in which both dermatological and skin care treatments can be helpful. Clients with rosacea should be referred to a dermatologist for diagnosis and management. If the right medication is given, along with the correct skin care treatment, flare-ups can be avoided and the condition can be stabilised.
>
> It is important to avoid products that are harsh, abrasive, fragranced and heavy, and to avoid excessive extraction, steam and very stimulating massage.
>
> Clients should be educated on avoiding known triggers, such as heat, spicy foods and alcohol.

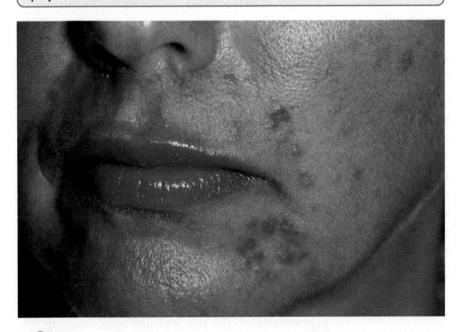

▲ Rosacea

Sebaceous cyst

This type of cyst, which develops from a sebaceous gland, is a round nodular lesion with a smooth, shiny surface. They are usually found on the face, neck, scalp and back. They are situated in the dermis and vary in size from 5 to 50 mm. The cause is unknown.

> ### In practice
> A client who presents with a sebaceous cyst should be referred to their medical practitioner, who may recommend that it is removed surgically.

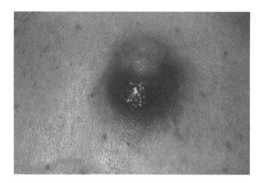

▲ Sebaceous cyst

Seborrhoea

Seborrhoea is defined as an excessive secretion of sebum by the sebaceous glands.

In this condition the glands appear enlarged and the skin appears greasy, especially on the nose and the centre zone of the face. It can resemble acne in that there may be swellings and breakout.

One of the main differences between acne and seborrhoea is that in seborrhoea the increased oil production is often accompanied by scaly, greasy, thickened skin, especially on the scalp.

Seborrhoea is common where there is a high incidence of sebaceous glands (for instance on the scalp and the sides of the nose). Seborrhoea can occur at any age, but is common in infancy (when it is called 'cradle cap') and at puberty due to glandular disturbances.

> ### In practice
> Depending on severity, clients with seborrhoea may need to be referred to their medical practitioner for topical medication to clear the condition.

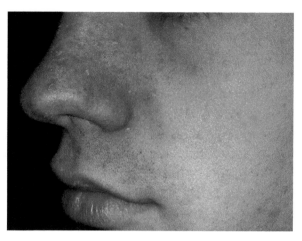

▲ Seborrhoea

Disorders of the sweat glands

Hyperhidrosis

This is the excessive production of sweat, affecting the hands, feet and underarms.

> ### KEY FACT
> Botox injections may be administered to help control hyperhidrosis.

Pigmentation disorders

> ### In practice
> Pigmentation disorders do not necessarily contraindicate all treatments, although care does need to be taken to avoid overstimulation and irritation of the skin, which could further exacerbate an existing condition.
>
> Also, clients should be educated about adequate protection of their skin from sunlight, to avoid further skin damage.

Albinism

This condition is caused by an inherited absence of pigmentation in the skin, hair and eyes, resulting in white hair, very pale skin and pink eyes. The pink colour is produced by the underlying blood vessels, which are normally masked by skin pigment. Other clinical signs of this condition include poor eyesight and sensitivity to light.

▲ Albinism

Chloasma

Chloasma is a pigmentation disorder with irregular areas of increased pigmentation, usually on the face. It commonly occurs during pregnancy and sometimes when taking the contraceptive pill due to stimulation of melanin by the female hormone oestrogen.

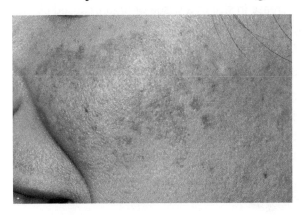

▲ Chloasma

Dermatosis papulosa nigra

Dermatosis papulosa nigra (DPN) is a unique benign skin condition that is common among people with black skin. It is characterised by multiple, small, hyperpigmented, asymptomatic papules.

Small, dark bumps most commonly affect the face, neck, chest and back. The cause of DPN is uncertain. There is a strong genetic basis for the disorder, and often the lesions can be seen in several members of the same family. Under the microscope, the lesions are revealed as a type of harmless keratosis.

DPN is not a skin cancer, and it will not turn into a skin cancer. The condition is chronic, with new lesions appearing over time.

No treatment is necessary other than to alleviate cosmetic concerns. In certain circumstances, if the lesions are symptomatic (painful, inflamed, itchy or catch on clothing) they can be treated via a minor surgical procedure.

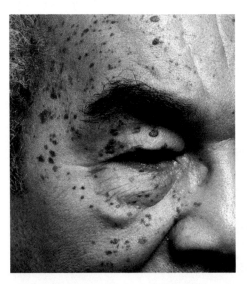

▲ Dermatosis papulosa nigra (DPN)

As black skin is thicker than white skin, it is prone to congestion and comedones. Black skin generally ages at a much slower rate than white skin, mainly due to the extra protection afforded by the melanin. A disadvantage of having more melanin is that it makes the skin more 'reactive'. This means that almost any stimulus, such as a rash, scratch or inflammation may trigger the production of excess melanin, resulting in dark marks or patches on the skin. This is known as **post-inflammatory hyperpigmentation**.

Occasionally some black skins develop a decrease in melanin, or **post-inflammatory hypopigmentation** in response to skin trauma. In either case (hypo- or hyperpigmentation), the light or dark areas may be disfiguring and may take months or years to fade. The increased thickness of the horny layer of black skin can lead to dehydration, which causes increased skin shedding. This can create a grey 'ashen' effect as the loose cells build up on the skin.

Ephelides

Ephelides is also known as freckles. These are small, harmless pigmented areas of skin. They appear where there is excessive production of the pigment melanin (after exposure to sunlight).

Lentigo

These are also known as 'liver spots'. They are flat dark patches of pigmentation found mainly in elderly people on skin that has been exposed to light.

Vitiligo

Vitiligo is a patchy lack of pigmentation in areas of the skin where the basal cell layer of the epidermis no longer produces melanin. The cause is unknown.

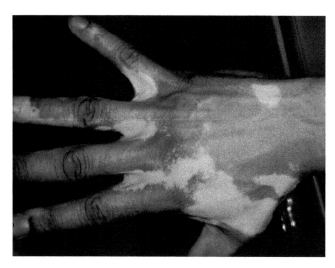

▲ Vitiligo

Naevus

A naevus is a birthmark or other clearly defined malformation of the skin. There are several different types of naevi:

- **Portwine stain** – also known as a 'deep capillary naevus'. These are present at birth and may vary in colour from pale pink to deep purple. They have an irregular shape but are not raised above the skin's surface. Usually, they are found on the face but may also appear on other areas of the body.
- **Spider naevi** – a collection of dilated capillaries radiating from a central papule. These often appear during pregnancy or after 'picking a spot'.

- **Strawberry naevus** – a red raised lump above the skin's surface. These usually develop before or shortly after a baby is born, but disappear spontaneously before the child reaches the age of ten.

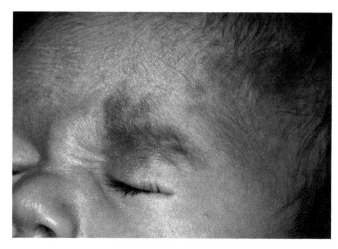

▲ Portwine stain

Hypertrophic disorders

Hypertrophic disorders refer to conditions that result in an increase in size of a tissue or organ, caused by an enlargement of the cells. Hypertrophic disorders of the skin are caused by an enlargement of the skin cells.

Hyperkeratosis

Keratoses are generally defined as a build-up of cells.

Hyperkeratosis is a rare skin disorder in which there is a gross thickening of the skin due to mass of keratinocytes that builds up to a horny overgrowth of skin cells.

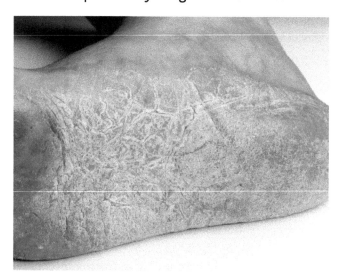

▲ Hyperkeratosis

In practice

Hyperkeratosis is a common problem for people with black skin. Because of the increased cell turnover, black skins desquamate dead skin cells more readily. The accumulation of dead skin cells on the skin's surface can give black skins an ashen grey appearance.

Care needs to be taken during treatment to avoid exfoliating too harshly to avoid irritation and sensitivity.

Skin cancer

Identifying skin cancers

It is important to be aware of the typical characteristics of different skin cancers. Cancers can present as:

- an open sore of any size that bleeds, oozes, or crusts and that remain open for three or more weeks
- a persistent non-healing sore
- a reddish patch or irritated area that doesn't go away, and fails to responds to moisturisers or treatment creams
- a smooth growth with a distinct rolled border and an indented centre
- a shiny bump or nodule with a smooth surface that can be pink, red, white, black, brown or purple in colour
- a white patch of skin that has a smooth, scar-like texture; the area of white stands out from the surrounding skin and can appear clear and taut.

In practice

There is an 'ABCD' rule to help in the identification of skin cancer:

- **Asymmetry** – one part of the lesion is unlike the rest
- **Border** – there is an irregular, scalloped border around the lesion
- **Colour** – colour varies from one area to another, and may appear with shades of tan, brown, black, white, red or blue
- **Diameter** – the area is generally larger than 6 mm across.

In practice

Any client who presents with an abnormal growth, undiagnosed lump or suspicious bump on the skin should be referred to a medical practitioner.

Basal cell carcinoma

This is a common form of skin cancer that originates in the basal cell layer of the epidermis. It is often found on the face and other sun-exposed areas (especially in fair-skinned people).

The most common presentation of basal cell carcinoma is a pearl-like bump, which may be pink or slightly flesh coloured, often with small capillaries running through it.

Superficial basal cell carcinomas appear red, flat and scaly and may be misdiagnosed as other conditions, such as eczema.

Basal cell carcinomas rarely spread to other tissues or organs, and although not life threatening they can produce unpleasant scarring if not detected early.

Malignant melanoma

A malignant melanoma is a deeply pigmented mole-like structure which is life threatening if not recognised and treated promptly. It presents as a blue–black module which increases in size, shape and colour, and is most commonly found on the head, neck and trunk. Overexposure to strong sunlight is a major cause and its incidence is increased in young people with fair skins.

Melanomas can occur in an existing mole or they may arise from normal skin. As they spread very quickly early detection is essential.

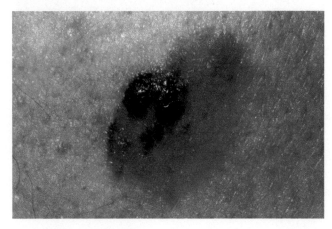

▲ Malignant melanoma

Rodent ulcer

This is a malignant tumour which starts off as a slow-growing pearly nodule, often at the site of a previous skin injury. As the nodule enlarges, the centre ulcerates and will not heal. The centre becomes depressed and the rolled edges become translucent, revealing many tiny blood vessels. Rodent ulcers do not disappear and if left untreated may invade the underlying bone. This is the most common form of skin cancer.

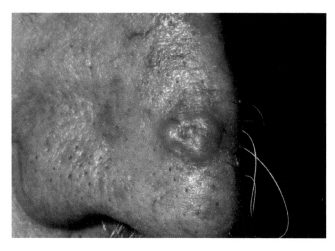

▲ Rodent ulcer

Squamous cell carcinoma

This is a malignant tumour which arises from the prickle cell layer of the epidermis. It is hard and warty, and eventually develops a heaped-up 'cauliflower' appearance. It is most frequently seen in elderly people. Unlike basal cell carcinomas, squamous cell carcinomas can spread to other organs, or deeply within the skin.

Fortunately, 90% of squamous cell carcinomas are detected and removed before they spread.

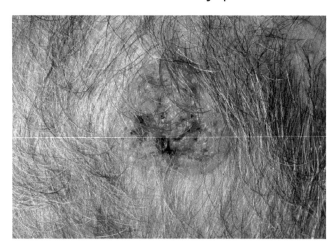

▲ Squamous cell carcinoma

Inflammatory skin conditions

> ### In practice
>
> In the case of an inflammatory skin condition, care should be taken to avoid any form of stimulation (through product or treatment method) that may escalate the inflammation.
>
> If there is severe inflammation and the skin is broken, or there are any signs of infection, treatment would be avoided and the client should be referred to their medical practitioner.

Contact dermatitis

The term *dermatitis* literally means 'inflammation of the skin'. Contact dermatitis is caused by a primary irritant which makes the skin red, dry and inflamed. Substances which are likely to cause this reaction include acids, alkalis, solvents, perfumes, lanolin, detergent and nickel. Affected areas are prone to skin infection.

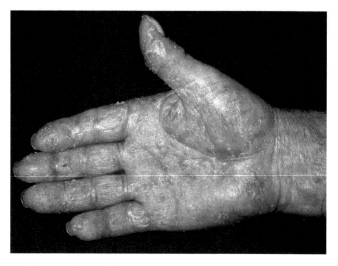

▲ Contact dermatitis

Eczema

This is a mild to chronic inflammatory skin condition characterised by itchiness, redness and the presence of small blisters that may be dry or weep if the surface is scratched. It can cause scaly and thickened skin, mainly at flexures such as the cubital area of the elbows and the back of the knees. Eczema is not contagious. Internal and external influences cause eczema in people with a genetic predisposition.

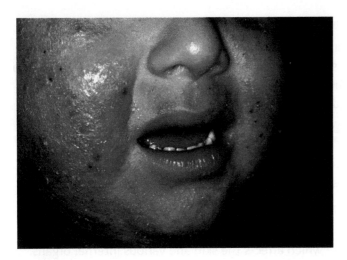

▲ Eczema

Psoriasis

Psoriasis is a genetic chronic inflammatory skin disease associated with a malfunction of the immune system, which causes skin cells to reproduce too quickly. A normal skin cell matures and falls off the body's surface in 28 to 30 days. However, skin affected by psoriasis takes only three to four days to mature and move to the surface. Instead of shedding, the cells pile up and form lesions. The skin also becomes very red due to increased blood flow.

Psoriasis may be recognised by the development of well-defined red plaques, varying in size and shape, and covered by white or silvery scales. Any area of the body may be affected by psoriasis but the most commonly affected sites are the face, elbows, knees, chest and abdomen. It can also affect the scalp, joints and nails.

Psoriasis is aggravated by stress and trauma, but is improved by exposure to sunlight.

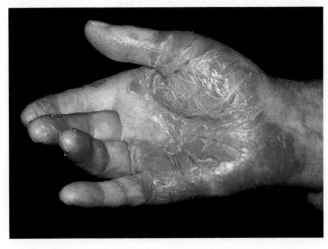

▲ Psoriasis

Seborrheic dermatitis

This is a mild to chronic inflammatory disease of the hairy areas that are well supplied with sebaceous glands. Common sites are the scalp, face, axilla and in the groin. The skin may have a grey tinge or a dirty yellow colour. Clinical signs include slight redness, scaling and dandruff in the eyebrows.

Autoimmune disorders of the skin

In autoimmune conditions the immune system mistakenly recognises a part of the body as foreign and attacks it.

Systemic lupus erythematosus

Systemic lupus erythematosus (SLE) is a chronic inflammatory disease of connective tissue which affects the skin and various internal organs. It is an autoimmune disease that can be diagnosed by the presence of abnormal antibodies in the bloodstream.

It is characterised by a red scaly rash on the nose and cheeks. Other symptoms include joint pain, hair loss and swelling of the feet and fingers.

Discoid lupus erythematosus (DLE) is a form of the disease that primarily affects the skin. Round, firm lesions with red raised bumps form around the hair follicles. These are called discoids.

All forms of lupus are aggravated by sun exposure.

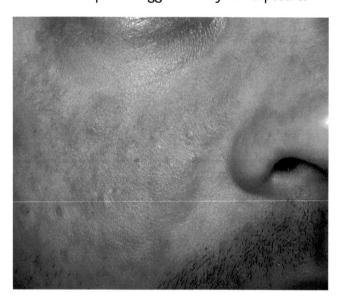

▲ Systemic lupus erythematosus

In practice

A client with lupus should be referred to their medical practitioner. Lupus is not contagious and, following appropriate medical advice on the client's condition, skin care treatments may be offered.

If a skin care service is considered suitable, stimulating products and/or treatments should be avoided. This condition has characteristics in common with a sensitive skin type and should be treated accordingly.

Interrelationships with other systems

Skin, hair and nails

The skin, hair and nails link to the following body systems.

Skeletal

Vitamin D is produced by the skin when exposed to ultraviolet light. Vitamin D is needed in bone formation and maintenance.

Muscular

Muscles provide a supportive function to the skin. Muscles lie directly under the skin and contribute to the skin's tone and elasticity.

Circulatory

Blood clots at the site of an injury. In the case of an external injury, a scab forms on the surface of the skin. This allows the skin to heal and protects underlying structures from any further damage and from infection.

Respiratory

Oxygen that is absorbed into the lungs from inhaled air is delivered to the cells of the skin, hair and nails to aid their renewal.

Nervous

There are numerous sensory nerve endings in the skin that respond to touch, temperature, pain and pressure.

Endocrine

Melanocyte stimulating hormone (MSH) secreted by the central lobe of the pituitary stimulates the production of melanin in the basal cell layer of the skin. The sex (gonadotrophic) hormones influence skin and hair growth during puberty, pregnancy and the menopause.

Digestive

Adipose (fatty) tissue is stored in the subcutaneous layer of the skin when caloric consumption in the daily diet is in excess of daily needs.

Renal

Water is lost from the skin as sweat. The kidneys regulate fluid balance in the body to prevent the skin (and other organs) from becoming dehydrated.

Key words

Acid mantle: a very fine, slightly acidic film on the surface of skin which acts as a barrier to bacteria, viruses and other potential contaminants

Anagen: the active growing phase of the hair life cycle

Apocrine gland: a type of sweat gland found in the genital and underarm regions

Carbuncle: a cluster of boils

Catagen: the transitional stage of hair growth from active to resting

Collagen: protein found in the dermis of the skin

Connective tissue sheath: the part of the hair structure that surrounds the follicle and sebaceous gland, supplies the hair follicle with nerves and blood

Corneocyte: dead skin cell of the stratum corneum

Cortex: the middle layer of a hair, contains the pigment melanin

Cuticle (hair): the outer layer of the hair, which gives the hair its elasticity

Cuticle (nail): the fold of overlapping skin that surrounds the base of the nail, providing a protective seal for the matrix

Dermal papilla: a raised elevation at the base of a hair bulb, contains blood supply

Dermis: the deeper thicker layer of the skin

Desmosome: a cell structure specialised for cell-to-cell adhesion

Desquamation: the shedding of dead skin cells

Eccrine gland: a simple coiled, tubular sweat gland that opens directly onto the surface of the skin

Elastin: protein in the skin (dermis) which gives it its elasticity

Epidermis: the outer thinner layer of the skin

Eponychium: dead cuticle that adheres to the base of the nail, near the lanula

Erector/arrector pili muscle: a small, smooth, weak muscle that attaches to the base of a hair follicle, and makes the hair stand erect in response to cold

Extracellular matrix (ECM): the support system of the dermis, made up of collagen, elastin and glycosaminoglycans (GAGs)

Free edge: part of the nail plate that extends beyond the nail bed and fingertip

Glycosaminoglycans (GAGs): water-binding molecules found in the dermis which give the skin its plumpness

Hair: appendage of the skin which grows from a sac-like depression called a hair follicle

Hair follicle: a sac from which the hair grows

Hair shaft: part of the hair lying above the surface of the skin

Hair bulb: the enlarged part at the base of the hair root

Hair root: the part found below the surface of the skin

Hyaluronic acid: a glycosaminoglycan that exists naturally in the dermis layer of the skin

Hydrolipidic film: an emulsion of fat and water, a film of which normally covers the skin

Hyponychium: the thickened layer of epidermis beneath the free edge of a nail

Inner root sheath: part of the hair follicle that is located between the outer root sheath and the hair shaft; shapes and contours the hair

Keratin: a key structural protein material making up hair, nails and the outer layer of skin; also the protein that protects epithelial cells from damage or stress

Keratinisation: the process cells undergo when they change from living cells with a nucleus to dead cells, filled with keratin and without a nucleus

Keratinocyte: the predominant cell of the epidermis, which serves as a barrier between an organism and its environment

Langerhans cells: special defence cells in the epidermis that set up an immune response to foreign bodies

Lanugo (hair): fine soft hair found on a foetus

Lunula: lightly coloured semicircular area of the nail, commonly called the half moon, which lies between the matrix and the nail plate

Mast cells: cells found in the reticular layer of dermis that secrete histamine during an allergic reaction

Matrix (hair): area of mitotic activity of hair cells, located at the lower part of hair bulb

Matrix (nail): area of nail where the living cells are produced, situated immediately below the cuticle

Mechanoreceptors: sensory receptors in the skin used to detect sensations such as pressure, vibrations and texture

Medulla: the inner layer of hair, which determines sheen and colour of hair

Melanin: a dark brown to black pigment occurring in the hair and skin: responsible for tanning of skin that is exposed to sunlight

Melanocyte: cells present in the epidermis and hair follicles that produce melanin

Melanosome: a melanin-producing granule in a melanocyte

Nail bed: part of the skin on which the nail plate rests

Nail groove: deep ridges under the sides of the nail

Nail mantle/proximal nail fold: a deep fold of skin above the matrix, protecting the nail root from physical damage

Nail plate: the main visible part of the nail which rests on the nail bed and ends at the free edge

Nail wall: a fold of skin overlapping the sides of the nail

Nociceptors: pain receptors in the skin which detect pain that is caused by mechanical, thermal or chemical stimuli

Outer root sheath: part of the hair structure that forms the follicle wall and provides a permanent source of hair germ cells

Papillary layer: the uppermost layer of the dermis

Perionychium: the part of the cuticle that outlines the nail plate

Phagocytic cells: white blood cells that destroy bacteria and other foreign matter found in the dermis

Reticular layer: the lower layer of the dermis

Sebaceous glands: sac-like secreting pouches found all over the body, except for the soles of the feet and palms of the hands

Sebum: oily substance produced by sebaceous glands, which lubricates the hair and skin

Sphingolipids: fats in the stratum spinosum layer of the epidermis that have an important role in the retention of moisture in the skin

Stem cells: a type of cell found in the stratum germinativum (basal cell layer) of the epidermis, involved in the process of skin renewal

Stratum corneum: the most superficial outer layer of the epidermis, consisting of dead skin cells

Stratum germinativum: the deepest of the five layers of the epidermis

Stratum granulosum: the layer of epidermis linking the living cells of epidermis to the dead cells above

Stratum lucidum: the epidermal layer below the most superficial layer (stratum corneum), which consists of small, tightly packed transparent cells that permit light to pass through

Stratum spinosum: the binding and transitional layer between the stratum granulosum and the stratum germinativum

Subcutaneous layer: a thick layer of connective tissue found below the dermis

Telogen: the resting stage of hair growth

Terminal hair: coarse, pigmented hair found on the scalp, underarms, eyebrows, pubic regions, arms and legs

Thermoreceptors: sensory receptors in the skin used to detect sensations related to temperature

Vellus (hair): soft, downy hair found all over the face and body, except for the palms of the hands, soles of the feet, eyelids and lips

Vitreous membrane: the basement membrane of the outer root sheath that separates the outer root sheath from the connective tissue sheath

Revision summary

Skin, hair and nails

- The skin and the appendages derived from it (hair, glands and nails) make up one of the largest organs of the body, known as the integumentary system.
- Functions of the skin include **protection**, **regulation of body temperature**, **sensation**, **excretion**, **storage**, **absorption** and **vitamin D production**.
- The principal parts of the skin are the outer **epidermis** and the inner **dermis**. Beneath the dermis lies the **subcutaneous** layer.
- The **epidermis** is the most superficial part and consists of five layers, from deepest to superficial:
  - **stratum germinativum (basal cell layer)** is the deepest layer and is concerned with cellular regeneration
  - **stratum spinosum (prickle cell layer)** is the next layer up, responsible for cellular transport of the melanosome
  - **stratum granulosum (granular layer)** responsible for keratinisation, cellular change and lipid formation
  - **stratum lucidum (clear layer)** responsible for lipid release
  - **stratum corneum (horny layer)** responsible for acting as a skin barrier defence.
- **Cell regeneration** occurs continuously in the basal cell layer and produces all other layers.
- It takes approximately a month for a new cell to complete its journey from the basal cell layer where it is reproduced, to the granular layer where it is keratinised, to the horny layer where it is desquamated or shed.
- **Stem cells** are found in the stratum germinativum (basal cell layer) of the epidermis and are responsible for constant renewal (regeneration) of skin and for healing wounds.
- The **dermis** is the deeper layer of the skin and provides support, strength and elasticity.
- It has a superficial **papillary layer** and a deeper **reticular layer**:
  - superficial **papillary layer** – consists of adipose connective tissue, dermal papillae, nerve endings and a network of blood and lymphatic capillaries
  - deeper **reticular layer** – consists of tough fibrous connective tissue and contains collagen, elastin and reticular fibres.

- The dermis layer of skin has three crucial components: **collagen**, **elastin** and **glycosaminoglycans (GAGs)**, which all form the bulk of an important support system called the **extracellular matrix (ECM)**.

- The ECM gives the dermis shape, structure and support, providing the **structural scaffolding** and maintaining the tissue architecture.

- Once the skin is broken, the process of wound repair is set in motion by four overlapping phases:
  - **haemostasis** occurs when the platelets clot and form a plug around the site of injury
  - the **inflammatory** phase is when bacteria and cell debris are removed from the wound by white blood cells
  - the **proliferation** phase occurs when the wound contracts and reduces in size, and new connective tissue forms to replace what was there before
  - **maturation** is the final phase, when the tissues are remodelled and scar tissue is formed.

- Appendages of the skin include the **hair**, **glands (sebaceous and sweat)** and **nails**.

- The **hair** is a dead keratinised structure, which grows out of a hair follicle and is divided into three parts: **hair shaft**, **root** and **bulb**.

- The role of a hair is protection.

- Internally, the hair has three layers from the outer to inner layer: **cuticle**, **cortex** and **medulla**.

- The **matrix** in the **hair bulb** is the hair's area of mitotic activity.

- There are three main types of hair in the body: lanugo, vellus and terminal.

- Each hair has its own **hair growth cycle**.

- **Anagen** is the active growing stage; **catagen** is the transitional stage from active to resting; and **telogen** is the short resting stage.

- Nails are made up of mainly **keratin** and are a modification of the horny and clear layers of the epidermis.

- The two main functions of the nail are protection for the fingers and toes, and manipulation of objects.

- Parts of the nail's anatomical structure include:

  - the **matrix**, which is the living, growing area of the nail
  - the **nail plate**, which protects the nail bed from damage
  - the **nail bed**, which provides nourishment and protection for the nail
  - the **cuticle**, which protects the matrix and nail bed from infection
  - the **nail mantle/proximal nail fold**, which protects the matrix from physical damage
  - the **nail wall**, which protects the edges of the nail plate from damage
  - the **lunula**, which is the area where the cells start to keratinise
  - the **nail groove**, which guides the growth of the nail up the fingers
  - the **free edge** which protects the fingertips.

- **Nail growth** occurs from the **nail matrix** by cell division.

- As new cells are produced in the matrix, older cells are pushed forwards and are hardened by **keratinisation** to form the hardened nail plate.

- Other structures of the skin are the erector pili muscle and the glands.
  - The **erector pili muscle** is the weak muscle associated with hair, which contracts when the body feels cold or when experiencing emotions such as fright or anxiety.
  - **Sebaceous glands** are also known as oil glands. They have ducts and are attached to hair follicles.
  - They secrete **sebum**, which is mildly antibacterial and antifungal, to lubricate the hair and the epidermis.
  - **Sweat glands** are located in the dermis and secrete sweat. There are two types of sweat glands, **eccrine** and **apocrine**:
    - **Eccrine glands** are the most numerous and are found in largest concentration in the palms of the hands, and soles of the feet.
    - **Apocrine glands** are attached to the hair follicles and are located in the axilla and groin.

- Factors affecting the skin include diet, water intake, sleep, stress and tension, exercise, alcohol, smoking, medication, chemicals, climate, environment, hormones and age.

Test your knowledge questions

Multiple choice questions

1 In which of these layers are epidermal cells constantly reproduced?
  a stratum corneum
  b stratum granulosum
  c stratum germinativum
  d stratum lucidum

2 Desquamation occurs in which layer of the epidermis?
  a stratum lucidum
  b stratum germinativum
  c stratum spinosum
  d stratum corneum

3 Which is the correct order of the layers of the epidermis, from deepest to most superficial?
  a stratum germinativum, stratum spinosum, stratum lucidum, stratum granulosum, stratum corneum
  b stratum germinativum, stratum spinosum, stratum granulosum, stratum lucidum, stratum corneum
  c stratum germinativum, stratum corneum, stratum lucidum, stratum granulosum, stratum corneum
  d stratum germinativum, stratum granulosum, stratum lucidum, stratum corneum

4 Epidermal stem cells are found in which layer of the epidermis?
  a stratum germinativum
  b stratum spinosum
  c stratum corneum
  d stratum granulosum

5 Which is the thickest layer of the epidermis?
  a stratum spinosum
  b stratum granulosum
  c stratum corneum
  d stratum germinativum

6 The erector pili muscle affects:
  a the hair
  b the sebaceous gland
  c motor nerves
  d the capillary network.

7 What is the function of the sebaceous gland?
  a to secrete sweat
  b to secrete sebum
  c to regulate temperature
  d to insulate

8 Which of the following is responsible for protecting the deeper layers of the skin from ultraviolet damage?
  a keratin
  b melanin
  c sebum
  d carotene

9 Which type of sweat gland is widely distributed throughout the body?
  a apocrine
  b eccrine
  c adipose
  d sebaceous

10 What do Merkel's disks, a type of cutaneous receptor, detect?
  a deep pressure, fast vibrations
  b sustained touch and pressure
  c changes in texture, slow vibrations
  d pain arising from mechanical stimuli

Exam-style questions

11 Name the cell types that make up 95% of cells in the epidermis. 1 mark

12 Which layer of the epidermis is also known as the Malpighian layer? 1 mark

13 a Explain the term *keratinisation*. 1 mark
  b In which layer of the epidermis does keratinisation begin? 1 mark

14 Describe the acid mantle of the skin. 2 marks

15 Briefly describe the stages of skin renewal. 4 marks

16 Which layer of the epidermis is responsible for acting as a skin barrier defence? 1 mark

4 The skeletal system

Introduction

The skeleton is made up of 206 individual bones, which collectively form a strong framework for the body. Bones provide support and protection. They must, however, be linked together in order to allow movement.

Joints provide the links between the bones of the skeletal system. At joints, ligaments hold bones together, offering stability while allowing flexibility. Movement at joints is carried out by associated muscles and tendons.

OBJECTIVES

By the end of this chapter you will understand:

- the functions of the skeleton
- the structure of bone
- the growth and development of bone
- the different types of bone in the body
- the names and positions of the bones of the skeleton
- the different types of joints and their range of movement
- the importance of good posture
- postural defects
- common pathologies of the skeletal system
- the interrelationships between the skeletal and other body systems.

Functions of the skeleton

The study of the structure and function of bones is known as **osteology**. Before learning the individual bones of the skeleton, it is important to understand the functions of the skeleton as a whole.

Protection of vital organs and delicate tissue

The skeleton surrounds vital organs and tissues with a tough and resilient covering. For example, the rib cage protects the heart and lungs, and the vertebral column protects the spinal cord.

Attachments for muscles and tendons

Muscles and tendons are anchored to bones, which provide strong sites of attachment.

Movement

This happens as a result of the co-ordinated action of muscles on bones and joints. Bones are, therefore, **levers** for muscles.

Support

The skeleton bears the weight of all other tissues. Without it we would be unable to stand up. The bones of the vertebral column, pelvis, feet and legs, for example, all support the weight of the body.

Shape

The bones of the skeleton give shape to structures such as the skull, thorax and limbs.

Formation of blood cells

Blood cells develop in red bone marrow, which is found in cancellous bone tissue.

Mineral reservoir

The skeleton acts as a storage depot for important minerals such as calcium, which can be released when needed for essential metabolic processes such as muscle contraction and the conduction of nerve impulses.

The structure of bone

Bone is one of the hardest types of connective tissue in the body and, when fully developed, is composed of:

- water
- protein in the form of collagen fibres, which give bone its tensile strength (resistance to stretching and tearing)
- mineral salts (calcium and phosphate), along with the inorganic mineral hydroxyapatite, which give bone its hardness.

Bone tissue is a type of living tissue that is made from special cells called osteoblasts.

There are two main types of bone tissue:

1 compact

2 cancellous.

All bones have both types of tissue, in different amounts depending on the type of bone.

Compact (dense) bone

This is the hard portion of the bone that makes up the main shaft of the long bones and the outer layer of other bones. It protects inner cancellous (spongy) bone and allows the skeleton to provide a firm framework.

The bone cells in this type of bone (osteocytes) are located in concentric rings (called lamellae) around a central Haversian canal. These canals are small tubes which form a network in bone through which nerves, blood and lymphatic vessels pass.

Cancellous (spongy) bone

Cancellous bone is more porous and lighter in weight than compact bone. It has an open sponge-like appearance. It is found at the ends of long bones and in the centre of other bones. It does not have a Haversian system but consists of a web-like arrangement of spaces that are filled with red bone marrow and separated by the thin processes of bone. Blood vessels run through every layer of cancellous bone, conveying nutrients and oxygen.

Bone marrow

Bones contain two types of marrow – red and yellow:

1   Red marrow manufactures red blood cells. It is found at the end of long bones and at the centre of the bones of the thorax and pelvis.

2   Yellow marrow is found chiefly in the central cavities of long bones. Yellow bone marrow is mainly a fatty tissue.

Structure of a long bone

A long bone consists of several sections:

- **diaphysis** – the long central shaft
- **epiphysis** – the larger rounded ends of long bones
- **epiphyseal cartilage** – the site of bone elongation during the growing years, which is located between the diaphysis and epiphysis. After we stop growing (between 18 and 25 years of age), this cartilage is replaced by compact bone
- **medullary canal/cavity** – the hollow centre of the bone shaft, which contains both red and yellow bone marrow
- **periosteum** – except for the ends that form joints, bones are covered with a thin membrane of connective tissue called the periosteum. The outer layer of the periosteum is extremely dense and contains a large number of blood vessels. The inner layer contains osteoblasts and fewer blood

vessels. The periosteum provides attachment for muscles, tendons and ligaments

- **hyaline/articular cartilage** – smooth hyaline cartilage covers the articular surfaces of the shaft endings of the long bones, where they form a joint with another bone. This is a firm but elastic type of cartilage which provides shock absorbtion to the joint and has no neural or vascular supply.

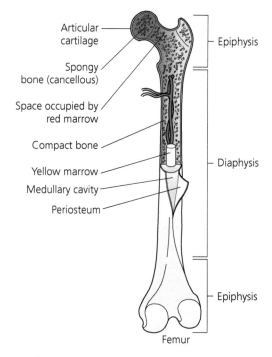

▲ Structure of a long bone

The development of bone

The process of bone development is called **ossification**. Fetal bones are made of cartilage rods. These are changed into bone as the child develops and grows. Ossification begins in the embryo near the end of the second month and is not complete until about the twenty-fifth year of life.

Ossification

Ossification takes place in three stages:

1 The cartilage-forming cells, called chondrocytes, enlarge and arrange themselves in rows, to give a structure like that of the bone they will eventually form.

2 Calcium salts are laid down by bone-building cells called osteoblasts.

3 A second set of cells called osteoclasts, known as cartilage-destroying cells, bring about an antagonistic action, enabling the absorption of any unwanted bone.

A fine balance of osteoblast and osteoclast activity helps to maintain the formation of normal bone.

Osteocytes are mature bone cells that maintain bone throughout life.

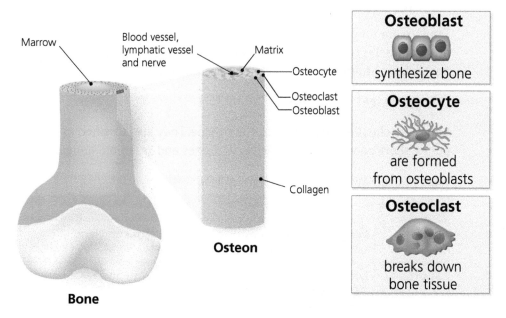

▲ Internal structure of bone

Cartilage

Cartilage is a dense connective tissue that consists of collagen and elastin fibres embedded in a strong gel-like substance. It is a flexible and durable tissue, providing cushioning and absorbing shock, thereby preventing direct transmission of damaging forces to bones.

There are three types of cartilage:

1 **hyaline** – covers the articular bone surfaces

2 **fibrous** – a strong and rigid type of cartilage found between the discs of the spine

3 **elastic** – a very flexible type of cartilage found in the auditory canal of the ear.

Cartilage has no blood supply and, therefore, does not repair or renew itself as easily as bone.

Ligaments

Ligaments are dense, strong and flexible bands of white fibrous connective tissue that link bones together at joints. They are inelastic but flexible, stabilising the joint and allowing the bones to move freely within a safe range.

Tendons

Tendons are tough, white and fibrous cords of connective tissue that attach muscles to the periosteum (fibrous covering) of a bone. Tendons enable bones to move when skeletal muscles contract.

Types of bone

Bones are classified according to their shape. They are classified as long bones, short bones, flat bones, irregular bones and sesamoid bones.

Table 4.1 Overview of the different types of bone

Bone type	Characteristics	Examples
Long	Weight-bearing bones, designed to provide structural support	Arms and legs
Short	Look like blocks Allow a wider range of movement than larger bones	Wrist and ankle bones
Flat	Designed for protection	Skull, scapula, ribs, sternum, pelvic bones
Irregular	Have a variety of shapes Usually have projections that muscles, tendons and ligaments can attach to	Vertebral column, some facial bones
Sesamoid	Small rounded bone embedded in a tendon	Kneecap/patella

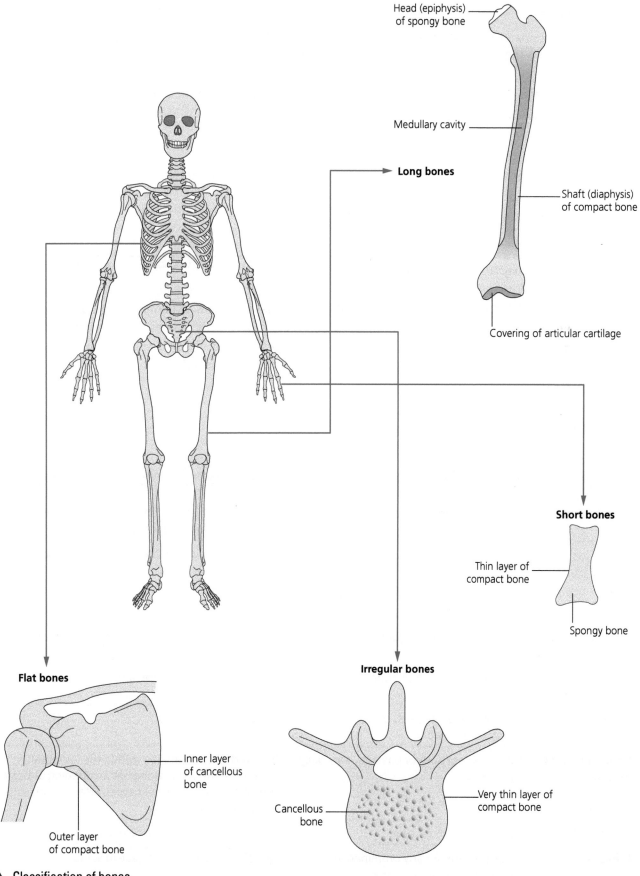

Head (epiphysis) of spongy bone

Medullary cavity

Long bones

Shaft (diaphysis) of compact bone

Covering of articular cartilage

Short bones

Thin layer of compact bone

Spongy bone

Flat bones

Inner layer of cancellous bone

Outer layer of compact bone

Irregular bones

Cancellous bone

Very thin layer of compact bone

▲ Classification of bones

KEY FACT
The smallest bones in the body are in the ear; the hammer, anvil and stirrup.

Long bones

Long bones are hard, dense bones that provide strength, structure and mobility. The thigh bone (femur) is an example of a long bone. All limb bones are long bones, except the wrist and ankle bones.

Long bones have a long shaft (diaphysis) and one or more endings or swellings (epiphysis).

Short bones

Short bones are generally cube shaped with roughly equal lengths and widths. Their primary function is to provide support and stability with little to no movement.

The bones of the wrist and the ankle are examples of short bones.

Flat bones

Flat bones are plate-like structures with broad surfaces. Examples include the ribs and the scapulae.

Irregular bones

Irregular bones don't fit into the three other classifications (flat, long or short bones). Irregular bones have a variety of shapes. Examples include the vertebrae, hips and some of the bones found in the skull.

Sesamoid bones

These are small rounded bones that are embedded in a tendon. The largest sesamoid bone is the patella, which is embedded in the quadriceps femoris tendon.

The surfaces of bones

The surfaces of bones are not always smooth, and they have all kinds of bumps, lumps, dips and ridges. These have specific names that are often used in the descriptions of bones and attachment points for muscle (Table 4.2).

Table 4.2 Types of bone surface

	Description	Example
Acetabulum	Concave surface of a pelvis	The head of the femur meets with the pelvis at the acetabulum, forming the hip joint
Crest	Large ridge of bone	The iliac crest (ilium of the pelvis in the pelvic girdle)
Condyle	Round (knuckle-shaped) prominence or expansion at the end of a bone, most often part of a joint	Condyles of the femur
Depression – also known as fossa(e)	A hollow, usually in a bone	Mandibular fossa of the temporal bone in the skull
Epicondyle	Smaller expansion of bone or projection over a condyle	Medial epicondyle of the humerus
Facet	Small, shallow depression, articulating with another bone	Vertebral articular facet
Foramen (plural foramina)	An opening, hole, or passage, especially in a bone	Intervertebral foramina – within the vertebral column (spine), each bone has an opening at both top and bottom to allow nerves, arteries and veins to pass through
Head	Rounded end of a bone	Head of the fibula (bone of lower leg)
Process	General term for any prominence or prolongation from a bone	Spinal processes of vertebrae in spine

	Description	Example
Protuberance	Knob-like protrusion of a bone	Occipital protuberance of the skull
Spine	Sharp, slender projection of a bone	Spine of the scapula
Trochanter	Large, blunt bump-like projection	Greater trochanter of the femur
Tubercle	Rounded projection of bone, usually blunt and irregular	Lesser tubercle of the humerus
Tuberosity	Large, rounded rough projection of bone, usually serving as the attachment point of muscles or ligaments	Deltoid tuberosity

The bones of the skeleton

The skeletal system is divided into two parts:

1 the **axial skeleton** – made up of 80 bones, this forms the main axis or central core of the body

2 the **appendicular skeleton** – made up of 126 bones, this describes the appendages (limbs) and the places where they attach to the axial skeleton.

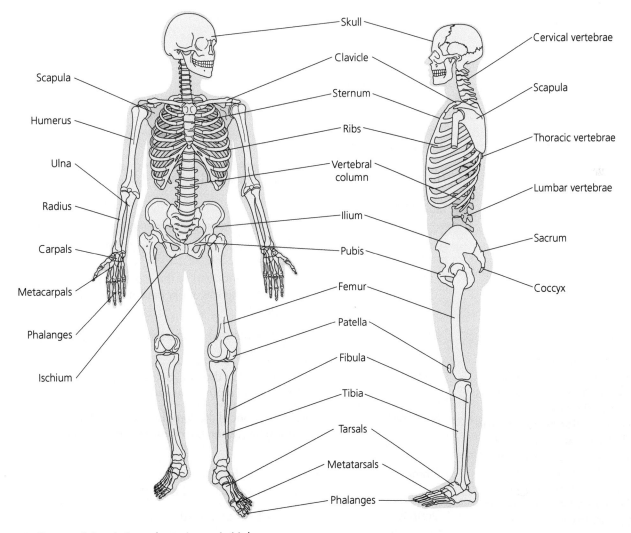

▲ Bones of the skeleton (anterior and side)

The axial skeleton

As the central core of the body, the axial skeleton consists of the following parts:

- skull
- vertebral column
- sternum
- ribs.

The bones of the skull

The skull (cranium) rests on the upper end of the vertebral column, weighs around 5 kg and consists of 22 bones. Eight bones make up the skull or cranium and 14 bones form the facial skeleton. The skull encloses and protects the brain and provides surface attachment points for various muscles of the face, jaw and neck.

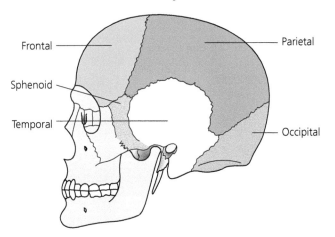

▲ Bones of the skull

The eight bones of the skull are described in Table 4.3.

Table 4.3 The bones of the skull

Name and number of bone(s)	Position
Frontal × 1	Forms the anterior part of the roof of the skull, the forehead and the upper part of the orbits (eye sockets)
Parietal × 2 (pa-ry-it-tal)	Form the upper sides of the skull and the back of the roof of the skull
Temporal × 2	Form the sides of the skull below the parietal bones and above and around the ears
Sphenoid × 1	Located in front of the temporal bone and serves as a bridge between the cranium and the facial bones
Ethmoid × 1	Forms part of the wall of the orbit, the roof of the nasal cavity and part of the nasal septum
Occipital × 1 (ox-sip-it-tal)	Forms the back of the skull

KEY FACT

There are many openings present in the bones of the skull. These holes are passages for the blood vessels and nerves that enter and leave the cranial cavity. An example is the large opening at the base of the skull called the foramen magnum through which the spinal cord and blood vessels pass to and from the brain.

Sutures

A suture is a type of joint between the bones of the skull in which the bones are held tightly together by fibrous tissue.

The four major sutures of the skull are:

- **coronal suture** – joins the frontal bone to the parietal bones
- **sagittal suture** – joins the two parietal bones to one another
- **lambdoid suture** – separates the parietal bones from the occipital bones
- **squamous suture** – separates the parietal bone from the temporal bone.

The bones of the face

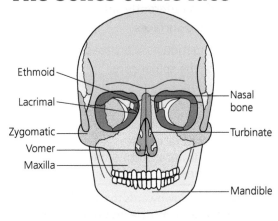

Ethmoid

Lacrimal

Zygomatic

Vomer

Maxilla

Nasal bone

Turbinate

Mandible

▲ Bones of the face

There are 14 facial bones in total. These occur mainly in pairs, one on either side of the face, as shown in Table 4.4.

Table 4.4 The bones of the face

Name and number of bone(s)	Position
Maxilla × 2	These are the largest bones of the face Form the upper jaw and support the upper teeth
Mandible × 1	This is the only movable bone of the skull Forms the lower jaw and supports the lower teeth The mandible is the largest and heaviest bone in the skull
Zygomatic × 2 (zi-go-mat-ik)	These are the most prominent of the facial bones Form the cheekbones

Name and number of bone(s)	Position
Nasal × 2	These small bones form the bridge of the nose
Lacrimal × 2	These are the smallest of the facial bones Located close to the medial part of the orbital cavity
Turbinate × 2	These are layers of bone located either side of the outer walls of the nasal cavities
Vomer × 1	This is a single bone at the back of the nasal septum
Palatine × 2	These are L-shaped bones which form the anterior part of the roof of the mouth

The hyoid bone

Although technically not a facial bone, the hyoid bone is a U-shaped structure located in the anterior neck. It lies at the base of the mandible, where it acts as a site of attachment for the anterior neck muscles.

KEY FACT

A cleft palate occurs when the palatine bones do not fuse during foetal development. Consequently, with the palatine unconnected, an opening exists between the roof of the mouth and the nasal cavity.

The sinuses

The sinuses are four pairs of air-containing spaces in the skull and face.

Their functions are to lighten the head, provide mucus and act as a resonance chamber for sound. The pairs of sinuses are named according to the closest facial bones:

- **frontal** sinuses – located in the forehead, above the eyes and nasal bridge
- **ethmoidal** sinuses – located behind the eyes and in the deeper recesses of the skull
- **sphenoidal** sinuses – located behind the ethmoid sinuses

- **maxillary** sinuses – the largest sinuses are located on either side of the nostrils in the cheekbone area.

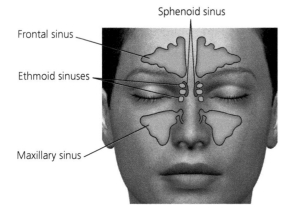

Sphenoid sinus

Frontal sinus

Ethmoid sinuses

Maxillary sinus

▲ Paranasal sinuses

The vertebral column

The vertebral column lies at the posterior of the skeleton, extending from the skull to the pelvis, providing a central axis to the body. It consists of 33 individual irregular bones called vertebrae. However, the bones of the base of the vertebral column, the sacrum and coccyx, are fused to give 24 movable bones in all.

The bones of the vertebral column are described in Table 4.5.

Functions of the vertebral column

The vertebral column:

- provides a strong and slightly flexible axis to the skeleton
- provides a surface for the attachment of muscle groups, by way of its differently shaped vertebrae with their roughened surfaces
- has a protective function as it encases the delicate nerve pathways of the spinal cord.

Table 4.5 The bones of the vertebral column

Vertebrae	Number	Position	Description
Cervical	7	Vertebrae of the neck	Smallest vertebrae in the vertebral column The top two vertebrae, C1 (the atlas) and C2 (the axis) allow the head and neck to move freely
Thoracic	12	Vertebrae of the mid spine, lie in the thorax where they articulate with the ribs	These vertebrae lie flatter and downwards to allow for muscular attachment of the large muscle groups of the back They can be easily felt as you run your fingers down the spine
Lumbar	5	Found in the lower back	These are much larger in size than the vertebrae above them as they are designed to support more body weight These vertebrae can be felt on the lower back due to their large shape and width
Sacral (sacrum)	5	Lies between the pelvic bones	This is a very flat triangular-shaped bone, consisting of five bones which are fused together A characteristic feature of the sacrum is the eight sacral holes, through which nerves and blood vessels penetrate
Coccygeal (coccyx)	4	Base of spine below the sacrum	These are made up of four bones which are fused together and are sometimes referred to as the coccyx (tail bone)

KEY FACT

The intervertebral discs are pads of fibrocartilage that lie between the vertebrae. These give the vertebrae a certain degree of flexibility and also act as shock absorbers, cushioning against mechanical stress.

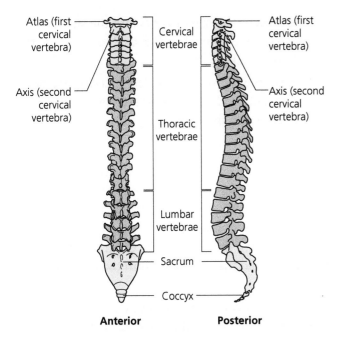

▲ Bones of the vertebral column

The thoracic cavity

This is the area of the body that is enclosed by the ribs, providing protection for the heart and lungs.

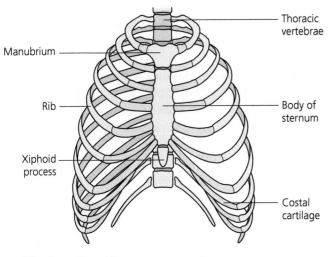

▲ The thoracic cavity

Essential parts that make up the skeleton of this cavity include:

● the sternum
● the ribs
● 12 thoracic vertebrae.

The sternum

This is commonly referred to as the breast bone. It is a flat bone lying just beneath the skin in the centre of the chest. The sternum is divided into three parts:

1 **the manubrium**, the top section, which articulates with the clavicle and the first rib

2 **the gladiolus**, the main and longest part located in the middle, which articulates with the costal cartilages that link the ribs to the sternum

3 **the xiphoid process**, the bottom section, which provides a point of attachment for the muscles of the diaphragm and the abdominal wall.

The ribs

There are 12 pairs of ribs. They articulate posteriorly with the thoracic vertebrae. Anteriorly, the first 10 pairs attach to the sternum via the costal cartilages, the first seven attach directly (known as the true ribs) and the remaining three attach indirectly (known as the false ribs). The last two ribs have no anterior attachment and are called the floating ribs.

The appendicular skeleton

The appendicular skeleton makes up and supports the body's appendages. It consists of the following parts:

● the shoulder girdle
● bones of the upper limbs
● bones of the lower limbs
● bones of the pelvic girdle.

The shoulder girdle

The shoulder girdle connects the upper limbs with the thorax and consists of four bones – two **scapulae** (singular: scapula) and two **clavicles**.

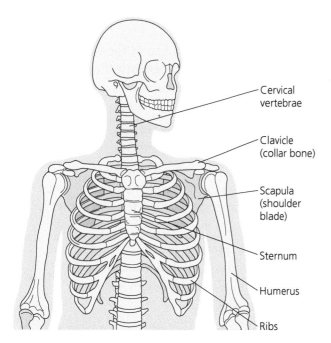

▲ Bones of the neck, chest and shoulder girdle

Labels:
Cervical vertebrae
Clavicle (collar bone)
Scapula (shoulder blade)
Sternum
Humerus
Ribs

The scapula

The scapula is a large flat bone, triangular in outline, which forms the posterior part of the shoulder girdle. It is located between the second and the seventh rib.

The scapula articulates with the clavicle and the humerus, and serves as a point of muscle attachment which connects the shoulder girdle with the trunk and upper limbs.

The scapula has several distinct features:

- The **acromion process** is a large bony projection on the superior end of the scapula. It is an important landmark of the skeletal system and a muscle attachment point that is essential to the function of the shoulder joint. The acromion also forms the acromioclavicular (ac) joint with the clavicle.

- The **spine** of the scapula is a prominent projection of bone that extends across the top of the dorsal surface of the scapula.

- The **glenoid cavity** is the shoulder structure that serves as the 'socket' of the ball-and-socket joint.

The head of the humerus fits into the shallow glenoid cavity, providing the 'ball' to complete the ball-and-socket joint.

- The **coracoid process** is on the anterior side of the scapula and serves as an attachment point for ligaments and muscles.

The clavicle

The clavicle is a long, slender bone with a double curve. It forms the anterior portion of the shoulder girdle. At its medial end it articulates with the top part of the sternum and at its lateral end it articulates with the scapula. The clavicle acts as a brace to hold the arm away from the top of the thorax.

The clavicle provides the only bony link between the shoulder girdle and the axial skeleton. The arrangement of bones and the muscle attached to the scapula and the clavicle allow for a considerable amount of movement of the shoulder and the upper limbs.

Bones of the upper limb

The upper limb consists of the bones described in Table 4.6.

Table 4.6 The bones of the upper limb

Bone(s)	Description
Humerus	Long bone forming the upper arm
Radius	Long bone of the forearm (thumb side)
Ulna	Long bone of the forearm (little-finger side)
Carpals	Eight bones forming the wrist
Metacarpals	Five long bones forming the palm of the hand
Phalanges (fal-ann-g-ees)	Fourteen bones forming the fingers and thumb

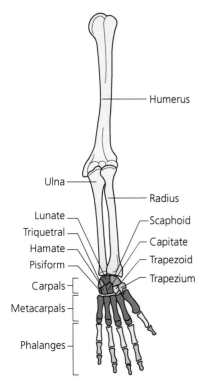

▲ Bones of the upper limb

Upper arm

The **humerus** is the long bone of the upper arm. The head of the humerus articulates with the scapula, forming the shoulder joint. The distal end of the bone articulates with the radius and ulna to form the elbow joint.

Forearm

The **radius** and **ulna** are the long bones of the forearm. The two bones are bound together by a fibrous ring. This allows a rotating movement in which the bones pass over each other. The ulna is the bone of the little-finger side and is the longer of the two forearm bones. The radius is situated on the thumb side of the forearm and is shorter than the ulna. The joint between the ulna and the radius permits a movement called pronation. This is when the radius moves obliquely across the ulna so that the thumb side of the hand is closest to the body. The movement called supination takes the thumb side of the hand to the lateral side. The radius and the ulna articulate with the humerus at the elbow and the carpal bones at the wrist.

The wrist and hand

The wrist consists of eight small bones of irregular size which are collectively called **carpals**. They fit closely together and are held in place by ligaments. The carpals are arranged in two groups of four. Those of the upper row articulate with the ulna and the radius, and those of the lower row articulate with the metacarpals.

> **KEY FACT**
>
> The bones of the upper row nearest the forearm are called the scaphoid, lunate, triquetral and pisiform.
>
> The bones of the lower row are called the trapezium, trapezoid, capitate and hamate.

- **Metacarpals** – there are five long metacarpal bones in the palm of the hand. Their proximal ends articulate with the wrist bones and the distal ends articulate with the finger bones.
- **Phalanges** – there are 14 phalanges. These are the finger bones; two are in the thumb or pollex, and three are in each of the other digits.

The lower limb

The lower limb consists of the bones shown in Table 4.7.

Table 4.7 The bones of the lower limb

Bone(s)	Description
Femur	Long bone forming the thigh
Patella	Bone forming the kneecap
Tibia	Long bone of the lower leg (anterior, medial side)
Fibula	Long bone of the lower leg (lateral side)
Tarsals	Seven bones forming the ankle
Metatarsals	Five bones forming the dorsal surface of the foot
Phalanges (fal-ann-g-ees)	Fourteen bones forming the toes

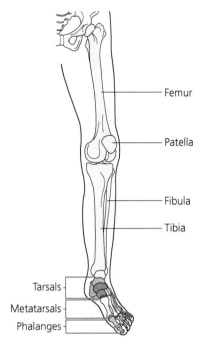

▲ Bones of the lower limb

The thigh

The **femur** is the bone of the thigh. It is the longest bone in the body and has a shaft and two swellings at each end. The proximal swelling has a rounded head like a ball, which fits into the socket of the pelvis to form the hip joint. Below the neck are swellings called trochanters which are sites for muscle attachment. The distal ends of the femur articulate with the patella, or kneecap.

The **patella** is located anterior to the knee joint. Its main function is to provide stabilisation, to cushion the hinge joint at the knee and protect the knee by shielding it from impact.

The lower leg

The **tibia** and **fibula** are the long bones of the lower leg.

- The tibia is situated on the anterior and medial side of the lower leg. It has a large head where it joins the knee joint and the shaft leads down to form part of the ankle. The tibia is the larger of the two bones of the lower leg and thus carries the weight of the body.

- The fibula is situated on the lateral side of the tibia in the lower leg and is the shorter and thinner of the two bones. The end of the fibula forms part of the ankle on the lateral side.

The foot

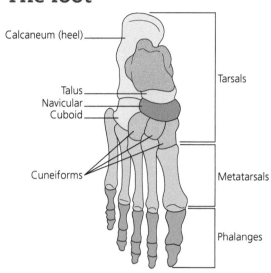

▲ Bones of the foot

There are seven bones in the foot which are collectively called the **tarsals**. Each tarsal is an irregular bone that slides over the next bone to collectively provide motion. The individual tarsals are listed:

- **Talus** – the main tarsal. This articulates with the tibia and fibula to form the ankle joint. The talus is significant in that it bears the weight of the entire body when standing or walking.

- **Calcaneum** – also known as the heel bone. It is the largest and most posterior tarsal bone. The calcaneum is an important site for attachment of muscles of the calf.

- **Cuboid** – this bone is situated between the fourth and fifth metatarsals and the calcaneum on the lateral (outer) border of the foot.

- **Cuneiforms** – there are three cuneiform bones which are located between the navicular bone and the first three metatarsal bones. They are numbered from I through to III (the most medial being I, the middle being II and the most lateral being III).

- **Navicular** – the navicular bone is situated between the talus bone and the three cuneiforms.

- **Metatarsals** – there are five metatarsals which form the dorsal surface of the foot.

- **Phalanges** – 14 phalanges form the toes, two of which are in the hallux, or big toe, and three of which form each of the other digits.

Arches of the foot

The bones of the feet form arches which are designed to support body weight and to provide leverage when walking. The arches of the foot are maintained by ligaments and muscles. These give the foot resilience when running or walking. The arches of the foot are the:

- **medial longitudinal arch** which runs along the medial side of the foot from the calcaneum bone to the end of the metatarsals
- **lateral longitudinal arch** which runs along the lateral side of the foot from the calcaneum bone to the end of the metatarsals
- **transverse arch** which runs between the medial and lateral aspect of the foot and is formed by the navicular, three cuneiforms and the bases of the five metatarsals.

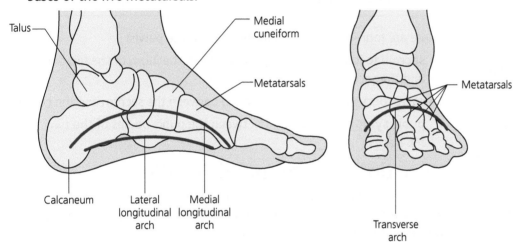

▲ Arches of the feet

The pelvic girdle

The pelvic girdle consists of two hip bones (also known as innominate bones) which are joined together at the back by the sacrum and at the front by the symphysis pubis.

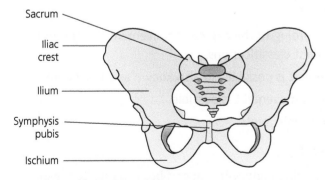

▲ Bones of the pelvic girdle

Each hip bone consists of three separate bones which are fused together, as described in Table 4.8.

Table 4.8 The bones of the hip

Pelvic bone	Position	Description
Ilium	Forms the superior (upper) part of the pelvic girdle	The largest and most superior pelvic bone in the pelvic girdle The iliac crest is the upper border and is an important site of attachment for muscles of the anterior and posterior abdominal walls
Ischium	Forms the inferior (lower) and posterior (back) part of pelvic girdle	The ischial tuberosity is a bony protrusion which is the part of the ischium that you sit on Receives the weight of the body when sitting and provides muscle attachments for the muscles, such as the hamstrings and the adductors
Pubis	Collective name for the two pubic bones in the most anterior (forward) portion of the pelvis	Two pubic bones resemble a wishbone and are linked via a piece of cartilage called the symphysis pubis The pubic bones provide attachment sites for some of the abdominal muscles and fascia

Activity

Write down the names of the individual bones of the skeleton on separate pieces of card or paper. Turn them face down and mix them up. One by one, turn over each card and place it into the correct category:

- Bones of the upper limb
- Bones of the lower limb
- Bones of the face
- Bones of the skull
- Bones of the vertebral column
- Bones of the pelvic girdle

There are four joints or articulations within the pelvis:

- **sacroiliac joints (×2)** – located between the ilium of the hip bones, and the sacrum
- **sacrococcygeal symphysis** – found between the sacrum and the coccyx
- **pubic symphysis** – located between the pubis bodies of the two hip bones.

Functions of the pelvic girdle

The pelvic girdle supports the vertebral column, bearing the body's weight, and offers protection by encasing delicate organs such as the uterus and bladder.

The joints of the body

A joint is formed where two or more bones (or sections of cartilage) meet and is otherwise known as an **articulation**. Where a bone acts as a lever in a movement, the joint is the fulcrum, or the support which steadies the movement and allows the bone to move in certain directions.

Types of joint

Joints are classified according to the degree of movement they permit. There are three main joint classifications:

1 **fibrous** – no movement is possible (so also known as a fixed joint)

2 **cartilaginous** – slight movement is possible

3 **synovial** – freely movable joints.

1 Fibrous joints

Fibrous joints are immovable joints with tough fibrous tissue between the bones. Often the edges of the bones are dovetailed together, as in the sutures of the skull. Some examples of fibrous joints include the joints between the teeth, and between the maxilla and mandible of the jaw.

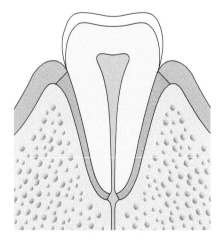
▲ A fibrous joint

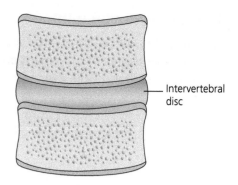

Intervertebral disc

▲ A cartilaginous joint

2 Cartilaginous joints

Cartilaginous joints are slightly movable joints which have a pad of fibrocartilage between the ends of the bones that make up the joint. The pads act as shock absorbers. Some examples of cartilaginous joints are those between the vertebrae of the spine and at the symphysis pubis, between the pubis bones.

3 Synovial joints

Synovial joints are freely movable joints which have a more complex structure than the fibrous or cartilaginous joints.

Before looking at the different types of synovial joints, it is important to have an understanding of the general structure of a synovial joint.

The general structure of a synovial joint

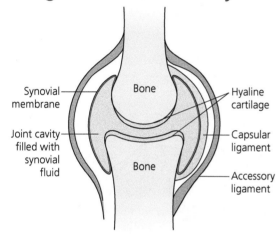

Synovial membrane

Joint cavity filled with synovial fluid

Bone

Bone

Hyaline cartilage

Capsular ligament

Accessory ligament

▲ A synovial joint

- A synovial joint has a space between the articulating bones. This is known as the synovial cavity.

- The surface of the articulating bones is covered by hyaline cartilage, which provides a hard-wearing surface that enables the bones to move against one another with the minimum of friction.

- The synovial cavity and the cartilage are encased within a fibrous capsule that helps to hold the bones together, enclosing the joint. The joint capsule is reinforced by tough sheets of connective tissue called ligaments – these bind the articular ends of bones together.

- The joint capsule is reinforced enough to resist dislocation but is flexible enough to allow movement.

- The inner layer of the joint capsule is formed by the synovial membrane which secretes a sticky, oily fluid called synovial fluid. This fluid lubricates the joint and nourishes the hyaline cartilage.

- As the hyaline cartilage does not have a direct blood supply, it relies on the synovial fluid to deliver oxygen and nutrients, and to remove waste from the joint via the synovial membrane.

Types of synovial joints

Synovial joints are classified into six different types according to their shape and the movements possible at each one. The degree of movement possible at each synovial joint is dependent on the type of synovial joint and its articulations (Table 4.9).

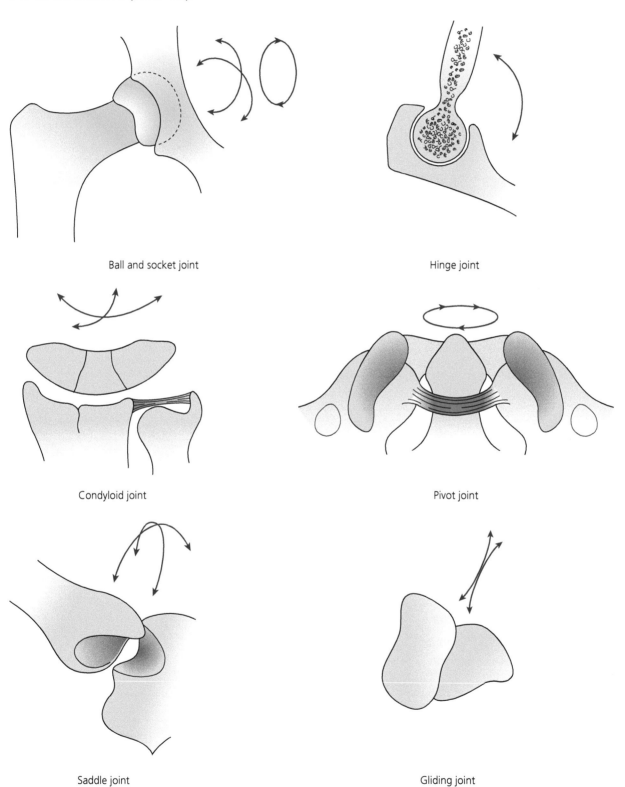

Ball and socket joint

Hinge joint

Condyloid joint

Pivot joint

Saddle joint

Gliding joint

▲ Types of synovial joint

Table 4.9 Types of synovial joints

Type of synovial joint	Description	Movement	Examples
Ball-and-socket	Formed when the rounded head of one bone fits into a cup-shaped cavity of another bone	Allows movement in many directions around a central point: flexion, extension, adduction, abduction, rotation and circumduction	Hip and shoulder joints
Hinge	Where the rounded surface of one bone fits the hollow surface of another bone	Movement is only possible in one direction Allows flexion and extension	Knee and elbow joints Joints between the phalanges
Condyloid	The joint surfaces are shaped so that the concave surface of one bone can slide over the convex surface of another bone in two directions	Although a condyloid joint allows movement in two directions, one movement dominates Movements include: flexion, extension, adduction and abduction	Wrist joint Joint between the metacarpals and phalanges (metacarpophalangeal joints)
Gliding	Often referred to as synovial plane joints as these occur where two flat surfaces of bone slide against one another	Allow only a gliding motion in various planes (side to side, and back and forth)	Joints between vertebrae and sacroiliac joint
Pivot	Occurs where a process of bone rotates in a socket One component is shaped like a ring and the other component is shaped so that it can rotate within the ring	Only permits rotation	Joint between the first and second cervical vertebrae (atlas and axis) and joint at the proximal ends of the radius and the ulna
Saddle	Shaped like a saddle Articulating surfaces of bone have both rounded and hollow surfaces so that the surface of one bone fits the complementary surface of the other	Movements include: flexion, extension, adduction, abduction and a small degree of axial rotation	Thumb joint

 Activity

Work with a partner to test your knowledge of joints.

Ask your partner to demonstrate the angular movements possible at the following joints:

- ball-and-socket
- pivot.

Glossary of angular movements possible at joints

Flexion	Bending of a body part at a joint so that the angle between the bones is decreased	 ▲ Flexion
Extension	Straightening of a body part at a joint so that the angle between the bones is increased	 ▲ Extension
Dorsiflexion	Upward movement of the foot so that feet point upwards	 ▲ Dorsiflexion
Plantar flexion	Downward movement of the foot so that feet face downwards towards the ground	 ▲ Plantar flexion
Adduction	Movement of a limb towards the midline	 ▲ Adduction

Abduction	Movement of a limb away from the midline	▲ Abduction
Rotation	Movement of a bone around an axis (180°)	▲ Rotation
Circumduction	A circular movement of a joint (360°)	▲ Circumduction
Supination	Turning the hand so that the palm is facing upwards	▲ Supination
Pronation	Turning the hand so that the palm is facing downwards	▲ Pronation

Eversion	Soles of the feet face outwards	
		▲ Eversion
Inversion	Soles of the feet face inwards	
		▲ Inversion

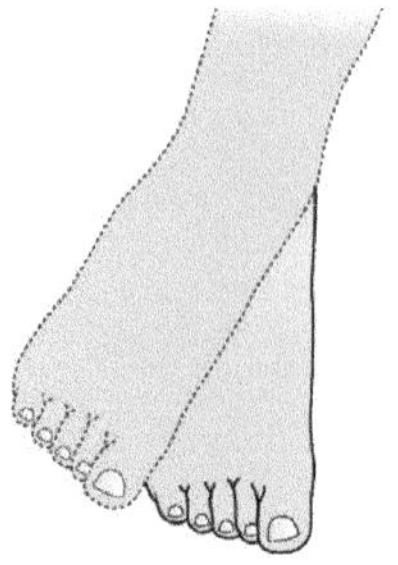

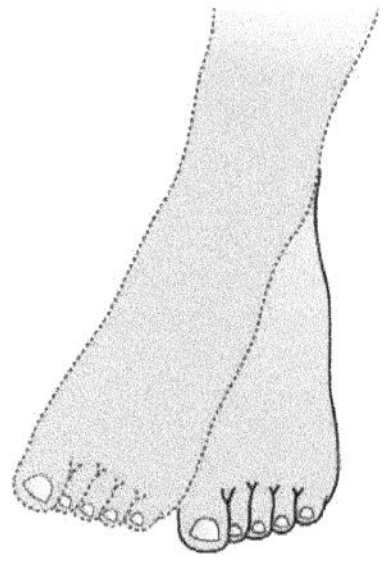

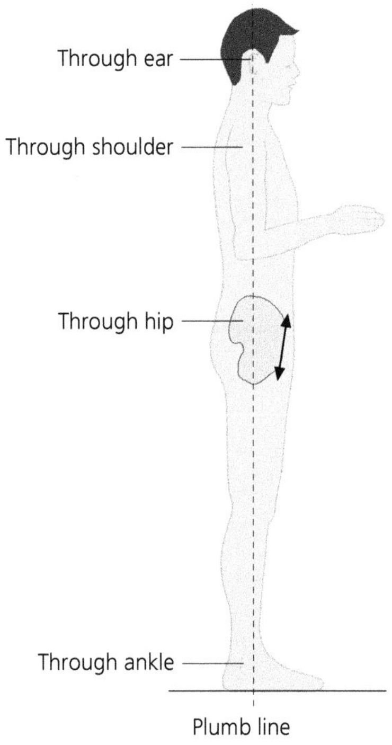

Posture

Posture describes body alignment and balance. It relies on the strength and tone of the body's muscles as they work against gravity.

Good posture

Good posture is when the maximum efficiency of the body is maintained with the minimum effort.

When evaluating posture, an imaginary line is drawn vertically through the body. This is called the centre of gravity line. From the front or back this line should divide the body into two symmetrical halves.

In good standing posture the following are observed:

- with feet together, the ankles and knees touch
- hips are the same height
- shoulders are level
- the sternum and vertebral column run down the centre of the body in line with the centre of gravity line
- head is erect and not tilted to one side.

Posture varies considerably in individuals and is influenced by factors such as body frame size, heredity, occupation, habits and personality. Additional factors which may also affect posture include clothing, shoes and furniture.

The importance of good posture

Good posture is important as it:

- allows a full range of movement
- improves physical appearance

Through ear
Through shoulder
Through hip
Through ankle
Plumb line

▲ Good posture

- keeps muscle action to a minimum, thereby conserving energy and reducing fatigue
- reduces susceptibility to injury
- helps the body's systems to function efficiently.

Poor posture

Poor posture:

- produces alterations in body function and movement
- wastes energy
- increases fatigue
- increases the risk of backache and headaches
- impairs breathing
- increases the risk of muscular, ligament or joint injury
- affects circulation
- affects digestion
- gives a poor physical appearance.

Postural defects

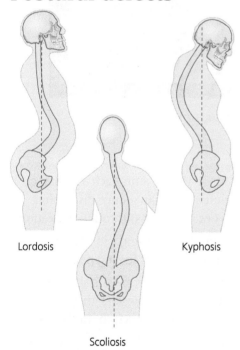

Lordosis

Kyphosis

Scoliosis

▲ Postural defects

Kyphosis (ky-fo-sis)

Kyphosis is an abnormally increased outward curvature of the thoracic spine. In this condition the back appears round as the shoulders point forwards and the head moves forwards. A tightening of the pectoral muscles is common in this condition.

Lordosis

Lodosis is an abnormally increased inward curvature of the lumbar spine. In this condition the pelvis tilts forwards and as the back is hollow, the abdomen and buttocks protrude and the knees may be hyperextended. Lordosis can cause tightening of the back muscles and weakening of the abdominal muscles. The anterior tilt of the pelvis can lead to hamstring problems. Lordosis is commonly exacerbated by increased weight gain or pregnancy.

Scoliosis (sko-lee-o-sis)

This is a lateral curvature of the vertebral column, either to the left or right side. Evident signs of this condition include unequal leg length, distortion of the rib cage, unequal position of the hips or shoulders and curvature of the spine (usually in the thoracic region).

KEY FACT
Poor posture or misalignment of the body is frequently found to be the cause of continued or chronic pain as the body makes compensatory changes which are habit forming.

Common pathologies of the skeletal system

In practice
When treating clients with a joint disorder, position the client so they are comfortable; remember that extra cushioning and support may be required.

Ankylosing spondylitis

This is a systemic joint disease characterised by inflammation of the intervertebral disc spaces, costovertebral and sacroiliac joints (costovertebral joints are those that connect the ribs to the thoracic spine – a plane synovial joint that only permits gliding). Fibrosis, calcification, ossification and stiffening of joints are common and the spine becomes rigid. Typically, clients with this condition complain of persistent or intermittent lower back pain. Kyphosis is present when the thoracic or cervical regions of the spine are affected and the weight of the head compresses the vertebrae and bends the spine forwards. This condition can cause muscular atrophy, loss of balance and falls. Typically this disease affects young male adults.

In practice
Avoid forcibly mobilising ankylosed joints and in the case of cervical spondylitis avoid hyperextending the neck.

Arthritis – gout

This is a joint disorder due to deposition of excessive uric acid crystals in the joint cavity. It affects the peripheral joints, commonly the metatarsophalangeal joint of the big toe. Kidneys can be affected. Other cartilage may be involved including the ear pinna.

Arthritis – osteoarthritis

This is a joint disease, also known as degenerative arthritis, which is characterised by the breakdown of articular cartilage, growth of bony spikes and swelling of the surrounding synovial membrane. It involves varying degrees of joint pain, stiffness, limitation of movement, joint instability and deformity. This form of arthritis is common in elderly people and takes a progressive course, usually affecting the weight-bearing joints – the hips, knees, lumbar and cervical vertebrae.

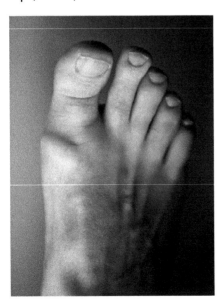

▲ Osteoarthritis

In practice
Passive and gentle friction movements (rubbing) around the joint may be beneficial if the client experiences minimal pain with the condition, but excessive movement may cause joint pain and damage.

Always ask the client to demonstrate the range of movement possible at each joint to guide you as to the limitations of treatment.

Arthritis – rheumatoid

This is a chronic inflammation of peripheral joints resulting in pain, stiffness and potential joint damage. It can cause severe disability. Joint swellings and rheumatoid nodules are tender.

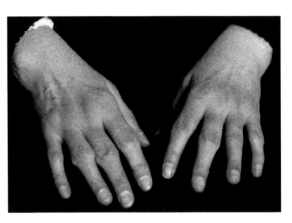

▲ Rheumatoid arthritis

In practice
Although therapeutic treatments such as massage cannot cure arthritis, they can reduce discomfort and help to prevent its progress through relaxation.

Take care when gently mobilising a joint and always ensure the client is not feeling pain.

Treatments of shorter duration are suitable. Note that, if the client is taking pain killers, they may be unable to give adequate feedback about their level of discomfort.

Bunion

This is a swelling of the joint between the big toe and the first metatarsal. Bunions are usually caused by ill-fitting shoes and are made worse by excessive pressure.

▲ Bunion

Bursitis

This condition is the inflammation of a bursa (small sac of fibrous tissue that is lined with synovial membrane and filled with synovial fluid). It usually results from injury or infection and produces pain, stiffness and tenderness of the joint adjacent to the bursa.

Dupuytren's contracture

This is the forward curvature of the fingers (usually the ring and little fingers) caused by contracture of the fibrous tissue in the palm and fingers.

Fracture

A fracture is a breakage of a bone, either complete or incomplete. There are six different types:

- **simple fracture** (also known as a closed fracture) – a clean break with little damage to surrounding tissues and no break in the overlying skin
- **compound fracture** – an open fracture where the broken ends of the bone protrude through the skin
- **comminuted fracture** – where the bone has splintered at the site of impact and smaller fragments of bone lie between the two main fragments
- **greenstick fracture** – only occurs in children and is a partial fracture in which one side of the bone is broken and the other side bends
- **impacted fracture** – where one fragment of bone is driven into another

- **complicated fracture** – occurs when a broken bone damages tissues and/or organs around it.

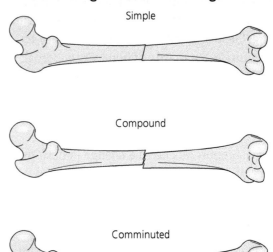
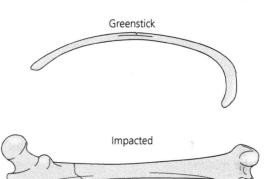

▲ Six types of fracture

Frozen shoulder (adhesive capsulitis)

This chronic condition causes pain, stiffness and reduced mobility (or locking) of the shoulder joint. This may follow an injury, stroke or myocardial infarction or may develop due to incorrect lifting or a sudden movement.

Hammer toe

A hammer toe is a deformity that causes a toe to bend or curl downwards instead of pointing forwards. This deformity can affect any toe on the foot, but most often affects the second or third one. Although a hammer toe may be present at birth, it usually develops over time due to arthritis, pressure from a bunion, a traumatic toe injury, an unusually high arch or wearing ill-fitting shoes, such as tight, pointed high-heels.

Osteoporosis

This condition causes brittle bones and is due to the fall in the level of the hormone oestrogen that occurs with ageing. Oestrogen affects the body's ability to deposit calcium in the matrix of bone. Osteoporosis can also result from prolonged use of steroids. Vulnerability to the condition can be inherited. Bones can break easily and vertebrae can collapse.

Women over the age of 50 are most at risk of developing osteoporosis due to a change in hormones (the decline of oestrogen) and the fact they have smaller, thinner bones than men.

In practice

Take care when handling clients with osteoporosis as they may have bone tenderness.

Avoid vigorous movements as there is a chance of spontaneous bone fracture and be aware that any movement may cause pain.

There is the potential for vertebral damage, so take particular care with the client's comfort.

Spina bifida

This is a congenital defect of the vertebral column in which the halves of the neural arch of a vertebra fail to fuse in the midline.

Sprain

A sprain is the injury to a ligament caused by overstretching or tearing. It occurs when the attachments to a joint are stressed beyond their normal capacity, resulting in pain and swelling. The ankle joint and lower back are most often sprained.

Stress

Stress can be defined as any factor which affects physical or emotional health. Examples of excessive physical/mechanical stress on the skeletal system include poor posture, stiff joints and repetitive strain injuries.

Synovitis

This is the inflammation of a synovial membrane in a joint.

Temporomandibular joint tension (TMJ syndrome)

This is a collection of symptoms and signs produced by disorders of the temporomandibular joint. It is characterised by bilateral or unilateral muscle tenderness and reduced motion. It presents with a dull aching pain around the joint, often radiating to the ear, face, neck or shoulder. The condition may start off with clicking sounds in the joint. There may be protrusion of the jaw or hypermobility and pain on opening the jaw. It slowly progresses to decreased mobility of the jaw, and locking of the jaw may occur. Causes include chewing gum, biting nails, biting off large chunks of food, habitual protrusion of the jaw, tension in the muscles of the neck and back, and clenching of the jaw. It may also be caused by injury, especially whiplash, or other trauma to the joint.

Whiplash

This condition is caused by damage to the muscles, ligaments, intervertebral discs or nerve tissues of the cervical region by sudden hyperextension and/or flexion of the neck. The most common cause is a road traffic accident when acceleration/deceleration causes a sudden stretch of the tissue around the cervical spine. It may also occur as a result of hard impact sports. It can present with pain and limitation of neck movements with muscle tenderness, which can start hours to days after the accident and may take months to recover.

In practice

Whiplash may last for a few months or even years; in the acute stages avoid manipulating and vigorously moving the neck.

Interrelationships with other systems

The skeletal system

The skeletal system links to the following body systems.

Skin

Vitamin D is produced in the skin. It has a role in helping bones absorb calcium in order to keep them strong and healthy.

Muscular

Muscles pull on bones at joints in order to effect movement.

Circulatory

Erythrocytes are produced in the bone marrow of long bones.

Digestive

Food that is ingested in the digestive system is broken down and vital nutrients, such as calcium and phosphorus, are carried in the blood to the bones.

Nervous

Skeletal muscles require stimulation from a nerve impulse in order to contract and produce movement.

Endocrine

Growth hormones produced by the pituitary gland are responsible for the growth rate of bones in childhood.

Key words

The skeletal system comprises two parts:

- The **axial skeleton** forms the main axis or central core of the body and consists of the bones of the skull, vertebral column, sternum and ribs.

- The **appendicular skeleton** is the part of the skeleton that supports and makes up the appendages or limbs. It consists of the shoulder girdle, bones of the upper limbs, lower limbs and bones of pelvic girdle.

Ball-and-socket joint: a type of synovial joint that allows multidirectional movement and rotation (an example is the hip joint)

Bone: a form of dense connective tissue that makes up the majority of the skeleton

Cancellous bone: a lightweight type of bone tissue with an open sponge-like appearance that is found at the ends of long bones or at the centre of other bones

Carpals: eight irregular shaped bones forming the wrist

Cartilage: flexible connective tissue found in the articulating surfaces of joints

Cartilagenous joint: a slightly movable joint which has a pad of fibrocartilage between the end of the bones making up the joint

Cervical vertebrae: vertebrae of the neck, made up of seven bones

Chondrocytes: cells found in cartilage connective tissue

Clavicle: a long slender bone that forms the anterior portion of the shoulder girdle

Coccyx (coccygeal vertebrae): the tail bone at the base of the spine, formed by four fused coccygeal vertebrae

Compact bone: type of hard bone tissue that makes up the main shaft of the long bones and the outer layer of other bones

Condyloid joint: a type of synovial joint in which the joint surfaces are shaped so that the concave surface of one bone can slide over the convex surface of another bone in two directions

Diaphysis: the central shaft of a long bone

Epiphyseal cartilage: the site of bone elongation, located between the diaphysis and epiphysis

Epiphysis: the larger rounded ends of long bones

Ethmoid: facial bones forming part of the wall of the orbit, the roof of the nasal cavity and part of the nasal septum

Femur: the long bone of the thigh

Fibrous joint: an immovable joint with tough fibrous tissue between the bones

Fibrous joint capsule: an envelope surrounding a synovial joint

Fibula: long bones of the lower leg (lateral side)

Frontal: the bone that forms the anterior part of the roof of the skull

Gliding joint: a type of synovial joint in which the opposed surfaces are flat or only slightly curved, so that the bones slide against each other in a simple and limited way

Hinge joint: a type of synovial joint that allows movement in only one plane, forwards and backwards (the elbow, for example)

Humerus: the long bone forming the upper arm

Hyoid: a U-shaped structure located in the anterior neck

Ilium: the largest and most superior pelvic bone in the pelvic girdle

Ischium: the bone forming the inferior (lower) and posterior (back) part of pelvic girdle

Joint: the point at which two or more bones (or cartilage) meet

Kyphosis: a postural defect in which there is an abnormally increased outward curvature of the thoracic spine

Lacrimal: smallest of the facial bones, located close to the medial part of the orbital cavity

Lateral longitudinal arch: one of the arches of the foot that runs along the lateral side of the foot from the calcaneum bone to the end of the metatarsals

Ligament: dense, strong flexible bands of white fibrous connective tissue that link bones together at a joint

Lordosis: a postural defect in which there is an abnormally increased inward curvature of the lumbar spine

Lower limb: part of the skeleton that includes the hip, knee and ankle joints, and the bones of the thigh, leg and foot

Lumbar vertebrae: vertebrae of the lower back (× 5)

Mandible: the bone in the face forming the lower jaw

Maxilla: bones of the face forming the upper jaw

Medial longitudinal arch: one of the arches of the foot that runs along the medial side of the foot from the calcaneum bone to the end of the metatarsals

Medullary canal/cavity: the hollow centre of the bone shaft containing both red and yellow bone marrow

Metacarpals: five long bones forming the palm of the hand

Metatarsals: five bones forming the dorsal (top) surface of the foot

Nasal: small bones that form the bridge of the nose

Occipital: the bone forming the back of the skull

Ossification: the process of bone development

Osteoblasts: cells that make bone

Osteoclasts: cells that break down bone tissue

Osteocytes: mature bone cells that maintain bone throughout life

Osteology: the study of the structure and function of bones

Palatine: L-shaped bones which form the anterior part of the roof of the mouth (× 2)

Parietal: two bones that form the upper sides of the skull and the back of the roof of the skull

Patella: the bone forming the kneecap

Pelvic girdle: a ring-like structure, located in the lower part of the trunk and connecting the axial skeleton to the lower limbs

Periosteum: the fibrous covering of a long bone

Phalanges: bones forming the fingers and thumb

Pivot joint: a type of synovial joint that permits rotation

Pubis: the collective name for the two pubic bones in the most anterior (front) portion of the pelvis

Radius: the long bone of the forearm (thumb side)

Ribs: a series of slender curved bones articulated in pairs to the spine (12 pairs in humans), protecting the thoracic cavity

Sacrum: a large, triangular bone at the base of the spine that is formed by the fusing of sacral vertebrae

Saddle joint: type of synovial joint where the articulating surfaces of bone have both rounded and hollow surfaces so that the surface of one bone fits the complementary surface of the other

Scapula: a large, flat bone, triangular in outline, which forms the posterior part of the shoulder girdle

Scoliosis: a postural defect which presents with a lateral curvature of the vertebral column, either to the left or right side

Shoulder girdle: connects the upper limbs with the thorax and consists of four bones – two scapulae and two clavicles

Sinuses: air-containing spaces in the skull and face

Sphenoid: one of the bones of the skull located in front of the temporal bone

Sternum: flat bone lying in the centre of the chest (also known as the breast bone)

Synovial joint: a freely movable joint

Synovial cavity: the space between the articulating bones of a synovial joint

Synovial fluid: a viscous fluid found in the cavities of synovial joints that helps to reduce friction between the articular cartilage of synovial joints during movement

Synovial membrane: specialised connective tissue that lines the inner surface of capsules of synovial joints

Tarsals: seven bones forming the ankle

Temporal: two bones that form the sides of the skull above and around the ears

Tendon: tough white fibrous cords of connective tissue that attach muscles to the bone

Thoracic vertebrae: vertebrae of the mid spine (× 12)

Tibia: long bone of the lower leg (anterior, medial side)

Transverse arch: one of the arches of the foot that runs between the medial and lateral aspect of the foot and is formed by the navicular, three cuneiforms and the bases of the five metatarsals

Turbinate: layers of bone located either side of the outer walls of the nasal cavities

Ulna: long bone of the forearm (little finger side)

Upper limb: part of the skeleton that includes the shoulder girdle, the arm, the forearm and the hand

Vomer: single facial bone at the back of the nasal septum

Zygomatic: facial bones that form the cheekbones

Revision summary

The skeletal system

- The functions of the skeleton are support, shape, protection, movement, attachment for muscles and tendons, formation of blood cells and mineral storage.
- **Bone** is the hardest type of connective tissue in the body.
- There are two types of bone tissue: **compact** (hard) and **cancellous** (spongy).
- The process of bone development is called **ossification** and is not complete until around the twenty-fifth year of life.
- **Cartilage** is a strong and rigid type of connective tissue that cushions and supports bone.
- A **ligament** binds bones to other bones.
- A **tendon** attaches a muscle to bone.
- Bones are classified according to shape: **long**, **short**, **flat**, **irregular** and **sesamoid**.
- The **axial skeleton** forms the main core of the body and consists of the skull, vertebral column, sternum and ribs.
  - There are eight bones of the skull: one **frontal**, two **parietal**, two **temporal**, one **sphenoid**, one **ethmoid** and one **occipital**.
  - There are 14 bones of the face: two **maxillae**, one **mandible**, two **zygomatic**, two **nasal**, two **lacrimal**, two **turbinate**, one **vomer** and two **palatine**.
  - There are 33 bones in the vertebral column: seven **cervical vertebrae** in the neck, 12 **thoracic vertebrae** in the mid spine, five **lumbar vertebrae** in the lower back, five **sacral vertebrae** (forming the sacrum) and four **coccygeal vertebrae** (forming the coccyx at the base of the spine).
  - The **thoracic cavity** protects vital organs in the chest and includes the **sternum**, 12 pairs of **ribs** and the 12 **thoracic vertebrae**.
- The **appendicular skeleton** supports and makes up the appendages or limbs and consists of the **shoulder girdle, bones of the upper and lower limbs** and **bones of the pelvic girdle**.
  - The shoulder girdle consists of two **scapulae** (posteriorly) and two **clavicle** bones (anteriorly).
  - The upper limb consists of the **humerus** in the upper arm, **radius** and **ulna** in the forearm, eight **carpals** in the wrist, five **metacarapals** in the palm and 14 **phalanges** in the fingers.
  - The lower limb consists of the **femur**, bone of the upper leg, **tibia** and **fibula** in the lower leg, seven **tarsals** in the ankle, five **metatarsals** in the dorsum of the foot and 14 **phalanges** in the toes.
  - There are three arches in the foot designed to support body weight and to provide leverage when walking – **medial longitudinal arch, lateral longtudinal arch** and **transverse arch**.
  - The pelvic girdle consists of the **ilium, ischium** and the **pubis**.
- A **joint** is a point of contact between two or more bones.
- **Ligaments** hold bones together at joints, provide flexibility and facilitate movement.
- Structurally joints are classified as **fibrous, cartilaginous** or **synovial**.
  - **Fibrous** joints are immovable, such as the sutures of the skull bones.
  - **Cartilaginous** joints are slightly movable, such as between the vertebrae of the spine.
  - **Synovial** joints are freely movable joints and there are several different types – **ball-and-socket** (hip), **hinge** (knee and elbow), **condyloid** (wrist), **gliding** (between the vertebrae), **pivot** (between the first and second cervical vertebrae), **saddle** (between the trapezium and metacarpal of the thumb).
  - Features of **synovial joints** include a **joint (synovial) cavity**, a **fibrous joint capsule** and a **synovial membrane** containing **synovial fluid**.

Test your knowledge questions

Multiple choice questions

1 How many bones make up the skeleton?
   a 208 bones
   b 206 bones
   c 106 bones
   d 80 bones

2 Which of the following statements is **false**?
   a The skeleton provides protection of vital organs.
   b The skeleton produces blood cells in red bone marrow.
   c The skeleton stores vitamins.
   d The skeleton provides support for the weight of the body.

3 Which skull bone forms the upper sides and the back of the roof of the skull?
   a sphenoid
   b occipital
   c parietal
   d temporal

4 How many bones are there in total in the skull?
   a 11
   b 14
   c 22
   d 24

5 Which is the largest bone in the face?
   a zygomatic
   b lacrimal
   c maxilla
   d turbinate

6 How many movable bones are there in the vertebral column?
   a 24
   b 33
   c 12
   d 9

7 How are the vertebrae classified?
   a as irregular bones
   b as short bones
   c as long bones
   d as sesamoid bones

8 The appendicular skeleton consists of:
   a the sternum, shoulder girdle and ribs
   b the shoulder girdle, bones of the lower and upper limbs and the pelvic girdle
   c the skull, vertebral column, sternum and ribs
   d the sternum, bones of the upper and lower limb, and the shoulder girdle.

9 Which of these makes up the main shaft of a long bone?
   a red bone marrow
   b compact bone
   c cancellous bone
   d chondrocyte cells

10 Which of the following statements is **true** in relation to cartilage?
   a Cartilage receives a generous blood supply.
   b Cartilage strengthens body structures.
   c Cartilage is completely flexible.
   d Cartilage cushions and absorbs shock.

Exam-style questions

11 State three functions of the skeletal system.
                                                    3 marks

12 Name the type of cell that makes bone tissue.
                                                    1 mark

13 Describe two differences between compact and cancellous bone.   2 marks

14 State the structural and functional difference between a ligament and a tendon.
                                                    2 marks

15 Name the parts that make up the axial skeleton.
                                                    4 marks

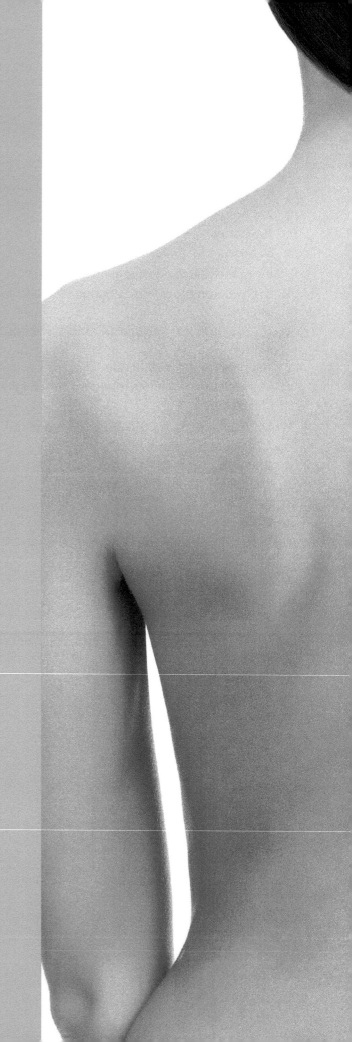

5 The muscular system

Introduction

The muscular system comprises over 600 individual muscles which are primarily concerned with movement and body co-ordination. There is an intimate relationship between muscle and bone, as both contribute to creating movement in the body. From Chapter 4, you know that bones and joints provide leverage for movement but it is, in fact, muscles that provide the force on the bones to effect movement. The key to learning the anatomical positions and actions of muscles is to first learn the individual positions of the bones. It is then a logical step to learn the muscle attachments in relation to each bone and what movements those muscles create.

OBJECTIVES

By the end of this chapter you will understand:

- the functions of the muscular system
- the structure and functions of the different types of muscle tissue
- how muscles contract
- the effects of temperature and increased circulation on muscle contraction
- the definition of the terms *muscle tone* and *muscle fatigue*
- the structure and function of the different types of attachments of muscles
- the position and action of the main superficial muscles of the face and body
- common pathologies of the muscular system
- the interrelationships between the muscular and other body systems.

The functions of the muscular system

Myology is the study of the muscular system, including the study of the structure, function and diseases of muscle.

The muscular system consists largely of skeletal muscle tissue, which covers the bones on the outside of the body, and connective tissue which attaches muscles to the bones of the skeleton. Muscles, along with connective tissue, help to give the body its contoured shape.

The muscular system has three main functions:

1 movement

2 maintaining posture

3 the production of heat.

1 Movement

Consider the action of picking up a pen that has dropped onto the floor. This seemingly simple action of retrieving the pen involves the co-ordinated action of several muscles pulling on bones at joints to create movement. Muscles are also involved in the movement of body fluids such as blood, lymph and urine. Heart muscle beats continuously and tirelessly throughout life.

2 Maintaining posture

Some fibres in a muscle resist movement and create slight tension in order to maintain body posture. This is essential, otherwise we would be unable to maintain normal body positions, such as sitting down or standing up.

3 The production of heat

As muscles create movement in the body they generate heat as a by-product, which helps to maintain our normal body temperature.

Muscle tissue

Muscle tissue makes up about 50% of your total body weight and is composed of:

- 20% protein
- 75% water
- 5% mineral salts, glycogen and fat.

There are three types of muscle tissue in the body:

1 **skeletal**, or **voluntary**, muscle tissue which is primarily attached to bone

2 **cardiac** muscle tissue which is found in the walls of the heart

3 **smooth (non-striated)** or **involuntary** muscle tissue which is found inside the digestive and renal tracts, as well as in the walls of blood vessels.

> **In practice**
>
> It is essential for therapists to have a good working knowledge of the muscular system as they primarily work on the muscles and their associated connective tissues. Together, these make up about half of the body's soft tissue mass.
>
> Having knowledge of the position, action and tone of muscles allows therapists to be more accurate in their treatment applications, ensuring effective results
>
> Understanding how muscles contribute to movement in the body helps therapists to appreciate how pathological disorders often result in muscle dysfunction.

All three types of muscle tissue differ in their structure and functions (Table 5.1) and the degree of control the nervous system has on them.

Table 5.1 Overview of the three types of muscle tissue

Muscle type	Description	Location	Function
Voluntary/ skeletal	Striped appearance with many nuclei Held together by connective tissue	Attached to bones, skin or other muscles	Facilitates movement of bones Moves blood and lymph Heat production Maintenance of posture
Cardiac	Striped appearance with branched structure Each cell has a single nucleus Has intercalated discs between cardiac muscle cells	Heart	Provides a consistent flow of blood throughout the body
Smooth/ involuntary	Non-striated and shaped like spindles Each cell has a single nucleus	In walls of stomach, intestines, bladder, uterus and in blood vessels	Move substances through the various tracts (digestive, genito-urinary)

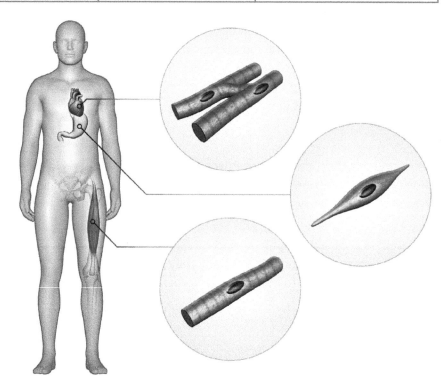

▲ Different types of muscle tissue in the human body

Voluntary (skeletal or striated) muscle tissue

Voluntary muscle tissue is made up of bands of elastic or contractile tissue bound together in bundles and enclosed by a connective tissue sheath which protects the muscle and helps to give it a contoured shape.

Voluntary or skeletal muscle tissue has very little intercellular tissue. It consists almost entirely of muscle fibres, held together by fibrous connective tissue and penetrated by numerous tiny blood vessels and nerves. The long slender fibres that make up muscle cells vary in size. Some are around 30 cm in length, whereas others are microscopic.

Each muscle fibre is enclosed in an individual wrapping of fine connective tissue called the endomysium. These are further wrapped together in bundles, known as fasciculi, and are covered by the perimysium (fibrous sheath). These bundles are then gathered to form the muscle belly (main part of the muscle) with its own sheath – the fascia epimysium.

Each striated muscle fibre is covered by a thin transparent extensible plasma membrane called the sarcolemma.

The relatively inelastic parts of muscles are tendons and these are usually made up from a continuation of the endomysium and perimysium.

Each muscle fibre is made up of even thinner fibres called myofibrils. These consist of long strands of microfilaments, made up of two different types of protein strands called actin and myosin. It is the arrangement of actin and myosin filaments which gives the skeletal muscle its striated or striped appearance when viewed under a microscope. Muscle fibre contraction results from a sliding movement within the myofibrils in which actin and myosin filaments merge.

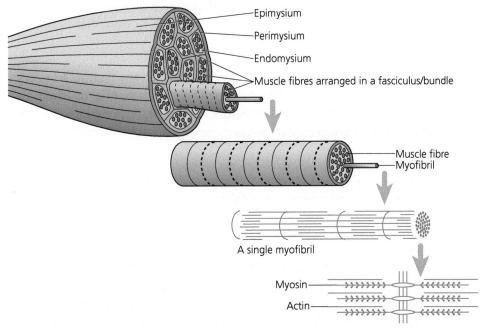

▲ The structure of voluntary muscle tissue

Types of muscle fibres

Most skeletal muscles are made up of a combination of fast twitch and slow twitch fibres.

Fast twitch fibres (white)

These have fast, strong reactions but tire quickly. They are well adapted for rapid movements and short bursts of activity. They have a rich blood supply and mainly use glucose as an energy store, which can be transferred into mechanical energy without oxygen.

Slow twitch fibres (red)

These fibres have greater endurance but do not produce as much force as fast twitch fibres. They are, therefore, suited to slower and more sustained movements and are relatively resistant to fatigue. Their energy comes from the breakdown of glucose by oxygen and they depend on a continuous supply of glucose for endurance. Slow twitch fibres have a good circulation (the red colour comes from both the circulation and from the presence of a red pigmented protein that stores oxygen).

During low-intensity work, such as walking, the body is working well below its maximal capacity and only slow twitch fibres are involved. As muscle intensity increases and exercise becomes more anaerobic, fast twitch fibres are activated. Whatever the intensity of movement, only a small number of fibres are used at any one time to prevent damage and injury to the tissues.

KEY FACT

Each person is born with a set number of muscle fibres which cannot be increased. An increase in the size of a muscle is due to exercise, which causes an increase in the individual fibres. However, with lack of use these shrink again as the muscle atrophies. It is interesting to note that men are more able to enlarge their muscles through exercise than women due to the effects of male hormones.

The way in which the bundles of fibres lie next to one another in a muscle will determine its shape. The contractile force of a muscle is partly attributable to the architecture of its fibres. Common muscle fibre arrangements are discussed below.

Parallel fibres

Muscles with parallel fibres can vary from short, flat muscles to spindle-shaped (fusiform) muscles to long straps.

Convergent

This is where the muscles fibres converge towards a single point for maximum concentration of the contraction. The direction of movement is determined by which sections of the muscle are activated. The muscle may be a triangular sheet (the pectoralis major muscle or the latissimus dorsi, for example). These muscles often cross joints that have a large range of possible movements. They provide a strong but steady pull, fine-tuning the angle of movement, thus balancing movement with continuing stability in the joint.

Pennate

This is where the fibres lie at an angle to the tendon and, therefore, also to the direction of pull. They have lots of short fibres, so the muscle pull is short but strong. They may be further classified as follows:

- **uni-pennate** – diagonal fibres attach to one side of the tendon only, such as the soleus

- **bi-pennate** – the fibres converge onto a central tendon from both sides, such as the rectus femoris
- **multipennate** – the muscle has several tendons of origin, such as the deltoid.

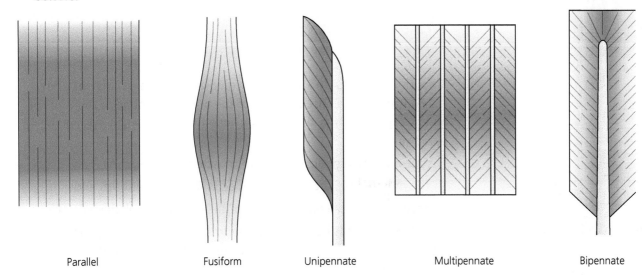

| Parallel | Fusiform | Unipennate | Multipennate | Bipennate |

▲ The six muscle shapes

Voluntary muscle works intimately with the nervous system so that it will only contract if a stimulus is applied to it via a motor nerve. Each muscle fibre receives its own nerve impulse so that fine and varied motions are possible. Voluntary muscles also have their own small stored supply of glycogen, which is converted to glucose and used as fuel for energy. Voluntary muscle tissue differs from other types of muscle tissue in that the muscles tire easily and need regular exercise.

Cardiac muscle

Cardiac muscle is a specialised type of involuntary muscle tissue found only in the walls of the heart. Forming the bulk of the wall of each heart chamber, cardiac muscle contracts rhythmically and continuously to provide the pumping action necessary to maintain a relatively consistent flow of blood throughout the body. Cardiac muscle resembles voluntary muscle tissue in that it is striated due to the actin and myosin filaments. However, it differs in two ways:

1 It is branched in structure.
2 It has intercalated discs between the cardiac muscle cells, forming strong junctions to assist in the rapid transmission of impulses throughout an entire section of the heart, rather than in bundles.

The contraction of the heart is automatic. The stimulus to contract is stimulated from a specialised area of muscle in the heart called the sinoatrial (SA) node which controls the heart rate.

KEY FACT

Every time a muscle is used, the muscle fibres shorten along their length. Therefore, tension often accumulates in lines in the longer muscles (particularly those with parallel fibres such as the paravertebral muscles). In muscles with shorter fibre pennates and convergent fibres, tension is often in knots rather than in lines.

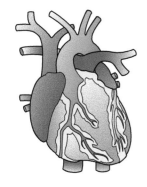

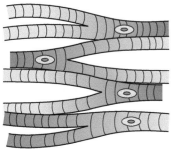

▲ Cardiac muscle tissue

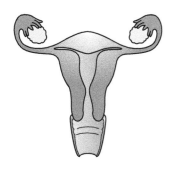

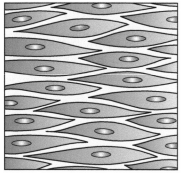

▲ Smooth/involuntary muscle tissue

As the heart has to alter its force of contraction under different conditions, its contraction is regulated not only by nerves but also by hormones, such as adrenaline which can speed up contractions.

Smooth/involuntary (non-striated) muscle

Smooth muscle is also known as involuntary muscle, as it is not under the control of the conscious part of the brain. It is found in the walls of hollow organs such as the stomach, intestines, bladder, uterus and in blood vessels.

The main characteristics of smooth muscle are that:

- the muscle cells are spindle shaped and tapered at both ends
- each muscle cell contains one centrally located oval-shaped nucleus.

Smooth muscle has no striations due to the different arrangement of the protein filaments actin and myosin, which are attached at their ends to the cell's plasma membrane.

The muscle fibres of smooth muscle are adapted for long, sustained contraction and, therefore, consume very little energy. One of the special features of smooth muscle is that it can stretch and shorten to a greater extent and still maintain its contractile function. Smooth muscle will contract or relax in response to nerve impulses, physical stretching or hormones but it is not under voluntary control.

Smooth muscle, like voluntary muscle, has muscle tone and this is important in areas such as the intestines where the walls have to maintain a steady pressure on the contents.

Muscle contraction

Muscle tissue has several characteristics which help contribute to the functioning of a muscle:

- **contractibility** – the capacity of the muscle to shorten and thicken
- **extensibility** – the ability to stretch when the muscle fibres relax
- **elasticity** – the ability to return to its original shape after contraction
- **irritability** – the response to stimuli provided by nerve impulses.

Muscles vary in the speed at which they contract. The muscle in your eyes will be moving very fast as you are reading this page, while the muscles in your limbs assisting you in turning the pages will be contracting at a moderate speed. The speed of a muscle contraction is, therefore, modified to meet the demands of the action concerned and the degree of nervous stimulus it has received.

Stimulus to contract

Skeletal or voluntary muscles contract as a result of nervous stimulus which they receive from the brain via a motor nerve. Each skeletal fibre is connected to the fibre of a nerve cell. Each nerve fibre ends in a motor point, which is the end portion of the nerve and is the part through which the stimulus to contract is given to the muscle fibre. A single motor nerve may transmit stimuli to one muscle fibre or to as many as 150, depending on the effect of the action required.

The site where the nerve fibre and muscle fibre meet is called a neuromuscular junction. In response to a nerve impulse, the end of the motor nerve fibre secretes a neuromtransmitter substance called acetylcholine, which diffuses across the junction and stimulates the muscles fibre to contract.

Cardiac and smooth muscle are innervated by the autonomic nervous system.

The contraction of voluntary muscle tissue

The functional characteristic of muscle is its ability to transform chemical energy into mechanical energy in order to exert force. Muscles exert force by contracting or making themselves shorter.

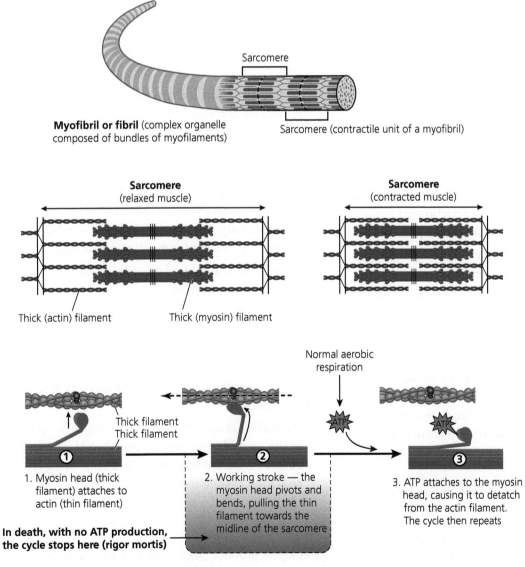

Sarcomere

Myofibril or fibril (complex organelle composed of bundles of myofilaments)

Sarcomere (contractile unit of a myofibril)

Sarcomere
(relaxed muscle)

Thick (actin) filament   Thick (myosin) filament

Sarcomere
(contracted muscle)

Normal aerobic respiration

Thick filament
Thick filament

ATP

ATP

1. Myosin head (thick filament) attaches to actin (thin filament)

2. Working stroke — the myosin head pivots and bends, pulling the thin filament towards the midline of the sarcomere

3. ATP attaches to the myosin head, causing it to detach from the actin filament. The cycle then repeats

In death, with no ATP production, the cycle stops here (rigor mortis)

▲ Mechanism of muscular contraction

The role of actin and myosin

A voluntary or skeletal muscle consists of many long cylindrical fibres. Each of these fibres is, in turn, filled with long bundles of even smaller fibres called myofibrils.

A myofibril resembles stacked blocks. In each block (or sarcomere) thick filaments containing the protein myosin overlap thin filaments containing the protein actin. Sarcomeres are the segments into which a fibril of striated muscle is divided. Sarcomeres are divided into dark Z lines with their centres known as H zones. As the muscle contracts, its sarcomeres shorten, reducing the distance between the Z lines and the width of the H zone. Muscle fibre contraction results from a sliding movement within the myofibrils in which the actin and myosin filaments merge. Actin and myosin affect contraction in the following way:

- During contraction, a sliding movement occurs within the contractile fibres (myofibrils) of the muscle in which the actin protein filaments move inwards towards the myosin and the two filaments merge.
- Cross-bridges of myosin filaments form linkages with actin filaments.
- This action causes the muscle fibres to shorten and thicken and then pull on their attachments (bones and joints) to effect the movement required.
- The attachment of myosin cross-bridges to actin requires the mineral calcium.
- The nerve impulses leading to contraction cause an increase in calcium ions within the muscle cell.
- During relaxation, the muscle fibres elongate and return to their original shape.

The force of muscle contraction depends on the number of fibres in a muscle which contract simultaneously. The more fibres that are involved, the stronger and more powerful the contraction will be.

KEY FACT

The basic contractile process is the same in cardiac, smooth and voluntary muscles, with movement being achieved through the action of the protein filaments, actin and myosin. However, since the requirements are different in terms of speed and force of contraction, the structure of cardiac and smooth muscles are slightly different to voluntary muscle tissue.

The energy needed for muscle contraction

A certain amount of energy is needed to effect the mechanical action of the muscle fibres. This is obtained principally from carbohydrate foods which are digested to glucose. Glucose is transported in the arterial blood supply; any that is not required immediately by the body is converted into glycogen and is stored in the liver and the muscles. Muscle glycogen, therefore, provides the glucose which is the fuel for muscle contraction. The process is as follows:

- During muscle contraction, glycogen is broken down to glucose. This undergoes oxidation, in which glucose combines with oxygen and releases energy. Oxygen is stored in the form of haemoglobin in the red blood cells and as myoglobin in the muscle cells.
- During oxidation, a chemical compound called adenosine triphosphate (ATP) is formed. Molecules of ATP are contained within voluntary muscle tissue and their function is to temporarily store energy produced from oxidation of glucose.
- When the muscle is stimulated to contract, ATP is converted to another chemical compound, adenosine diphosphate (ADP) with the release of the energy needed for muscle contraction.
- During the oxidation of glucose, a substance called pyruvic acid is formed.
- If plenty of oxygen is available to the body, as at rest or undertaking moderate exercise, then the pyruvic acid is broken down into waste products, carbon dioxide and water, which are excreted into the venous system. This is known as aerobic respiration.
- If insufficient oxygen is available to the body, as in the case of vigorous exercise, then the pyruvic acid is converted into lactic acid. This is known as anaerobic respiration.

Muscle fatigue

The waste product lactic acid, which diffuses into the bloodstream after vigorous exercise, causes the muscles to ache. This condition is known as muscle fatigue and is defined as the loss of the ability of a muscle to contract efficiently due to insufficient oxygen, exhaustion of energy supply and the accumulation of lactic acid.

The effects of increased circulation on muscle contraction

During exercise, muscles require more oxygen to cope with the increased demands made on the body. The body initiates certain circulatory and respiratory changes to meet the increased oxygen requirements of the muscles.

Circulatory changes that occur in the body during muscle contraction

During exercise, there is an increased rate of return of venous blood to the heart, owing to the more extensive movements of the diaphragm and the general contractions of the muscles, which compress the veins. With the increased heart rate and greater output from each heartbeat, a greater volume of blood is circulated around the body and lungs, which leads to an increase in the amount of oxygen in the blood.

More blood is distributed to the muscle and less to the intestines and skin to meet the needs of the exercising muscles. During exercise, a muscle may receive as much as 15 times its normal flow of blood.

Respiratory changes

The presence of lactic acid in the blood stimulates the respiratory centre in the brain to increase the rate and depth of breathing, producing panting. The rate and depth of breathing remains above normal for a while after strenuous exercise has ceased. Large amounts of oxygen are taken in to allow the cells of the muscles and the liver to dispose of the accumulated lactic acid by oxidising it and converting it back to glucose or glycogen. Lactic acid is formed in the tissues in amounts far greater than can be immediately disposed of by available oxygen. The extra oxygen needed to remove the accumulated lactic acid is called the oxygen debt, which must be repaid after the exercise is over.

KEY FACT

The conversion of lactic acid back into glucose is a relatively slow process and it may take several hours to repay the oxygen debt, depending on the extent of the exercise undertaken. This situation can be minimised by massaging muscles before and after an exercise schedule, which will increase the blood supply to the muscles and prevent formation of an excess of lactic acid.

The effects of temperature on muscle contraction

Exercising muscles produce heat, which is carried away from the muscles by the bloodstream and is distributed to the rest of the body. Exercise is, therefore, an effective way to increase body temperature. When muscle tissue is warm, the process of contraction occurs faster due to the acceleration of the chemical reactions and the increase in circulation. However, it is possible for heat cramps to occur in muscles that are exercising at high temperatures, as increased sweating causes loss of sodium in the body, leading to a reduction in the concentration of sodium ions in the blood supplying the muscle.

Cramp occurs when muscles overcontract and go into spasm. This is usually caused by an irritated nerve or an imbalance of mineral salts such as sodium in the body. Cramp most commonly affects the calf muscles or the soles of the feet. Cramp can be very painful as it is a sudden involuntary contraction of the muscle.

Treatment to relieve the pain of cramp includes stretching the affected muscle group and using soothing effluerage movements to help to relax the muscles. Conversely, as muscle tissue is cooled, the chemical reactions and circulation slow, causing the contraction to be slower. This causes an involuntary increase in muscle tone (known as shivering) that increases body temperature in response to cold.

Muscle tone

Even in a relaxed muscle, a few muscle fibres remain contracted to give the muscle a certain degree of firmness. At any given time a small number of motor units in a muscle are stimulated to contract, while the others remain relaxed. This causes tension in the muscle rather than full contraction and movement. The groups of motor units functioning in this way change periodically so that muscle tone is maintained without fatigue.

This state of partial contraction of a muscle is known as muscle tone and is important for maintaining body posture. Good muscle tone is apparent when muscles appear firm and rounded, whereas poor muscle tone may be recognised by muscles that are loose and flattened.

KEY FACT

An increase in the size and diameter of muscle fibres, usually caused by exercise and/ or weightlifting, leads to a condition called hypertrophy.

Muscles with less than the normal degree of tone are said to be flaccid, and when muscle tone is greater than normal the muscles become spastic and rigid.

KEY FACT

Muscle tone varies from person to person and largely depends on the amount of exercise undertaken. Muscles with good tone have a better blood supply as their blood vessels are not inhibited by fat.

Muscle attachments

In order to understand how skeletal muscles produce movement, it is helpful to know how muscles are attached to the rest of the body.

Tendons

Tendons are glistening white, tough fibrous bands or cords that link muscle to bone. They do not stretch or contract like muscles do. They are mechanically strong, as their primary role is to transmit the contractile force of the muscle to the bone. For this reason, tendons are relatively inflexible structures, designed to be strongest in the direction of tensile stress.

Despite their great strength, tendons are susceptible to excessive tensile stress injuries. A tendon's blood supply is limited, so it will not heal quickly or easily. Luckily complete tendon tears or ruptures are infrequent (the most common ruptured tendon is the Achilles tendon).

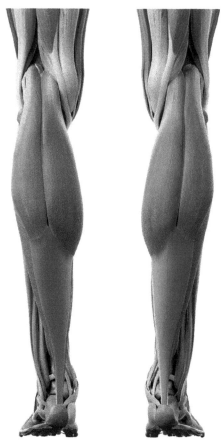

▲ The Achilles tendon

Ligaments

Ligaments are strong, fibrous and elastic tissues, usually cord-like in nature. Their attachments to various skeletal components help to maintain the bones in correct relationship to one another, stabilising joints.

They are found lying parallel to or closely interlaced with one another, which creates a shining white or silvery effect. A ligament is pliant and flexible, so as to allow good freedom of movement, but is also tough, strong and inextensible (does not stretch).

The orientation of a ligament's fibres (parallel arrangements complemented by transverse fibres) gives the ligament an ability to resist stress in several different planes. Ligaments also contain a greater concentration of elastin than a tendon. This allows the ligament a small degree of 'give' before it pulls taut. This small amount of give is important because it helps to prevent injury; if ligaments were as rigid and resistant to tensile stress as tendons, the frequency of ligament injuries would be much greater. The most common injury to a ligament is a sprain (torn ligament).

When torn, ligaments heal slowly due to the fact that they have a relatively poor blood supply compared to muscles and tendons.

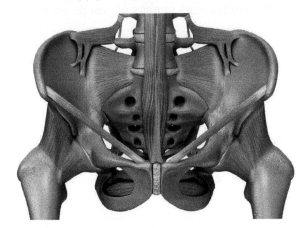

▲ Skeleton hip with ligaments

Fascia

Fascia consists of fibrous connective tissue that envelops certain muscles and forms partitions for others. Fascia is found all over the body – it packages, supports and encloses all the body's muscles and organs. It separates different muscle sets allowing them to glide smoothly beside each other. The fascial planes provide pathways for nerves, blood vessels and lymphatic vessels.

Fascia, therefore, plays a key role in maintaining the muscle health. When these cellophane-like sheets become adhered to neighbouring muscle tissue, efficiency and function can be significantly diminished. If the fascia becomes torn or overstressed, its subsequent loss of elasticity will cause chronic tissue congestion.

KEY FACT

One of the most problematic features of fascia is its response to prolonged immobilisation. If the body is held in one position for long periods of time, the fascia has a tendency to adapt to that position. This is especially the case when the fascia is held in a shortened position – it will structurally adapt to that position and resist attempts to return it to its normal length.

Origins and insertions

Muscle attachments are called origins and insertions. Generally, the end of the muscle closest to the centre of the body is referred to as the origin and the insertion is the furthest attachment.

Origins are often shorter and broader and attach over a larger area, while insertions are commonly longer and the fibres are more densely concentrated at a smaller bone area. The insertion is generally the most movable point and, therefore, the point at which the muscle work is done. Where a muscle divides into more than one attachment at one end or has a long line of attachments at one end, a number of actions are usually possible.

Muscle movement

In the co-ordination of movement, muscles work in pairs of groups. Muscles are classified by functions as:

- antagonists
- agonists (prime movers)
- fixators (stabilisers)
- synergists.

Although muscles are usually described as performing a particular action, they do not act alone. All movement is the result of co-operation between a large number of muscles, which are co-ordinated in the cerebellum in the brain for smooth and efficient actions.

Antagonists

Antagonists are two muscles (or two sets of muscles) that pull in opposite directions. They don't actually work *against* one another, but function in a reciprocal and complementary way, with one relaxing to allow the other to contract.

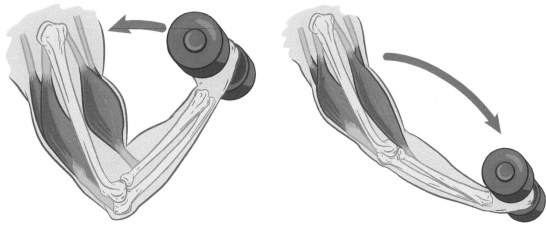

▲ Antagonistic muscles

Agonists/prime movers

Agonists are the main activating muscles. Note that the term is used in relation to a specific action. The roles are, therefore, relative to one another and are interchangeable. An example is the action of the biceps and triceps of the upper arm. Biceps are the agonist in flexion of the elbow joint and triceps are the antagonist. In relation to extending or straightening the elbow the roles are reversed.

Fixators

These are muscles that stabilise a bone to give a steady base from which the agonist works. For the biceps and triceps to flex and extend the elbow joint, muscles around the shoulder and upper back must control the position of the arm.

Synergists

This term refers to muscles on the same side of a joint that work together to perform the same movement. An example of this is flexing the elbow. The biceps actually work synergistically with the brachialis muscle that lies underneath them.

Muscle contraction in movement

Biomechanically, muscles do one of two things – stretch (relax) or shorten (contract). Muscular contractions can be isometric or isotonic. Isotonic contractions may be further classified as either concentric or eccentric. The opposite of contracting is stretching, which extends the muscles.

Isometric contraction

This is when the muscle works but without creating actual movements (*iso* means 'same' and *metric* means 'length'). Postural muscles work by isometric contraction.

Isotonic contraction

This term refers to a muscular contraction with constant force (*tonic* meaning 'tone' or 'tension') but where the muscle length changes. There are two types of isotonic contraction:

1 **Concentric contractions** (towards the centre) – this type of contraction occurs when the muscle shortens to move the attachment closer, such as when the biceps bends up the forearm.

2 **Eccentric contractions** (away from the centre) – this type of contraction occurs when a muscle is stretched as it tries to resist a force, pulling the bones of attachment away from one another, such as when tensing the biceps as someone pulls your forearm straight.

During many everyday actions, both isometric and isotonic contractions occur. For example, when standing the quadriceps muscle straightens the knee to keep you upright (isometric contraction), thereby preventing your knee from bending. When you sit down slowly, the muscle is stretched and an eccentric contraction controls the rate at which the knee bends to lower the body. If you then stand up, the muscle works concentrically to straighten the knee again.

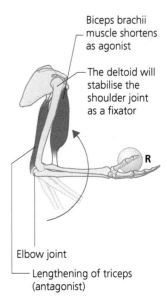

Biceps brachii muscle shortens as agonist

The deltoid will stabilise the shoulder joint as a fixator

R

Elbow joint

Lengthening of triceps (antagonist)

▲ Agonist and fixator

Study tip

Learning muscle names and positions can be daunting. It is helpful to break down the information into manageable chunks, so you learn a few muscles at a time.

The following tips may help you when studying muscles:

- There is often a clue in the name of the muscle as to where it is located in the body. For example, the tibialis anterior muscle is located alongside the tibia bone in the front of the lower leg.
- Picturing where the muscle is in your own body may help you to remember where it is located.
- Look for information that will help you remember the action for each muscle. See the key facts in the last column of the muscle tables on pages 143 to 145.
- If you know where the muscle is located and attached, you can work out its action by moving that body part and feeling the muscle contracting.

Muscles of the head and neck

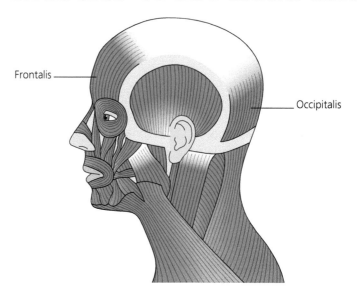

Frontalis

Occipitalis

▲ Frontalis and occipitalis

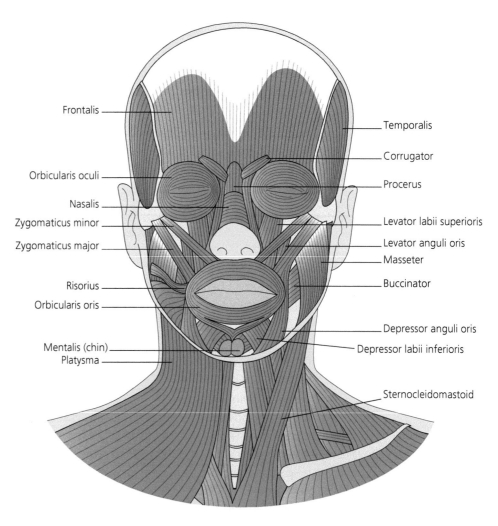

Frontalis

Temporalis

Corrugator

Orbicularis oculi

Procerus

Nasalis

Levator labii superioris

Zygomaticus minor

Levator anguli oris

Zygomaticus major

Masseter

Risorius

Buccinator

Orbicularis oris

Depressor anguli oris

Mentalis (chin)

Depressor labii inferioris

Platysma

Sternocleidomastoid

▲ Muscles of the head and neck

Table 5.2 Muscles of the head and neck

Name of muscle	Position	Attachments	Action(s)	Key facts
Frontalis (front-ta-lis)	Extends over the front of the skull and across the width of the forehead	Attaches to the skin of the eyebrows and the frontal bone at the hairline	Wrinkles the forehead and raises the eyebrows	The frontalis muscle is used when expressing surprise
Occipitalis (ok-sip-it-ta-lis)	Back of the skull	Attaches to the occipital bone and the skin of the scalp	Moves the scalp backwards	The occipitalis muscle is united to the frontalis muscle by a broad tendon called the epicranial aponeurosis which covers the skull like a cap
Temporalis (tem-po-ra-lis)	A fan-shaped muscle situated on the side of the skull above and in front of the ear	Attaches to the temporal bone and to the upper part of the mandible	Raises the lower jaw when chewing	Becomes overtight and painful in the condition known as TMJ syndrome Also becomes tightened with a tension headache
Orbicularis oculi (or-bik-you-la-ris ock-you-ly)	Circular muscle that surrounds the eye	Attached to the bones at the outer edge and the skin of the upper and lower eyelids at the inner edge	Closes the eye	The orbicularis oculi muscle is used when blinking or winking It also compresses the lacrimal gland, aiding the flow of tears
Orbicularis oris (or-bik-you-la-ris or-ris)	A circular muscle that surrounds the mouth	Its fibres attach to the maxilla, mandible, the lips and the buccinator muscle	Closes the mouth	The orbicularis oris muscle is used when shaping the lips for speech and when kissing Also contracts the lips when tense
Corrugator (kor-u-gay-tor)	Located between the eyebrows	Attached to the frontalis muscle and the inner edge of the eyebrow	Brings the eyebrows together	Used when frowning
Procerus (pro-ser-rus)	Located between the eyebrows	Attached to the nasal bones and the frontalis muscle	Draws the eyebrows inwards	Contraction creates a puzzled expression
Nasalis (nay-sa-lis)	Located at the sides of the nose	Attached to the maxillae bones and the nostrils	Dilates and compresses the nostrils	Used when blowing the nose
Zygomatic major and minor or zygomaticus (zi-go-mat-ik-us)	Lies in the cheek area	Extends from the zygomatic bone to the angle of the mouth	Draws the angle of the mouth upward and laterally	Used when laughing or smiling

Name of muscle	Position	Attachments	Action(s)	Key facts
Levator labii superioris (le-vay-tor lay-be-eye soo-pee-ri-o-ris)	Located towards the inner cheek beside the nose	Extends from the upper jaw to the skin of the corners of the mouth and the upper lip	Raises the upper lip and the corner of the mouth	Used to create a snarling expression
Levator anguli oris (le-vay-tor ang-you-lie o-ris) or caninus (kay-ni-nus)	Above the lip, located at an angle above the side of the mouth	Extends from the maxilla (upper jaw) to the angle of the mouth	Elevation of the angle of the mouth	Contraction can result in the teeth, especially the canine teeth, becoming visible
Depressor anguli oris (dee-pres-or ang-you-lie o-ris) or triangularis	Side of chin, extending down at an angle from the side of mouth	Extends from the mandible (lower jaw) to the angle of the mouth	Depression of the angle of the mouth	Contributes to the facial expression of sadness or uncertainty
Depressor labii inferioris (dee-pres-or lay-be-eye in-fee-ri-o-ris)	Side of chin, extending down from lower lip	Extends from the mandible to the midline of the lower lip	Depression of the lower lip	Contributes to the facial expressions of sorrow, doubt or irony
Risorius (ri-sor-ri-us)	Triangular-shaped muscle that lies horizontally on the cheek, above the buccinator muscles, joining at the corners of the mouth	Attached to the zygomatic bone at one end and the skin of the corner of the mouth at the other	Pulls the corner of the mouth sideways and upwards	Creates a grinning expression
Buccinator (buk-in-a-tor)	Main muscle of the cheek	Attached to both the upper and lower jaw, its fibres are directed forwards from the bones of the jaws to the angle of the mouth	Helps hold food in contact with the teeth when chewing and compresses the cheek	Used when blowing a balloon or blowing a trumpet. A common site for holding tension in the face
Mentalis (men-ta-lis)	Radiates from the lower lip over the centre of the chin	Attached to the lower jaw and the skin of the lower lip	Elevates the lower lip and wrinkles the skin of the chin	Used when expressing displesaure and when pouting
Masseter (ma-sa-ter)	Thick, flattened and superficial muscle at sides of jaw/cheek	Fibres extend downwards from the zygomatic arch to the mandible	Main muscle of mastication, raises the jaw and exerts pressure on the teeth when chewing	Tends to hold a lot of tension and can be felt just in front of the ear when the teeth are clenched

Name of muscle	Position	Attachments	Action(s)	Key facts
Sternocleidomastoid (ster-no-kli-do-mas-toyd)	Long muscle that lies obliquely across each side of the neck	Fibres extend upwards from the sternum and clavicle at one end to the mastoid process of the temporal bone (at the back of the ear)	When working together, they flex the neck (pull the chin down towards the chest) and when working individually, they rotate the head to the opposite side	Spasm of this muscle results in a condition known as torticollis or wryneck The only muscle that moves the head but is not attached to any vertebrae
Platysma (pla-tiz-ma)	Superficial muscle that covers the front of the neck	Extends from the chest (fascia covering the upper part of pectoralis major and deltoid) up either side of the neck to the chin	Depresses the lower jaw and lower lip	Used in yawning and when creating a pouting expression

The pterygoids

Lateral pterygoid (lat-er-al ter-i-goyds)

- **Position and attachments** – this muscle extends from the sphenoid bone to the mandible and temporomandibular joint.
- **Action** – protraction of the mandible.

Medial pterygoids (mee-dee-al ter-i-goyd)

- **Position and attachments** – this muscle extends from the sphenoid bone to the internal surface of the mandible.
- **Action** – elevation of the mandible.

KEY FACT

Tension in the lateral pterygoid may be associated with dysfunction of the temporomandibular joint (TMJ syndrome).

KEY FACT

Tension in the medial pterygoid may be associated with dysfunction of the temporomandibular joint (TMJ syndrome).

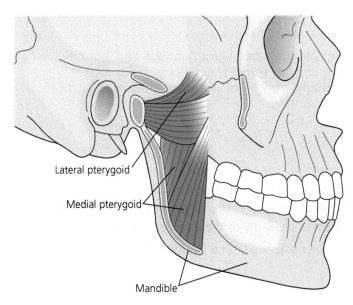

Lateral pterygoid

Medial pterygoid

Mandible

▲ Medial and lateral pterygoids

Muscles of the posterior of the neck

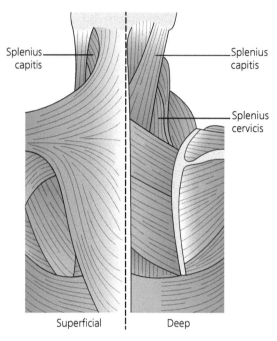

Splenius capitis

Splenius capitis

Splenius cervicis

Superficial | Deep

▲ Muscles of the posterior of the neck

Splenius capitis (splee-knee-us kap-i-tis)

- **Position and attachments** – this is a long posterior neck muscle that extends from the spinous processes of C7–T3 to the mastoid process of the temporal bone and the occipital bone.
- **Action** – extension of the head and neck, and lateral flexion of the head and neck.

> **KEY FACT**
>
> The splenius capitis muscle is shaped like a bandage and attaches onto the head. The right and left splenius captitis muscles form a V-shape. Due to this V-shape, they are sometimes referred to as the 'golf tee' muscles.

Splenius cervicus (splee-knee-us ser-vi-sis)

- **Position and attachments** – this is a long posterior neck muscle (with fibres slightly thinner and longer than splenius capitis) that extends from the spinous processes of T3–T6 to the transverse processes of C1–C3.
- **Action** – extension of the neck, and lateral flexion of the neck.

Muscles of the shoulder

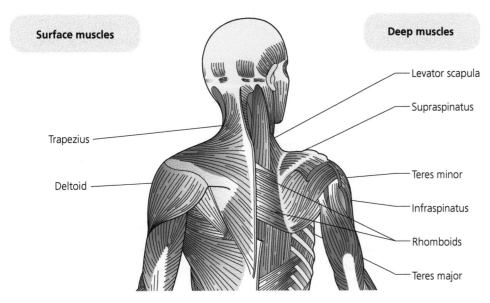

Surface muscles

Trapezius

Deltoid

Deep muscles

Levator scapula

Supraspinatus

Teres minor

Infraspinatus

Rhomboids

Teres major

▲ Muscles of the shoulder

Table 5.3 Muscles of the shoulder

Name of muscle	Position	Attachments	Action(s)	Key facts
Trapezius (tra-pee-zee-us)	Large, triangular-shaped muscle in the upper back	Extends horizontally from the base of the skull (occipital bone) and the cervical and thoracic vertebrae to the scapula Fibres are arranged in three groups – upper, middle and lower	The upper fibres raise the shoulder girdle, the middle fibres pull the scapula towards the vertebral column and the lower fibres draw the scapula and shoulder downward When the trapezius is fixed in position by other muscles, it can pull the head backwards or to one side	Commonly holds upper body tension, causing discomfort and restrictions in the neck and shoulder Carrying a bag on one shoulder can cause tightness in the upper fibres of trapezius
Levator scapula (le-vay-tpr skap-you-lee)	A strap-like muscle that runs almost vertically through the neck	Connects the cervical vertebrae to the scapula	Elevates and adducts the scapula	Due to its attachments, tension in the levator scapula can effect mobility of both the neck and the shoulder
Rhomboids (rom-boyds)	Fibres of these muscles lie between the scapulae	Attach to the upper thoracic vertebrae at one end and the medial border of the scapula at the other end	Adduct the scapula	Known as the 'Christmas tree' muscles because the fibres are arranged obliquely like Christmas tree branches Tension in the rhomboid muscles often results in aching and soreness between the scapulae

Name of muscle	Position	Attachments	Action(s)	Key facts
Supraspinatus (soo-pra-spy-nay-tus)	Located in the depression above the spine of the scapula	Attached to the spine of the scapula at one end and the humerus at the other	Abducts the humerus, assisting the deltoid	One of the rotator cuff muscles (along with infraspinatus, teres major and subscapularis) The only muscle of the rotator cuff that does not rotate the humerus
Infraspinatus (in-fra-spy-nay-tus)	Deep muscle covering the lower part of the scapula	Attaches to the middle two-thirds of the scapula below the spine of the scapula at one end and the top of the humerus at the other	Rotates humerus laterally (outwards)	One of the four rotator cuff muscles (along with supraspinatus, teres minor and subscapularis) Tension here can affect the range of mobility in the arm and the shoulder
Teres major (te-reez may-jor)	Deep, small muscle located between the lower border of the scapula and the humerus	Attaches to the bottom lateral edge of the scapula at one end and the back of the humerus (just below the shoulder joint) at the other	Adducts and medially (inwardly) rotates humerus	Sometimes referred to as the 'little helper' of the latissimus dorsi muscle because they run together between the scapula and the humerus
Teres minor (te-reez my-nor)	Deep, small muscle located above teres major	Attaches to the lateral edge of the scapula, above teres major at one end, and into the top of the posterior of the humerus at the other	Rotates humerus laterally (outwards)	One of the four rotator cuff muscles (along with supraspinatus, infraspinatus and subscapularis)
Subscapularis (sub-skap-u-la-ris)	Large muscle located beneath the scapula	Attaches to the inside surface of the scapula to the anterior of the top of the humerus	Rotates the humerus medially, draws the humerus forwards and down when the arm is raised	One of the four rotator cuff muscles (along with supraspinatus, infraspinatus and teres minor) Often implicated in frozen shoulder
Deltoid (del-toid)	Thick, triangular muscle that caps the top of the humerus and shoulder	Attaches to the clavicle and the spine of the scapula at one end, and to the side of the humerus at the other	Abducts arm, draws the arm backwards and forwards	Has anterior, lateral and posterior fibres that give the shoulder its characteristic shape

KEY FACT

Although the teres major and minor may appear similar by name, they wrap around the humerus in opposite directions and, therefore, have opposite rotary actions.

Muscles of the upper limbs

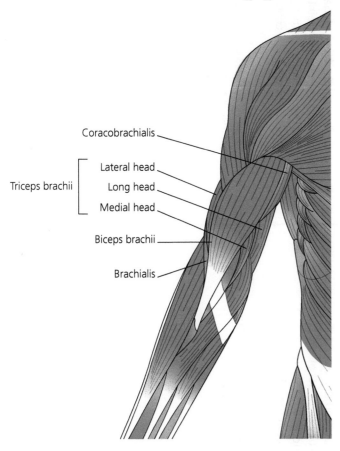

Coracobrachialis

Triceps brachii
- Lateral head
- Long head
- Medial head

Biceps brachii

Brachialis

▲ Muscles of the upper arm

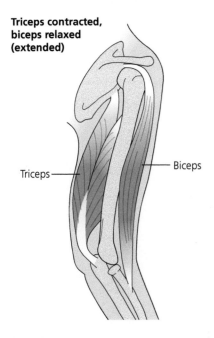

Triceps contracted, biceps relaxed (extended)

Triceps

Biceps

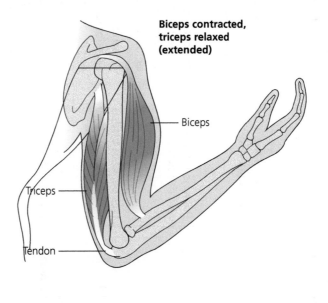

Biceps contracted, triceps relaxed (extended)

Biceps

Triceps

Tendon

▲ Biceps and triceps

Table 5.4 Muscles of the upper limbs

Name of muscle	Position	Attachments	Action(s)	Key facts
Coraco-brachialis (kor-a-ko-bra-key-al-is)	On the upper arm between the shoulder joint and elbow joint	Extends from the scapula to the middle of the humerus along its medial surface	Flexes and adducts the humerus	The name coraco-brachialis indicates that this muscle is related to the coracoid process (in the scapula) and the brachium (the arm)
Biceps (by-seps)	Anterior of the upper arm (humerus)	Attaches to the scapula at one end and the radius and flexor muscles of the forearm at the other	Flexes the forearm at the elbow and supinates the forearm	The action of this muscle is similar to the action of twisting a corkscrew (supination) and pulling it from a wine bottle (flexion)
Triceps (try-seps)	Posterior of the upper arm (humerus)	Attaches to the posterior of the humerus and outer edge of the scapula at one end, and to the ulna below the elbow at the other	Extends (straightens) the forearm	Also referred to as the 'boxer's muscle' as it is used when delivering a punch
Brachialis (bray-key-al-is)	Lies beneath biceps	Attaches to the distal half of the anterior surface of the humerus at one end and the ulna at the other	Flexes the forearm at the elbow	A strong and fairly large muscle, which accounts for much of the contour of the biceps muscle
Pronator teres (pro-nay-tor te-reez)	Crosses the anterior aspect of the elbow	Attaches to the distal end of the humerus and the upper aspect of the ulna at one end, and the lateral surface of the radius at the other	Pronates and flexes forearm	Due to the fact that the fibres of the pronator teres cross the elbow joint, irritation and inflammation of this muscle may contribute to the condition 'tennis elbow'
Supinator radii brevis (sue-pin-a-tor rade-ee brev-is)	Runs diagonally across forearm crossing the elbow joint to outer edge of radius	Attaches to the lateral aspect of the lower humerus and the radius	Supinates the forearm	Due to the fact that the fibres of the supinator cross the elbow joint, irritation and inflammation of this muscle may also contribute to tennis elbow

Flexors of the forearm

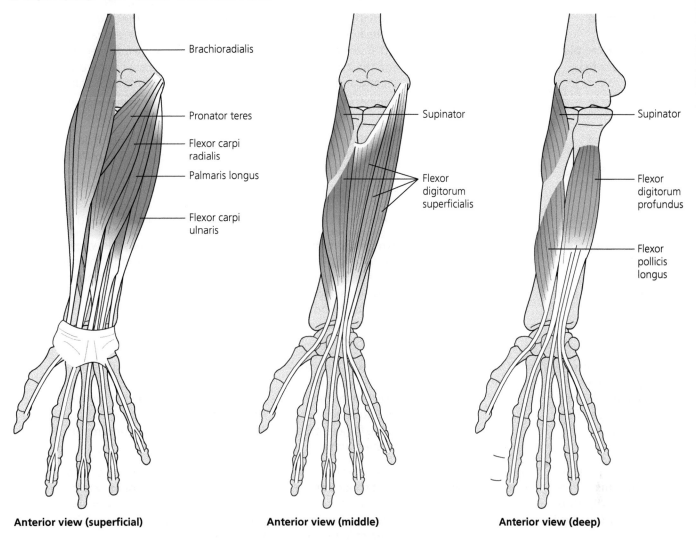

| Anterior view (superficial) | Anterior view (middle) | Anterior view (deep) |

▲ Muscles of the anterior of the forearm (flexors)

Table 5.5 Flexors of the forearm

Name of flexor	Position	Attachments	Action(s)	Key facts
Brachioradialis (bray-key-o-ray-dee-al-is)	Anterior of forearm; connects the humerus to the radius	Connects the humerus to the radius; attaches to the distal end of the humerus at one end and the radius at the other end	Flexes forearm at the elbow	Can be felt as the bulge on the radial side of the forearm Sometimes nicknamed the 'hitchhiker muscle' for its characteristic action of flexing the forearm in a position halfway between full pronation and full supination

Name of flexor	Position	Attachments	Action(s)	Key facts
Flexor carpi radialis (fleks-or kar-pie ray-dee-a-lis)	Extends along the radial side of the anterior of forearm	From the medial end of the humerus to radial side of forearm and the base of the second and third metacarpal	Flexes the wrist	Any of the flexor muscles in the foream can easily become inflamed due to excess pressure and overwork, such as from working on a keyboard for extended periods of time
Flexor carpi ulnaris (fleks-or kar-pie ul-na-ris)	Extends along the ulnar side of the anterior of the forearm	From the medial end of the humerus to the pisiform and hamate carpal bones and the base of the fifth metacarpal	Flexes the wrist	Pain in this muscle mimics the pain of a sprained wrist and is the most common source of pain in the wrist and hand on the (ulnar) little finger side of hand
Flexor carpi digitorum (fleks-or kar-pie dij-i-toe-rum)	In the forearm, extending down to the palm of the hands	Extends from the medial end of the humerus, the anterior of the ulna and radius to the anterior surfaces of second to fifth fingers	Flexes the fingers	Deep muscles whose tendons cross the palmar surface of each hand
Flexor digitorum superficialis (fleks-or dij-i-toe-rum soo-per-fish-ee- ar-lis)	In the forearm, wrist and hand	Originates near the elbow joint and inserts onto the middle bone of each of the fingers, except for the thumb	Helps flex four of the fingers (except the thumb) of each hand	Weakness in the flexor digitorum superficialis results in the muscle decreasing grip strength and wrist flexion strength
Flexor digitorum profundus (fleks-or dij-i-toe-rum pro-fun-dus)	Long, thin muscle located in the forearm, hand, and fingers	Originates from the front surface of the top part of the ulna, extends down the forearm to the tip of the phalanges, not including the thumb	Flexes the fingers, also playing a role in the flexion of the whole hand	Considered the most important muscle for separate movement of interphalangeal joints
Palmaris longus (pal-mar-is long-us)	Between the flexor carpi radialis and the flexor carpi ulnaris	From lower end of humerus to the wrist	Flexes wrist	Dysfunction can cause the ring and little finger to curl into the palm of the hand
Flexor pollicis longus (fleks-or)	Runs along medial anterior surface of the radius in the forearm down to the thumb	From below the elbow, on medial side of the radius to the base of distal phalanx of thumb	Flexes the thumb	Pain here is known as weeder's thumb because it is often seen in gardeners and horticulturists

KEY FACT

Finger flexion is required for grabbing items, such as a spoon or fork when you eat or a door knob when you open a door.

Extensors of the forearm

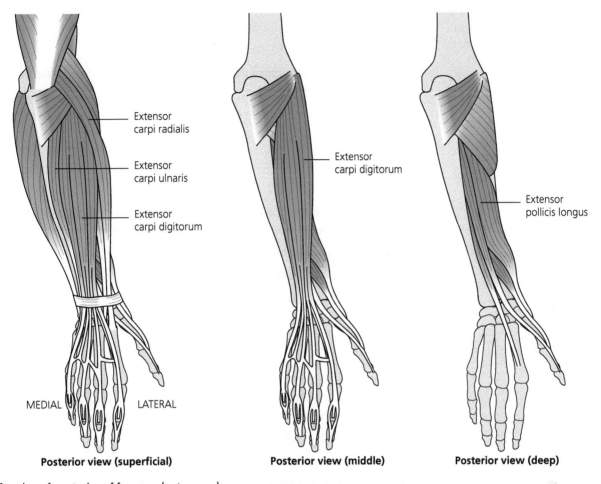

		Extensor carpi radialis			
		Extensor carpi ulnaris			
		Extensor carpi digitorum		Extensor carpi digitorum	Extensor pollicis longus
MEDIAL	LATERAL				
Posterior view (superficial)		**Posterior view (middle)**		**Posterior view (deep)**	

▲ Muscles of posterior of forearm (extensors)

Table 5.6 Extensors of the forearm

Name of muscle	Position	Attachments	Action(s)	Key facts
Extensor carpi radialis (eks-ten-sor kar-pie ray-dee-a-lis)	Extends along the radial side of the posterior of the forearm	From above the lateral end of the humerus to the posterior of the base of the second metacarpal	Extends the wrist	Can become easily inflamed due to excess pressure and overwork, for example from working on a keyboard for extended periods of time
Extensor carpi ulnaris (eks-ten-sor kar-pie ul-na-ris)	Extends along the ulnar side of the posterior of the forearm	From above the lateral end of the humerus to the ulna and the posterior side of the base of the fifth metacarpal	Extends the fingers	
Extensor carpi digitorum (eks-ten-sor dij-i-toe-rum)	Extends along the lateral side of the posterior of the forearm	From the lateral end of the humerus to the second and fifth phalanges	Extends the fingers	
Extensor pollicis longus (eks-ten-sor poll-ik-kus long-us)	Posterior surface of the forearm	Extends from the ulna to the tip of the thumb	Extends the thumb	

Muscles of the hand

Palmar aponeurosis (pal-mar app-an-nur-os-is)

- **Position and attachments** – a strong, triangular membrane overlying the tendons in the palm.
- **Action** – firm attachment to the skin of the palm to improve the grip, and it protects the underlying tendons.

Thenar eminence (thee-na emm-in-nen-s)

This is an eminence of soft tissue located on the radial side of the palm of the hand. There are three muscles of the thenar eminence:

1 abductor pollicis brevis
2 flexor pollicis brevis
3 opponens pollicis.

- **Action** – all three muscles move the thumb.

Hypothenar eminence (hi-po- thee-na emm-in-nen-s)

This is an eminence of soft tissue located on the ulnar side of the palm of the hand. There are three muscles of the hypothenar eminence:

1 abductor digiti minimi manus
2 flexor digiti minimi manus
3 opponens digiti minimi.

- **Action** – all three muscles move the little finger.

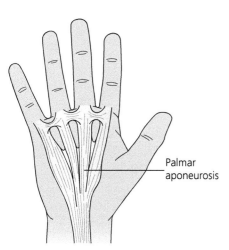

Palmar aponeurosis

▲ Palmar aponeurosis

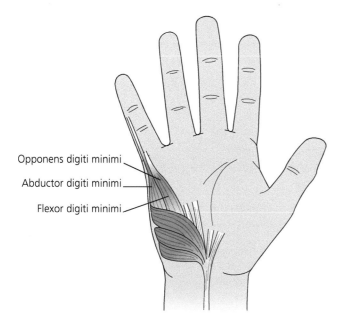

Opponens digiti minimi
Abductor digiti minimi
Flexor digiti minimi

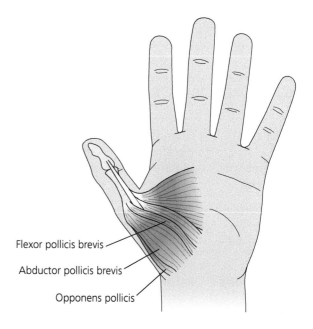

Flexor pollicis brevis
Abductor pollicis brevis
Opponens pollicis

▲ Hypothenar and thenar eminences

Muscles of the lower limb

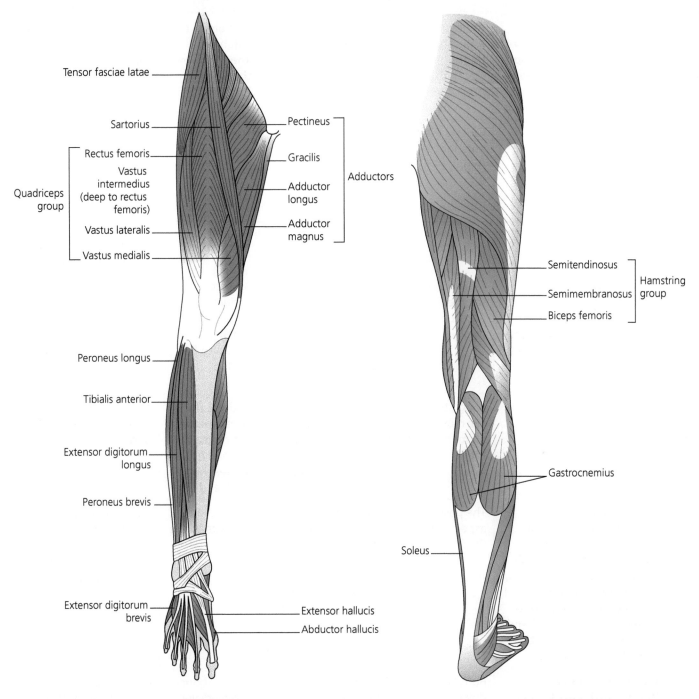

Tensor fasciae latae

Sartorius

Quadriceps group
- Rectus femoris
- Vastus intermedius (deep to rectus femoris)
- Vastus lateralis
- Vastus medialis

Pectineus

Gracilis

Adductor longus

Adductor magnus

Adductors

Peroneus longus

Tibialis anterior

Extensor digitorum longus

Peroneus brevis

Extensor digitorum brevis

Extensor hallucis

Abductor hallucis

Semitendinosus

Semimembranosus

Biceps femoris

Hamstring group

Gastrocnemius

Soleus

▲ Muscles of the anterior of the lower limb

▲ Muscles of the posterior of the lower limb

Table 5.7 Muscles of the lower limb

Name of muscle	Position	Attachments	Action(s)	Key facts
Quadriceps extensor (quad-ri-cep-s ex-ten-soo-r) The quadriceps is made up of four muscles: rectus femoris, vastus lateralis, vastus intermedius, vastus medialis	Anterior aspect of the thigh	Attached to the pelvic girdle (rectus femoris) and femur (vastus group) at one end to the patella and tibia at the other end	As a group they extend the knee and flex the hip	A group of strong muscles that are used for walking, kicking and raising the body from a sitting or squatting position
Sartorius (sar-tor-ee-us)	Crosses the anterior of thigh	Attached to the ilium of the pelvis, crosses the anterior of the thigh to the medial aspect of the tibia	Flexes the hip and knee and rotates the thigh laterally (turns it outwards)	Flexes both the hip and the knee as a result of its positioning Over contraction puts pressure on the knee and can lead to knee problems The longest muscle in the human body
Adductors (ad-duk-tors) A group of four muscles: adductor brevis, adductor longus, adductor magnus and pectineus	Situated on the medial aspect of the thigh	Attached to the lower part of the pelvic girdle at one end (pubic bones and the ischium) and the inside of the femur at the other end	As a group, adduct and laterally rotate the thigh They also flex the hip	The adductors are important muscles in the maintenance of posture Groin strain is associated with these muscles
Gracilis (gra-sil-is)	Long, strap-like muscle on medial of thigh	Attached to the lower edge of the pubic bone at one end and the upper part of the medial aspect of the tibia at the other end	Adducts thigh, flexes knee and hip, medially (inwardly) rotates the thigh and tibia	The second longest muscle in the human body
Hamstrings The hamstrings consist of three muscles – two situated on the inside of the thigh (semitendinosus, sem-ee-ten-da-noo-sis and semimembranosus, sem-ee-mem-braa-noo-sis) and one on the outside of the thigh (biceps femoris)	The posterior aspect of the thigh	Attaches to the lower part of the pelvis (ischium), and the lower part of the posterior of the femur to either side of the posterior of the tibia	Flex the knee and extend the hip	Contract powerfully when raising the body from a stooped position and when climbing stairs
Tensor fascia latae (ten-sor fash-ee-a la-tee)	Runs laterally down the side of the thigh	Attached to the outer edge of the ilium of the pelvis and runs via the long fascia lata tendon to the lateral aspect of the top of the tibia	Flexes, abducts and medially rotates thigh	Attached to a broad sheet of connective tissue (fascia lata tendon) which helps to strengthen the knee joint when walking and running

Name of muscle	Position	Attachments	Action(s)	Key facts
Gastrocnemius (gas-trok-nee-me-us)	A large, superficial calf muscle with two bellies (central portion of the muscle) on the posterior of the lower leg	Attached to the lower aspect of the posterior of the femur across the back of the knee and runs via the Achilles tendon at the ankle to the calcaneum at the back of the heel	Plantar flexes the foot and assists in knee flexion	Provides the push during fast walking and running
Soleus (so-lee-us)	Situated deep to the gastrocnemius in the calf	Attached to the tibia and fibula just below the back of the knee at one end and runs via the Achilles tendon to the calcaneum at the other end	Plantar flexes the foot	A thicker and flatter muscle than the gastrocnemius and accounts for the contours of the gastrocnemius being so visible
Tibialis anterior (tib-ee-a-lis an-tee-ri-or)	Anterior aspect of the lower leg	Attached to the outer side of the tibia at one end and the medial cuneiform and the base of the first metatarsal at the other end	Dorsiflexes and inverts the foot	If this muscle becomes weak, it can lead to rolling in of the lower leg due to the collapse of the medial longitudinal arch of the foot
Tibialis posterior (tib-ee-a-lis pos-tee-ri-or)	Found on the posterior aspect of the lower leg, very deeply situated in the calf	Attached to the back of the tibia and fibula at one end and to the navicular, third cuneiform and second, third and fourth metatarsals at the other end	Assists in plantar flexion and inverts the foot	Weakness in this muscle can cause the feet to turn out from the ankles rather than the knees, causing the muscle to stretch and the medial longitudinal arch of the foot to drop
Peroneus longus (pero-knee-us long-us)	Situated on the lateral aspect of the lower leg	Attaches the fibula to the underneath of the first (longus) and fifth metatarsal (brevis)	Plantar flexes and everts the foot	Going over on to the outside of the ankle, as in a trip or a fall, can sprain the peroneal muscles in the lower leg; if the injury is not treated properly it can affect future stability of the ankle joint
Peroneus brevis (pero-knee-us brev-is)	Lies under the peroneus longus muscle	Extends from the lateral surface of the fibula to the lateral aspect of the fifth metatarsal	Assists in plantar flexion and eversion of the foot	The shorter and smaller of the peroneus muscles

Name of muscle	Position	Attachments	Action(s)	Key facts
Peroneus tertius (pero-knee-us ter-she-us)	Located on the front of the lower leg	Extends from front surface of fibula to head of fifth metatarsal	Assists with dorsiflexion of the foot at the ankle	Pain in the lower leg, ankle, heel and foot, and weak ankles can all be a sign of peroneus tertius muscle dysfunction
Flexor digitorum longus (fleks-or dij-i-toe-rum long-us)	Located on the inner side of the leg by the tibia	Extends from the middle third of the posterior of the tibia to the plantar surface of the second to fifth toes	Flexes the toes, plantar flexes and inverts the foot	An indicator of flexor digitorum muscle dysfunction is a sharp pain radiating across the ball of the foot and a deep ache at rest
Flexor hallucis longus (fleks-or hal-oo-sis long-us)	Deep muscle found in the lower back region of the leg	Extends from the distal two-thirds of the posterior fibula to the plantar surface of the big toe	Flexes the big toe, plantar flexes and inverts the foot	An indicator of flexor hallucis longus muscle dysfunction is pain in the big toe and into the ball of the foot under the big toe
Extensor digitorum longus (eks-ten-sor dij-i-toe-rum long-us)	Situated along the outside of the lower leg, just behind the tibialis anterior	Extends from the proximal two-thirds of the anterior of the fibula to the dorsal surface of the second to fifth toes	Extends of the second to fifth toes, dorsiflexes and everts the foot	An indicator of extensor digitorum longus dysfunction is pain and numbness in the top of the foot, extending from the ankle to the bottom of the lower leg, and cramping in the foot.
Extensor digitorum brevis (eks-ten-sor dij-i-toe-rum brev-is)	Located on the top of the foot	Attached to tendons that extend to the toes	Controls the movements of the all toes except the smallest toe	An indicator of extensor digitorum brevis dysfunction is a condition known as 'drop foot' (a muscular weakness which makes it difficult to lift the fingers and toes)
Extensor hallicus longus (eks-ten-sor hal-oo-sis long-us)	A thin muscle, situated between the tibialis anterior and the extensor digitorum longus	Extends from the middle third of the anterior of the fibula to the dorsal surface of the big toe	Extends the big toe, dorsiflexes and inverts the foot	The flexor and extensor muscles of the lower leg can become weak due to excess pressure and overuse in walking and running
Abductor halluces (ab-duc-tor hal-ik-us)	Runs along the medial border of each foot	Originates from the medial part of the heel bone and that inserts into the first phalanx of the big toe	Moves the big toe away from the other toes	An indicator of abductor hallucis dysfunction is overpronation of the foot, and pain along the medial longitudinal arch of the foot

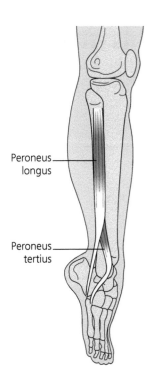
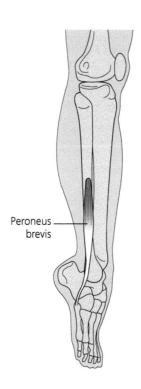

▲ The peroneal muscles

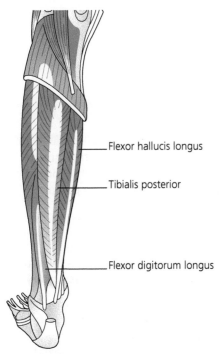

Flexor hallucis longus

Tibialis posterior

Flexor digitorum longus

▲ Tibialis posterior, flexor digitorum longus and flexor hallucis longus

Muscles of the pelvic floor

The **levator ani** and the **coccygeus** are the muscles that form the pelvic floor. These muscles support and elevate the organs of the pelvic cavity, such as the uterus and the bladder. They provide a counterbalance to increased intra-abdominal pressure, which would otherwise expel the contents of the bladder, rectum and the uterus.

During childbirth, these muscles can become weakened and need to be strengthened by pelvic floor exercises as soon as possible after the birth.

Muscles of the anterior aspect of the trunk

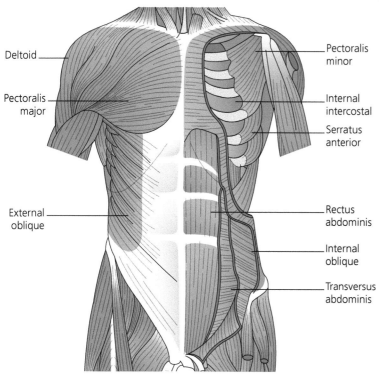

Deltoid

Pectoralis major

External oblique

Pectoralis minor

Internal intercostal

Serratus anterior

Rectus abdominis

Internal oblique

Transversus abdominis

▲ Muscles of the anterior of the trunk

Table 5.8 Muscles of the anterior aspect of the trunk

Name of muscle	Position	Attachments	Action(s)	Key facts
Pectoralis major (pek-to-ra-lis may-jor)	Thick, fan-shaped muscle covering the anterior surface of the upper chest	Attaches to the clavicle and the sternum at one end and to the humerus at the other end	Adducts arm, medially (inwardly) rotates arm	Tightness in this muscle can cause restrictions of the chest and postural disortions such as rounded shoulders
Pectoralis minor (pek-to-ra-lis my-nor)	A thin muscle that lies beneath the pectoralis major	Fibres attach laterally and upwards from the ribs at one end to the scapula at the other end	Draws the shoulder downwards and forwards	Involved in forced expiration and is therefore an accessory respiratory muscle
Serratus anterior (ser-at-tus an-tee-ri-or)	A broad, curved muscle located on the side of the chest/rib cage below the axilla	Attaches to the outer surface of the upper eighth or ninth rib at one end to the inner surface of the scapula, along the medial edge nearest the spine	Pulls the scapula downwards and forwards	Has a serrated appearance which comes from attaching onto separate ribs
External obliques (eks-turn-al o-bleek)	Laterally at the sides of the waist	Fibres slant downwards from the lower ribs to the pelvic girdle and the linea alba (tendon running from the bottom of the sternum to the pubic symphysis)	Flexes, rotates and side-bends the trunk, compresses the contents of the abdomen	The external oblique muscles are often referred to as the pocket muscles as their fibres run in the direction in which you put your hands in your pocket

→

Name of muscle	Position	Attachments	Action(s)	Key facts
Internal obliques (in-turn-al o-bleek)	A broad, thin sheet of muscle located beneath the external obliques	Fibres run up and forwards from the pelvic girdle to the lower ribs	Flexes, rotates and side-bends the trunk, compresses the contents of the abdomen	The fibres of the internal obliques are deeper and run at right angles to the external obliques
Rectus abdominis (rek-tus ab-dom-i-nis)	Long, strap-like muscle extending medially along the length of the abdomen	Attaches to the pubic bones at one end and the ribs and the sternum at the other	Flexes the vertebral column, flexes the trunk (as in a sit-up), compresses the abdominal cavity	The rectus abdominis has three fibrous bands that give the muscle a segmented appeearance and divides it into the so-called 'six pack'
Transversus abdominus (trans-ver-sus ab-dom-i-nis)	Large, deep muscle with fibres extending across the anterior of the abdominal cavity	Attaches to the inner surfaces of the ribs (last six) and iliac crest at one end and extends down to the pubis via the linea alba (a long tendon that extends from the bottom of the sternum to the pubic symphysis)	Compresses the abdominal contents and supports the organs of the abdominal cavity	Often called the corset muscle because it wraps around the abdomen like a corset

Muscles of respiration

Table 5.9 Muscles of respiration

Name of muscle	Position	Attachments	Action(s)	Key facts
Diaphragm (di-a-fram)	A large, dome-shaped muscle that separates the thorax from the abdomen	Attaches to the lower part of the sternum, lower six ribs and upper three lumbar vertebrae Fibres converge to meet on a central tendon in the abdominal cavity	On contraction the diaphragm flattens to expand the volume of the thoracic cavity to assist inspiration On relaxation and expiration it returns to its dome shape	Unusual in that it is under both unconscious control (from the brainstem) in the regulation of breathing and conscious control (in that we can choose to override the brainstem to hold our breath, sigh, sing or talk)
External intercostals (eks-turn-al in-ter-kos-tals)		Superficial muscles that occupy and attach to the space between the ribs (called external because they are positioned on the outside)	Help to elevate the rib cage during inhalation	Help to increase the depth of the thoracic cavity
Internal intercostals in-turn-al in-ter-kos-tals)	Lie deeper than the external intercostals (called internal because they are positioned on the inside)	Occupy and attach to the spaces between the ribs	Depress the rib cage, which helps to move air out of the lungs when exhaling	Help to increase the depth of the thoracic cavity

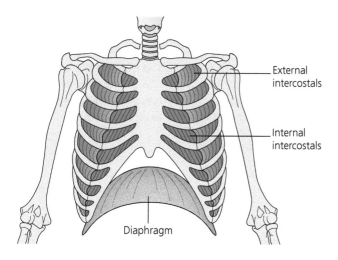

External
intercostals

Internal
intercostals

Diaphragm

▲ Muscles of respiration

Muscles of the posterior aspect of the trunk

Surface muscles

Deep muscles

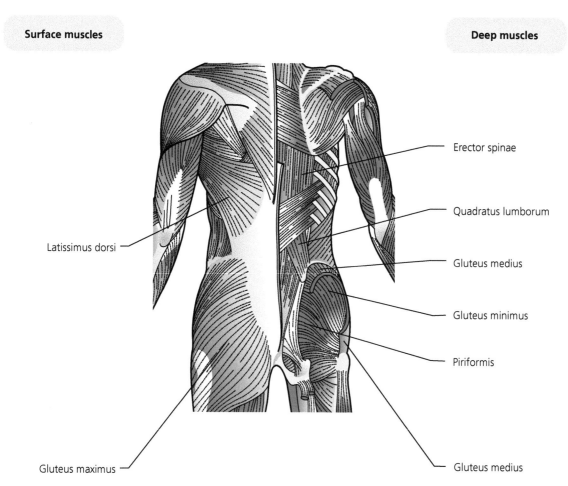

Erector spinae

Quadratus lumborum

Gluteus medius

Gluteus minimus

Piriformis

Latissimus dorsi

Gluteus maximus

Gluteus medius

▲ Muscles of the posterior of the trunk

Table 5.10 Muscles of the posterior aspect of the trunk

Name of muscle	Position	Attachments	Action(s)	Key facts
Erector spinae (ee-rek-tor spee-nee)	Made up of separate bands of muscle that lie in the groove between the vertebral colum and the ribs	Attaches to the sacrum and iliac crest at one end to the ribs, transverse and spinous processes of the vertebrae and the occipital bone at the other end of the ribs	Extends, laterally flexes and rotates the vertebral column	A very important postural muscle as it helps to extend the spine
Latissimus dorsi (la-tis-i-mus door-si)	Large sheet of muscle extending across the back of the thorax	Attaches to the posterior of the iliac crest and sacrum, lower six thoracic and five lumbar vertebrae at one end and the humerus at the other end	Extends, adducts and rotates the humerus medially	Often referred to as the 'swimmer's muscle' as it allows extension of the arm to propel the body in water Implicated in lower back pian due to its pelvic attachments
Quadratus lumborum (quad-dra-tus lum-bor-um)	Deep muscle located medially, either side of the lumbar vertebrae	Attaches to the top of the posterior of the iliac crest at one end to the twelfth rib and transverse processes of the first four lumbar vertebrae at the other end	Lateral flexion (side-bending) of the lumbar vertebrae	Excessive bending to the side can strain and injure this muscle
Gluteus maximus (gloo-tee-us max-i-mus)	A large muscle covering the buttock	Attaches to the back of the ilium along the sacroiliac joint at one end, and into the top of the femur at the other	Extends the hip, abducts and laterally rotates thigh	Sometimes referred to as the 'speedskater's muscle' as it is powerful in extending, abducting and laterally rotating the thigh Often implicated in postural defects such as lordosis (excess curvature in the lumbar spine)
Gluteus medius (gloo-tee-us meed-ee-us)	Situated on the outer surface of the pelvis, partly covered by gluteus maximus	Attaches to the outer surface of the ilium at one end and the outer surface of the femur at the other end	Abducts thigh, medially rotates thigh	When this muscle becomes tight it can create postural distortions, pulling and depressing the pelvis towards the thigh on that side, resulting in a 'functional short leg' and a compensatory scoliosis
Gluteus mimimus (gloo-tee-us min-i-mus)	Lies beneath the gluteus medius	Attachments are the same as for gluteus medius: outer surface of the ilium at one end to the outer surface of the femur at the other end	Abducts thigh, medially rotates thigh	When chronically tight, can contribute to postural conditions such as 'functional short leg' and compensatory scoliosis
Piriformis (pi-ri-for-mis)	Located deep in the buttock (behind the gluteus maximus)	Attaches to the anterior of the sacrum at one end and the top of the femur at the other	Lateral rotation and abduction of the hip	Largest lateral rotator of the hip If tight, it can restrict mobility in the hip

Deep pelvic muscles

Psoas (so-as)

- **Position and attachments** – this is a long, thick and deep pelvic muscle. It attaches to the anterior transverse processes of T12–L5 (twelfth thoracic to fifth lumbar vertebrae) to the inside of the top of the femur at the other end.
- **Action** – flexes the thigh.

Iliacus (i-lee-ak-us)

- **Position and attachments** – this is a large, fan-shaped muscle deeply situated in the pelvic girdle. It attaches to the iliac crest at one end and to the inside of the top of the femur at the other end.
- **Action** – flexes and laterally rotates the femur.

KEY FACT

The iliacus and psoas muscles are often considered as one unit and may be referred to as the iliopsoas. Both muscles are primary flexors of the thigh and, therefore, serve to advance the leg in walking.

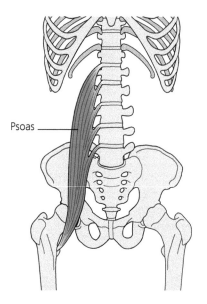
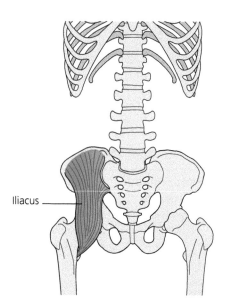

▲ Deep pelvic muscles – psoas and iliacus

Surface muscles

Deep muscles

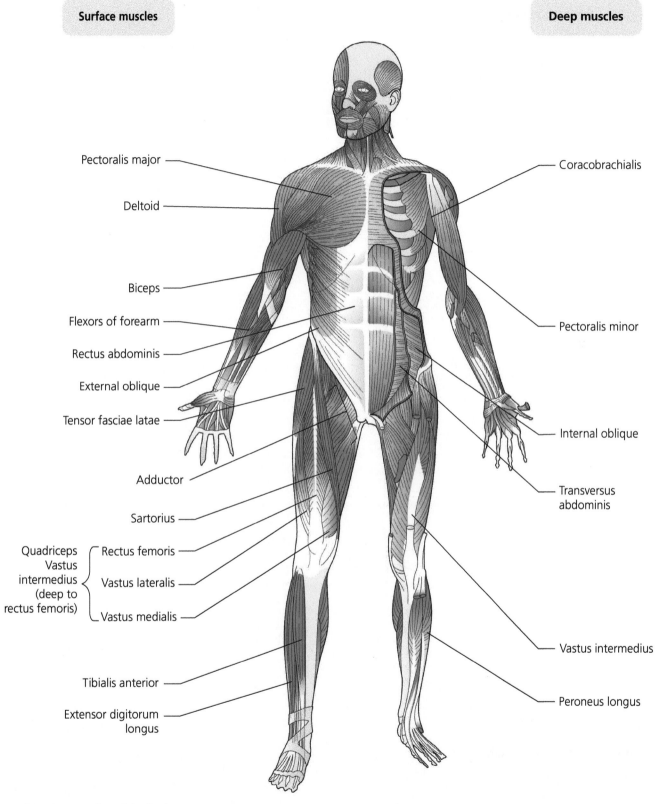

Pectoralis major

Deltoid

Biceps

Flexors of forearm

Rectus abdominis

External oblique

Tensor fasciae latae

Adductor

Sartorius

Quadriceps
Vastus
intermedius
(deep to
rectus femoris)

Rectus femoris

Vastus lateralis

Vastus medialis

Tibialis anterior

Extensor digitorum
longus

Coracobrachialis

Pectoralis minor

Internal oblique

Transversus
abdominis

Vastus intermedius

Peroneus longus

▲ Anterior muscles of the body

Surface muscles

Deep muscles

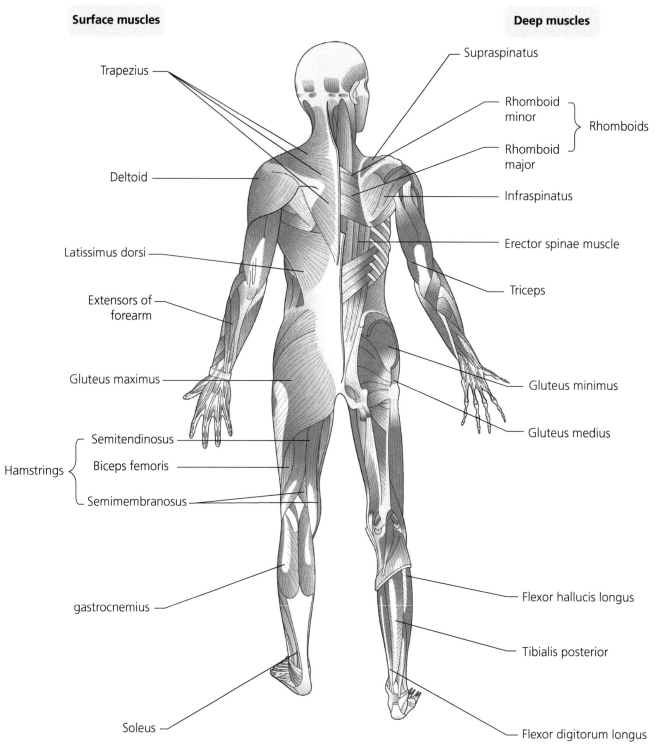

Supraspinatus

Trapezius

Rhomboid minor ⎫
⎬ Rhomboids
Rhomboid major ⎭

Deltoid

Infraspinatus

Latissimus dorsi

Erector spinae muscle

Extensors of forearm

Triceps

Gluteus maximus

Gluteus minimus

Gluteus medius

Semitendinosus

Hamstrings ⎰ Biceps femoris

Semimembranosus

Flexor hallucis longus

gastrocnemius

Tibialis posterior

Soleus

Flexor digitorum longus

▲ Posterior muscles of the body

Common pathologies of the muscular system

Atony

This is a state in which the muscles are floppy and lack their normal degree of elasticity.

Atrophy

This is the wasting of muscle tissue due to undernourishment, lack of use and diseases such as poliomyelitis.

Carpal tunnel syndrome

This syndrome is characterised by pain and numbness in the thumb or hand, resulting from pressure on the median nerve of the wrist. Pain and a pins-and-needles sensation may radiate to the elbow. It is known to cause severe pain at night and can cause muscle wasting of the hand. There is a higher risk of this condition in occupations requiring repetitive strains of the wrist. Those at risk include massage therapists and intensive computer users.

Fibromyalgia

This is a chronic condition in which the predominant symptoms include widespread musculoskeletal pain, lethargy and fatigue. Other characteristic features include a non-refreshing, interrupted sleep pattern which leaves the patient feeling more exhausted than later in the day.

Other recognised symptoms include early morning stiffness, pins-and-needles sensations, unexplained headaches, poor concentration, memory loss, low mood, urinary frequency, abdominal pain and irritable bowel syndrome. Anxiety and depression are also common.

Fibrositis

Fibrositis is an inflammatory condition of the fibrous connective tissues, especially in the muscle fascia (also known as muscular rheumatism).

Muscle cramp

This is an acute, painful contraction of a single muscle or group of muscles. Cramp is often associated with a mineral deficiency, an irritated nerve or muscle fatigue.

Muscle fatigue

This is the loss of the ability of a muscle to contract efficiently due to insufficient oxygen, exhaustion of energy supply and the accumulation of lactic acid.

Muscle spasm

This is an increase in muscle tension due to excessive motor nerve activity, resulting in a knot in the muscle.

 Activity

Discuss the main muscles used in the following exercises and sports activities:

- sit-ups
- side-bends
- press-ups
- squats
- boxing
- football.

Muscular atrophy

This is the wasting away of muscles due to poor nutrition, lack of use or a dysfunction of the motor nerve impulses.

Muscular dystrophy

This is a progressively crippling disease in which the muscles gradually weaken and atrophy. The cause is unknown.

Myositis

This condition is the inflammation of a skeletal muscle.

Rupture

A rupture is the tearing of a muscle fascia or tendon.

Shin splints

This is a soreness in the front of the lower leg due to straining of the flexor muscles used in walking.

Spasticity

This is characterised by an increase in muscle tone and stiffness. In severe cases, movements may become unco-ordinated and involve a nervous dysfunction.

Sprain

This is a complete or incomplete tear in the ligaments around a joint. It usually follows a sudden, sharp twist to the joint, which stretches the ligaments and ruptures some or all of its fibres. Sprains commonly occur in the ankle, wrist and the back where there is localised pain, swelling and loss of mobility.

Strain

A strain is an injury that is caused by excessive stretching or working of a muscle or tendon, resulting in a partial or complete tear. Symptoms include pain, swelling, tenderness and stiffness in the affected area. Muscle strains are common in the lower back and the neck.

Stress

Stress is excessive muscular tension resulting in tight, painful muscles and restricted joint movements.

Tendinitis

This is the inflammation of a tendon, accompanied by pain and swelling.

Tennis elbow

This condition is the inflammation of the tendons (tendinistis) that attach the extensor muscles of the forearm at the elbow joint.

Torticollis

This is a condition in which the neck muscles (sternomastoids) contract involuntarily. It is commonly called 'wryneck'.

Interrelationships with other systems

The muscular system

The muscular system is linked to the following body systems.

Cells and tissues/histology

There are three types of muscle tissue in the body – skeletal or voluntary muscle, smooth muscle and cardiac muscle. Fascia, tendons and ligaments are all made from connective tissue and serve a function in muscle attachments.

Skeletal system

Bones and joints provide the leverage in a movement and the muscles provide the pull on the bone to effect the movement.

Circulatory system

The circulatory system is responsible for delivering oxygen, glucose and water to the working muscles.

It also transports waste products, such as carbon dioxide and lactic acid, away from the muscles.

Respiratory system

The respiratory system provides the working muscles with vital oxygen, which is transported in the blood to be combined with glucose to release energy.

Nervous system

Muscles rely on nervous stimulation in order to function. Skeletal muscles are moved as a result of nervous stimulus, which they receive from the brain via a motor nerve.

Digestive system

The energy needed for muscle contraction is obtained principally from carbohydrate digestion. Carbohydrates are broken down into glucose. Any glucose that is not required immediately by the body is converted into glycogen and stored in the liver and muscle.

Key words

Actin: one of proteins (along with myosin) that is involved in contraction of muscle fibres

Agonist/prime mover: a muscle whose contraction moves a part of the body directly

Antagonists: when two muscles or sets of muscles pull in opposite directions to each other

Cardiac muscle tissue: specialised type of involuntary muscle tissue found only in the walls of the heart

Concentric contraction: type of contraction which causes the muscle to shorten as it contracts

Eccentric contraction: the opposite of concentric contraction; occurs when the muscle lengthens as it contracts

Endomysium: a fine connective tissue sheath surrounding a muscle fibre

Epimysium: a sheath of fibrous elastic tissue surrounding a muscle

Fascia: a sheet of connective tissue, primarily collagen, beneath the skin that attaches, stabilises, encloses, and separates muscles and other internal organs

Fasciculi: a small bundle of muscle fibres wrapped by a layer of connective tissue

Fixators: muscles that stabilise a bone to give a steady base from which the agonist works

Glycogen: a substance found in the liver and muscles that stores glucose, and provides the energy for muscle contraction

Insertion: the most movable part of the muscle during contraction

Isometric contraction: a type of muscle contraction in which the muscle works without actual movements, and there is no change in the length of the contracting muscle

Isotonic contraction: a type of muscle contraction in which the muscle changes length as it contracts and causes movement of a body part

Lactic acid: a type of acid produced in the muscle tissues during strenuous exercise

Ligament: a short band of tough, flexible fibrous connective tissue which connects two bones or cartilages or holds together a joint

Linea alba: a long tendon that extends from the bottom of the sternum to the pubic symphysis

Motor nerve: a nerve carrying impulses from the brain or spinal cord to a muscle or gland

Motor point: a point at which a motor nerve enters a muscle

Muscle fatigue: the loss of ability of a muscle to contract due to insufficient oxygen, exhaustion of energy and the accumulation of lactic acid

Muscle tone: state of partial contraction of a muscle

Myofibrils: the elongated contractile threads found in striated muscle cells

Myosin: one of the proteins (along with actin) involved in contraction of muscle fibres

Neurotransmitter: a chemical released from a nerve cell which transmits an impulse from a nerve cell to a muscle

Origin: the fixed attachment site of a muscle that does not move during contraction

Perimysium: the sheath of connective tissue surrounding a bundle of muscle fibres

Sarcolemma: thin, transparent, extensible membrane covering every striated muscle fibre

Sarcomere: one of the segments into which a fibril of striated muscle is divided

Skeletal/voluntary muscle tissue: a form of striated muscle tissue which is under 'voluntary' control. Most skeletal muscles are attached to bones by tendons.

Smooth/involuntary muscle: a specialised type of involuntary muscle tissue found only in the walls of the heart

Synergists: term referring to muscles on the same side of a joint that work together to perform the same movement

Tendon: a flexible but inelastic cord of strong fibrous collagen tissue attaching a muscle to a bone

Revision summary

The muscular system

- The muscular system is comprised mainly of **skeletal** or **voluntary muscle tissue** that is primarily attached to bones.

- The other types of muscle tissue are **cardiac muscle tissue**, which is found in the wall of the heart, and **smooth muscle tissue**, which is located in the wall of the stomach and small intestines.

- Through contraction, muscle performs three important functions – movement, maintaining posture and heat production.

- **Voluntary** or **skeletal muscle tissue** consists of muscle fibres held together by fibrous connective tissue and penetrated by numerous tiny blood vessels and nerves.

- Voluntary muscle tissue is made up of bands of elastic or contractile tissue bound together in bundles and enclosed by a connective tissue sheath.

- Each muscle fibre is enclosed in an individual wrapping of connective tissue called the **endomysium**.

- The muscle fibres are wrapped together in bundles, known as **fasciculi**, and are covered by the **perimysium** (fibrous sheath). These are gathered together to form the muscle belly (main part of the muscle) with its own sheath, the fascia **epimysium**.

- Each skeletal muscle fibre is made up of thin fibres called **myofibrils** which consist of two different types of protein strands called **actin** and **myosin**. This gives skeletal muscle its striated, or striped, appearance.

- Skeletal muscle is moved as a result of nervous stimulus received from the brain via a motor nerve.

- Each nerve fibre ends in a **motor point**, the end portion of the nerve and is the part through which the stimulus is given to contract.

- The muscle cells in smooth or involuntary muscle are spindle shaped and tapered at both ends with each muscle cell containing one centrally located oval-shaped nucleus.

- **Smooth muscle** contracts or relaxes in response to nerve impulses, stretching or hormones.

- **Cardiac muscle** is found only in the heart and, like skeletal muscle, it is striated. However, it is branched in structure and has intercalated discs between each muscle cell.
- The contraction of cardiac muscle is regulated by nerves and hormones.
- During muscular contraction, a sliding movement occurs within the contractile fibres (**myofibrils**).
- The **actin** filaments move in towards the **myosin** and cause the muscle fibres to shorten and thicken.
- During relaxation, the muscle fibres elongate and return to their original shape.
- The energy needed for muscle contraction comes from **glucose** (stored as **glycogen** in the liver and the muscles) and oxygen.
- If insufficient oxygen is available to a working muscle a waste product called **lactic acid** forms which can cause a muscle to ache.
- The term *muscle fatigue* is defined as the loss of ability of a muscle to contract efficiently due to insufficient oxygen, exhaustion of glucose and the accumulation of lactic acid.
- During exercise, the circulatory and respiratory systems adjust to cope with the increased oxygen demands of the body. More blood is distributed to the working muscles and the rate and depth of breathing is increased.
- When muscle tissue is warm, muscle contraction occurs faster due to the increase in circulation and acceleration of chemical reactions.
- Conversely, when muscle tissue is cool, the chemical reactions and circulation slow down.
- The term *muscle tone* is the state of partial contraction of a muscle to help maintain body posture.
  - Good muscle tone can be recognised by the muscles appearing firm and rounded.
  - Poor muscle tone may be recognised by the muscles appearing loose and flattened.
- **Tendons** are tough bands of white fibrous tissue that link muscle to bone. Unlike muscle, they are inelastic and therefore do not stretch.
- **Ligaments** are strong, fibrous, elastic tissues that link bones together and, therefore, stabilise joints.
- **Fascia** consists of fibrous connective tissue that envelops a muscle and provides a pathway for nerves, blood vessels and lymphatic vessels. It, therefore, plays a key role in maintaining the health of a muscle.
- Muscle attachments are known as **origin** and **insertion**.
  - The **origin** is the end of the muscle closest to the centre of the body and the **insertion** is the furthest attachment.
  - The **insertion** is generally the most movable point and is the point at which the muscle work is done.
- In movement co-ordination muscles work in pairs or groups.
- Muscles are classified by function as **agonists** (prime movers), **antagonists**, **synergists** and **fixators** (stabilisers).
  - **Antagonists** are two muscles or sets of muscles pulling in opposite directions to each other, with one relaxing to allow the other to contract.
  - The **agonist/prime mover** is known as the main activating muscle.
  - **Synergist** refers to muscles on the same side of a joint that work together to perform the same movements.
- Muscular contractions can be **isometric** or **isotonic**.
  - **Isometric contraction** is when the muscle works without actual movements (postural muscles).
  - **Isotonic contraction** is when the muscles force is considered to be constant but the muscle length changes.
  - There are two types of isotonic contraction – **concentric contractions** (towards the centre) and **eccentric contractions** (away from the centre).

Test your knowledge

Multiple choice questions

1 Which of the following is **not** a function of the muscular system?
 a movement
 b exchanging of gases
 c production of heat
 d maintaining posture

2 Which of the following applies a stimulus to cause a voluntary muscle to contract?
 a sensory nerve
 b motor nerve
 c mixed nerve
 d none of the above

3 A tendon attaches:
 a muscle to bone
 b muscle to ligament
 c bone to bone
 d none of the above.

4 Where would you **not** find involuntary muscle tissue?
 a stomach
 b bladder
 c brain
 d heart

5 Which of these provides the fuel for muscle contraction?
 a ATP
 b glucose
 c pyruvic acid
 d actin and myosin

6 Which of the following statements is **true**?
 a On contraction voluntary muscle fibres elongate.
 b The attachment of myosin to actin requires the mineral sodium.
 c The merging of actin and myosin filaments causes muscle fibres to shorten and thicken on contraction.
 d The force of muscle contraction depends on where the muscle fibres are located.

7 What causes muscle fatigue?
 a insufficient oxygen
 b exhaustion of energy supply
 c accumulation of lactic acid
 d all of the above

8 Which term describes the state of continuous partial contraction of muscles?
 a atrophy
 b muscle tone
 c hypertrophy
 d none of the above

9 Which of the following statements is **false**?
 a During exercise, there is an increase in return of venous blood to the heart.
 b During exercise, a muscle may receive as much as 15 times its normal flow of blood.
 c The presence of lactic acid in the blood stimulates the respiratory centre in the brain, decreasing the rate and depth of breath.
 d The rate and depth of breathing remains above normal for a while after strenuous exercise has ceased.

10 Which of the following statements is **true**?
 a Muscles with less than the normal degree of tone are said to be spastic.
 b Muscles that appear firm and rounded indicate good muscle tone.
 c An increase in the size and diameter of muscles fibres leads to a condition called atrophy.
 d Poor muscle tone is a cause of muscle cramps.

Exam-style questions

11 State three functions of the muscular system. 3 marks

12 Briefly describe the following types of muscle tissue and where they may be found in the body:
 a cardiac muscle 2 marks
 b voluntary/striated muscle 2 marks
 c involuntary muscle. 2 marks

13 State the four characteristics of muscle tissue that contribute to its function. 4 marks

14 Define the term *muscle tone*. 1 mark

15 State one action of each of the following muscles:
 a Brachioradialis 1 mark
 b Extensor carpi radialis 1 mark
 c Quadriceps 1 mark
 d Sartorius 1 mark
 e Tibialis anterior 1 mark
 f Flexor digitorum longus. 1 mark

6 The cardiovascular system

Introduction

The cardiovascular system is the body's transport system and comprises blood, blood vessels and the heart. Blood provides the fluid environment for our body's cells and is transported in specialised tubes called blood vessels. The heart acts like a pump which keeps the blood circulating constantly around the body.

OBJECTIVES

By the end of this chapter you will understand:

- the composition and functions of blood
- the structural and functional significance of the different types of blood cells
- the structural and functional differences between the different blood vessels
- the major blood vessels of the heart
- the pulmonary and systemic blood circulation
- the main arteries of the head, neck and body
- the main veins of the head, neck and body
- blood pressure and the pulse rate
- common pathologies of the circulatory system
- the interrelationships between the cardiovascular and other body systems.

In practice

It is essential for therapists to have a good working knowledge of the cardiovascular system in order to be able to understand the physiological effects of treatments. Treatments such as massage help to improve circulation by assisting the venous flow back to the heart. By enhancing blood flow, delivery of oxygen and nutrients to the tissues is improved and the removal of waste products is hastened.

Blood

Blood is the fluid tissue or medium in which all materials are transported to and from individual cells in the body. Blood is, therefore, the chief transport system of the body.

The composition of blood

The percentage composition of blood

Blood is 55% plasma, a clear, pale yellow and slightly alkaline fluid. The other 45% of blood is made up of the blood cells: erythrocytes (red blood cells), leucocytes (white blood cells) and thrombocytes (platelets).

Plasma consists of:

- 91% water
- 9% dissolved blood proteins, waste, digested food materials, mineral salts and hormones.

Blood types

Although all blood has the same basic elements, not all blood is alike. Blood types are determined by the presence or absence of certain antigens, which are substances that can trigger an immune response if they are foreign to the body.

Safe blood transfusions depend on careful blood typing and cross-matching, since some antigens can trigger the immune system to attack transfused blood.

The ABO blood group system

There are four major blood groups which are determined by the presence or absence of two antigens – A and B – on the surface of red blood cells:

- **Group A** – has only the A antigen on red cells (and B antibody in the plasma)
- **Group B** – has only the B antigen on red cells (and A antibody in the plasma)
- **Group AB** – has both A and B antigens on red cells (but neither A nor B antibody in the plasma)
- **Group O** – has neither A nor B antigens on red cells (but both A and B antibody are in the plasma)

Source: www.redcrossblood.org

There are very specific ways in which blood types must be matched for a safe blood transfusion (Table 6.1).

Table 6.1 Blood type matches

Blood group	Can receive blood from	Can donate blood to
A	A and O	A and AB
B	B and O	B and AB
AB	Any of these groups	AB
O	O only	Any of these groups

In addition to the A and B antigens, there is a third antigen called the rhesus (Rh) factor, which can be either present (+) or absent (−). In general, Rh-negative blood is given to Rh-negative patients; Rh-positive blood or Rh-negative blood may be given to Rh-positive patients.

- The universal red cell donor has type O rhesus negative blood type.
- The universal plasma donor has type AB blood type.

The functions of the blood

Blood has four main functions:

1 transport
2 defence
3 regulation
4 clotting.

1 Transport

Blood is the primary transport medium for a variety of substances that travel throughout the body.

- Oxygen is carried from the lungs to the cells of the body by red blood cells.

- Carbon dioxide is carried from the body's cells to the lungs.
- Nutrients such as glucose, amino acids, vitamins and minerals are carried from the small intestine to the cells of the body.
- Cellular wastes such as water, carbon dioxide, lactic acid and urea are carried in the blood to be excreted.
- Hormones, which are internal secretions that help to control important body processes, are transported by the blood to target organs.

> **KEY FACT**
> Red blood cells are called erythrocytes and they contain the red protein pigment haemoglobin, which combines with oxygen to form oxyhaemoglobin. The pigment haemoglobin assists the function of the erythrocyte in transporting oxygen from the lungs to the body's cells and in carrying carbon dioxide away.

2 Defence

White blood cells are collectively called leucocytes and they play a major role in combating disease and fighting infection.

> **KEY FACT**
> Some white blood cells are known as phagocytes as they have the ability to ingest micro-organisms which invade the body and cause disease. Other specialised white blood cells, called lymphocytes, produce antibodies to protect the body against infection.

3 Regulation

Blood helps to regulate heat in the body by absorbing large quantities of thermal energy produced by the liver and the muscles. This is transported around the body to help maintain a constant internal temperature. Blood also helps to regulate the body's pH balance.

4 Clotting

Clotting is an effective mechanism for controlling blood loss from blood vessels when they become damaged, as in the case of a cut. Specialised blood cells called thrombocytes, or platelets, form a clot around the damaged area to prevent blood loss and to stop the entry of bacteria.

The process of blood clotting

Blood clotting or coagulation is a process that can stop bleeding. When a blood vessel is injured, there is a rapid, localised response to help arrest the bleeding and prevent excessive blood loss. This is known as haemostasis.

Haemostasis (heem-o-stay-sis)

Haemostasis is the process in which bleeding is stopped. It involves three steps.

1 **Vasoconstriction**: narrowing of damaged blood vessels to reduce blood loss. This is caused by contraction of the smooth muscle in the wall of vessels.

2 **Activation of platelets**: activated platelets stick to each other and to collagen fibres in the broken walls of blood vessels, forming a platelet plug that temporarily blocks blood flow. The platelets also release chemicals that attract other platelets and stimulate further vasoconstriction.

3 **Formation of a blood clot**: the clot contains fibres that trap the platelets and is stronger and longer-lasting than the initial platelet plug. Clot formation is summarised below.

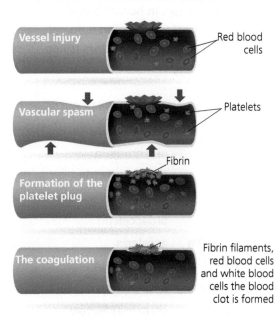

▲ The stages of haemostasis

Summary of the blood-clotting process

The blood-clotting process is complex and involves a long sequence of chemical reactions. However, the process can be summarised in three steps:

1 The damaged tissue releases thromboplastin (a plasma protein that helps with blood coagulation) and a prothrombin activator complex, which converts a blood protein called prothrombin into another protein called thrombin.

2 Thrombin converts a soluble blood protein called fibrinogen into an insoluble protein called fibrin.

3 Fibrin exists as solid fibres which form a tight mesh over the wound. The mesh traps platelets and other blood cells and forms the blood clot.

KEY FACT

Prothrombin and fibrinogen are always present in our blood, but they aren't activated until a prothrombin activator is made in response to injury.

Clotting factors

Clotting factors are proteins in the blood that control bleeding. Many different clotting factors work together in a series of chemical reactions to stop bleeding.

There are 12 clotting factors in human blood and tissues, which are designated by roman numerals. There are 13 numerals but only 12 factors (since factor VI was subsequently found to be part of another factor).

Most clotting factors are manufactured in the liver.

- Clotting factor I — fibrinogen
- Clotting factor II — prothrombin
- Clotting factor II — thromboplastin
- Clotting factor IV — calcium
- Clotting factor V — proaccelerin
- Clotting factor VI/Va — accelerin
- Clotting factor VII — proconvertin
- Clotting factor VIII — antihaemophilic factor A
- Clotting factor IX — christmas factor/ antihaemophilic factor B
- Clotting factor X — Stuart–Prower factor
- Clotting factor XI — plasma thromboplastin component
- Clotting factor XII — Hageman factor
- Clotting factor XIII — fibrin stabilising factor

Blood cells

There are three types of blood cells:

1 **Erythrocytes** – red blood cells

2 **leucocytes** – white blood cells

3 **thrombocytes** – platelets.

Table 6.2 Overview of the three types of blood cells

Type of blood cell	Description	Function
Erythrocyte (err-rith-ro-sytes)	Disc-shaped structures Non-nucleated Red in colour due to protein haemoglobin	Transport the gases of respiration
Leucocytes (loo-co-sytes)	Largest of all the blood cells White due to lack of haemoglobin	Protect the body against infection and disease
Thrombocytes or platelets	Granular disc-shaped, small fragments of cells	Blood clotting

1 Erythrocytes

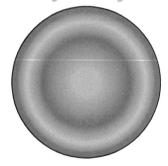

▲ An erythrocyte

Erythrocytes are disc-shaped structures that make up more than 90% of the formed elements in blood. They are made in red bone marrow and contain the iron–protein compound haemoglobin.

Old and worn-out erythrocytes are destroyed in the liver and the spleen. The haemoglobin is broken down and the iron within it is retained for further haemoglobin synthesis. Erythrocytes have a life span of only about four months and, therefore, have to be continually replaced.

The function of erythrocytes is to transport the gases of respiration (they transport oxygen to the cells and carry carbon dioxide away from the cells).

2 Leucocytes

Leucocytes are the largest of all the blood cells and appear white due to their lack of haemoglobin. They have a nucleus and are generally more numerous than erythrocytes.

Leucocytes usually only survive for a few hours, but in a healthy body some can live for months or even years. The main function of leucocytes is to protect the body against infection and disease via a process known as phagocytosis, which means to engulf and digest microbes, dead cells and tissue.

There are two main categories of leucocytes: granulocytes and agranulocytes.

- **Granulocytes** – these account for about 75% of white blood cells and can be further divided into:
  - **Neutrophils** – one of the first immune cell types to travel to the site of an infection. Neutrophils help fight infection by ingesting micro-organisms and releasing enzymes to kill them. A neutrophil is a type of white blood cell, a type of granulocyte, and a type of phagocyte.
  - **Eosonophils** – a type of immune cell that has granules (small particles) with enzymes that are released during infections, allergic reactions and asthma. An eosinophil is a type of white blood cell and a type of granulocyte.
  - **Basophils** – a type of immune cell that has granules (small particles) with enzymes that are released during allergic reactions and asthma. A basophil is a type of white blood cell and a type of granulocyte.
- **Agranulocytes** – these can be divided into lymphocytes, which account for about 20% of all white blood cells, and monocytes, which account for about 5% of white blood cells.
  - **Lymphocytes** – a type of white blood cell that is crucial to our immune system. Lymphocytes recognise antigens, produce antibodies, and destroy cells that could cause damage. There are two main types of lymphocytes:

- The **B-cells** produce antibodies that attack invading bacteria, viruses and toxins.
- The **T-cells** destroy the body's own cells that have been taken over by viruses or become cancerous.
- **Monocytes** – can develop into two types of cell:
  - **Dendritic cells** are antigen-presenting cells which are able to mark out cells with foreign antigens that need to be destroyed by lymphocytes.
  - **Macrophages** are large scavenging phagocytes that clean up areas of infection. Monocytes increase in number during chronic infections. They are larger and live longer than neutrophils.

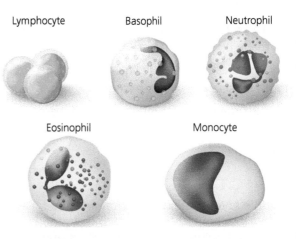

▲ White blood cells

3 Thrombocytes

▲ Thrombocytes

Thrombocytes are also known as platelets. These are small fragments of cells and are the smallest cellular elements of the blood. They are formed in bone marrow and are disc-shaped with no nucleus.

Thrombocytes normally have a short life span of just five to nine days.

They are very significant in the blood-clotting process as they initiate the chemical reaction that leads to the formation of a blood clot. Platelets stop the loss of blood from a damaged blood vessel (see page 175).

Blood vessels

Blood flows round the body due to the pumping action of the heart and is carried in vessels called arteries, veins and capillaries.

- **arteries** – carry blood away from the heart
- **veins** – carry blood towards the heart
- **capillaries** – unite arterioles and venules, forming a network in the tissues.

Blood vessel walls

Blood vessels have to withstand the pressure of the blood as it is pumped continuously by the heart. To resist this pressure, the walls of the blood vessels are constructed of three layers, known as tunics.

1 **Tunica adventitia** (or tunica externa) is the outer layer made up of fibrous tissue.
2 **Tunica media** is the middle layer made up of smooth muscle and elastic tissue.
3 **Tunica intima** (or tunica interna) is the innermost layer made up of squamous epithelium (endothelial cells).

KEY FACT

The middle layer (tunica media) of arteries contains more smooth muscle than is found in veins, thus allowing arteries to constrict and dilate to adjust the volume of blood supplied to the tissues.

Lumens

Blood vessels such as arteries, veins and capillaries have a central void called a lumen, which is the space through which the blood flows.

Veins are generally larger in diameter, carry a greater volume of blood and have thinner walls in proportion to the size of the lumen. The tunica media is smaller in relation to the lumen than in arteries.

Veins, therefore, have a wide lumen to accommodate the slow-flowing blood under low pressure.

Arteries have thicker walls in proportion to their narrow lumen and carry blood under higher pressure than veins.

KEY FACTS

Arteries:

- have thick muscular and elastic walls to withstand pressure
- are generally deep-seated, except where they cross over a pulse spot
- have no valves, except at the base of the pulmonary artery where they leave the heart
- have a narrow lumen to carry blood under high pressure
- carry blood away from the heart, and carry oxygenated blood (except the pulmonary artery to the lungs)
- carry blood under high pressure
- give rise to small blood vessels called arterioles, which deliver blood to the capillaries.

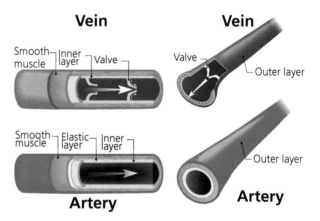

▲ The structure of an artery and a vein

KEY FACTS

Veins:

- have thinner muscular walls than arteries
- are generally superficial, not deep-seated
- have valves at intervals to prevent the backflow of blood
- have a large lumen, allowing more blood to flow with less resistance
- carry blood towards the heart
- carry deoxygenated blood (except the pulmonary veins) from the lungs
- carry blood under low pressure
- form finer blood vessels called venules which continue from capillaries.

KEY FACTS

Capillaries:

- are superficial microscopic blood vessels that form part of the microcirculation

- have thin walls, only a single layer of cells thick, to enable the diffusion of dissolved substances to and from the tissues

- have no valves

- have a narrow lumen; this means that many capillaries can fit in a small space, increasing the surface area for diffusion

- carry blood under low pressure, but higher than in veins

- carry both oxygenated and deoxygenated blood as they exchange oxygen and carbon dioxide with tissues

- are responsible for supplying the cells and tissues with nutrients

- unite arterioles and venules, forming a network in the tissues.

KEY FACT

Both arteries and veins have three layers (external, middle and internal layers) but because an artery must contain the pressure of blood pumped from the heart, its walls are thicker and more elastic.

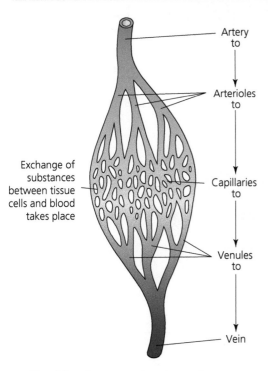

Exchange of substances between tissue cells and blood takes place

Artery to

Arterioles to

Capillaries to

Venules to

Vein

▲ Blood flow from an artery to a vein

KEY FACT

Oxygenated blood flowing through the arteries appears bright red in colour due to the colour of the oxygen-binding pigment, haemoglobin. As haemoglobin moves through capillaries in red blood cells, it offloads some oxygen and picks up carbon dioxide, changing colour in the process. This explains why blood flowing in veins appears darker.

Vasodilation and vasoconstriction

Capillaries have the ability to narrow (vasoconstriction) or widen (vasodilation), which allows for changes in body temperature.

- **Vasoconstriction** is the **narrowing** of the vessels, resulting in a decreased vascular diameter and conservation of heat in the blood, so the body keeps warm.

- **Vasodilation** is the **widening** of the vessels, resulting in an increased vascular diameter and loss of heat from the blood through radiation, cooling the body.

The heart

The heart is a hollow organ made of cardiac muscle tissue which lies in the thorax above the diaphragm and between the lungs.

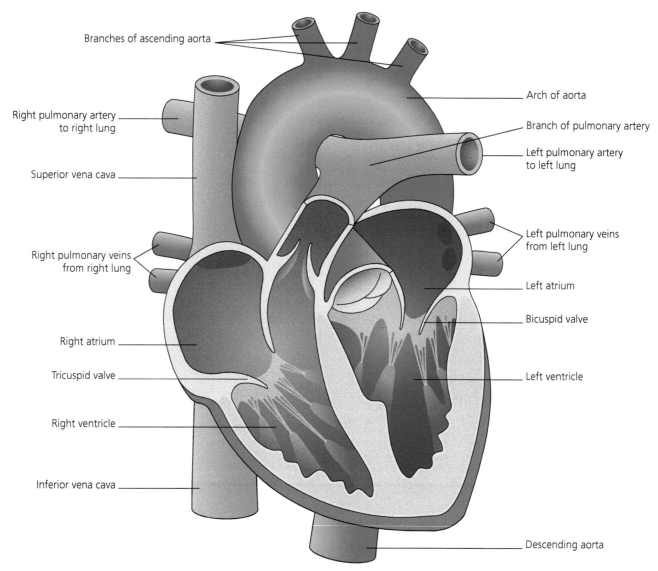

Branches of ascending aorta

Right pulmonary artery to right lung

Superior vena cava

Right pulmonary veins from right lung

Right atrium

Tricuspid valve

Right ventricle

Inferior vena cava

Arch of aorta

Branch of pulmonary artery

Left pulmonary artery to left lung

Left pulmonary veins from left lung

Left atrium

Bicuspid valve

Left ventricle

Descending aorta

▲ The structure of the heart

Composition of the heart

The heart is composed of three layers of tissue:

1 **Pericardium** (the outer layer) – this consists of an outer fibrous layer and an inner, double-layered bag of serous membrane enclosing a cavity that is filled with pericardial fluid. This fluid reduces friction as the heart moves as it beats inside the bag.

2 **Myocardium** (the middle layer) – this is a strong layer of cardiac muscle that makes up the bulk of the heart.

3 Endocardium (the inner layer) – this thin layer lines the heart's cavities and is continuous with the lining of the blood vessels.

The heart is divided into a right and left side by a partition called a **septum**, and each side is further divided into a thin-walled **atrium** above and a thick-walled **ventricle** below. The top chambers of the heart (the atria, plural) take in blood from the body from the large veins and pump it to the bottom chambers. The lower chambers, the ventricles, pump blood to the body's organs and tissues.

There are four sets of valves that regulate the flow of blood though the heart, as shown in Table 6.3.

Table 6.3 The valves of the heart

Valve	Location
1 Tricuspid valve	Between the right atrium and the right ventricle
2 Bicuspid or mitral valve	Between the left atrium and the left ventricle
3 Aortic valve	Between the left ventricle and the aorta
4 Pulmonary valve	Between the pulmonary artery and the right ventricle

The bicuspid and tricuspid valves (also known as the atrioventricular valves) help to maintain the direction of blood flow through the heart by allowing blood to flow into the ventricles while preventing it from returning to the atria.

The aortic and pulmonary valves are known as the semilunar valves. They control the blood flow out of the ventricles into the aorta and the pulmonary arteries, and prevent any backflow of blood into the ventricles. These valves open in response to pressure generated when the blood leaves the ventricles.

The heart muscle is supplied by the two coronary arteries (right and left) which originate from the base of the aorta.

KEY FACT

If either of the coronary arteries is unable to supply sufficient blood to the heart muscle, a heart attack occurs. The most common site of a heart attack is the anterior or inferior part of the left ventricle.

Blood flow through the heart

Blood moves into and out of the heart in a co-ordinated and precisely timed rhythm.

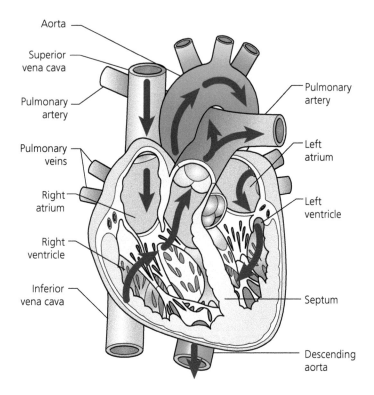

Aorta

Superior vena cava

Pulmonary artery

Pulmonary veins

Right atrium

Right ventricle

Inferior vena cava

Pulmonary artery

Left atrium

Left ventricle

Septum

Descending aorta

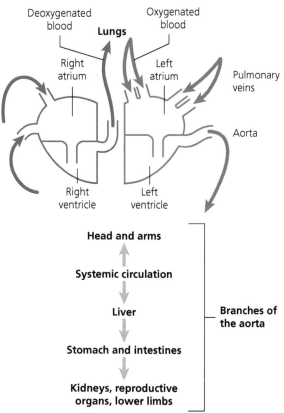

Deoxygenated blood

Lungs

Oxygenated blood

Right atrium

Left atrium

Pulmonary veins

Right ventricle

Left ventricle

Aorta

Head and arms

Systemic circulation

Liver

Stomach and intestines

Kidneys, reproductive organs, lower limbs

Branches of the aorta

▲ Blood flow through the heart

Blood flow through the heart can be considered in four stages:

- **Stage 1** – deoxygenated blood from the body enters the superior and inferior vena cavae and flows into the right atrium. When the right atrium is full, it empties through the tricuspid valve into the right ventricle.

- **Stage 2** – when the right ventricle is full, it contracts and pushes blood through the pulmonary valve into the pulmonary artery. The pulmonary artery divides into the right and left branch and takes blood to both lungs, where the blood becomes oxygenated. The four pulmonary veins leave the lungs carrying oxygen-rich blood back to the left atrium.

- **Stage 3** – this process takes place at the same time as the process described in stage 1. Oxygen-rich blood fills the left atrium. When full, the blood passes through to the left ventricle via the bicuspid or mitral valve.

- **Stage 4** – this process takes place at the same time as the process described in stage 2. When the left ventricle is full it contracts, forcing blood through the aortic valve into the aorta and to all parts of the body (except the lungs). The walls of the left ventricle are thicker than those of the right in order to provide the extra strength to push blood out of the heart and around the whole body.

Blood is transported in a double circuit which consists of two separate systems (**pulmanory** and general/**circulatory**) which are joined only at the heart.

> **Study tip**
>
> To study the blood flow through the heart, follow the arrows on the diagram on the previous page.
>
> Blue indicates the flow of deoxygenated blood and red shows the flow of oxygenated blood.
>
> Remember that although arteries carry oxygenated blood and veins carry deoxygenated blood, there is an exception to the rule: the pulmonary arteries carry deoxygenated blood and the pulmonary veins carry oxygenated blood.

The function of the heart

The function of the heart is to maintain a constant circulation of blood throughout the body. The heart acts as a pump and its action consists of a series of events known as the **cardiac cycle**.

The cardiac cycle

The cardiac cycle is the sequence of events between one heartbeat and the next, and is normally less than a second in duration.

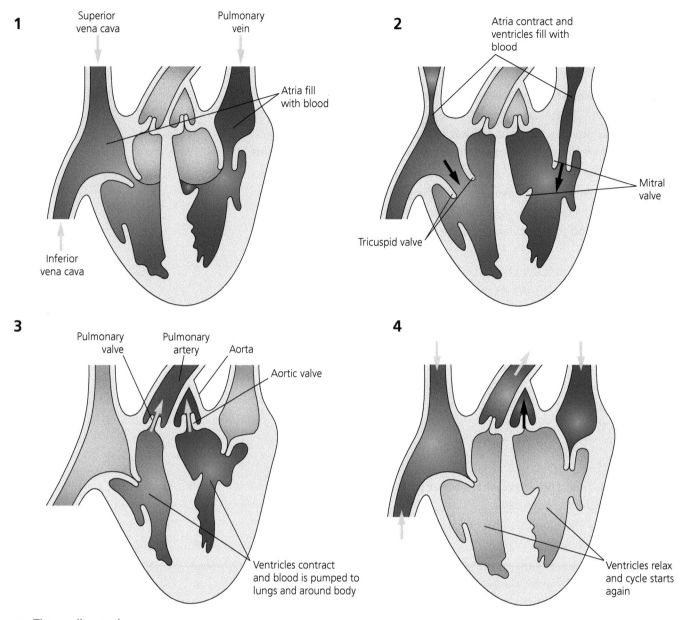

1
- Superior vena cava
- Pulmonary vein
- Atria fill with blood
- Inferior vena cava

2
- Atria contract and ventricles fill with blood
- Mitral valve
- Tricuspid valve

3
- Pulmonary valve
- Pulmonary artery
- Aorta
- Aortic valve
- Ventricles contract and blood is pumped to lungs and around body

4
- Ventricles relax and cycle starts again

▲ The cardiac cycle

- During a cardiac cycle, the atria contract simultaneously and force blood into the relaxed ventricles.
- The ventricles then contract very strongly and pump blood out through the aorta and the pulmonary artery.
- During ventricular contraction, the atria relax and fill up again with blood.

The heart rate is the number of cardiac cycles per minute. In an average healthy person this is likely to be between 60 and 70 cycles, or beats per minute.

> **KEY FACT**
> On average the heart beats 100 000 times a day. In an average lifetime, that is around 2.5 billion heartbeats!

The heart has its own built-in rhythm. The co-ordinated rhythm of the heart is initiated by the electrical system in the sinoatrial (SA) node, which sets the pace of the heart rate. The signal originates in the right atrium and travels to the left atrium, causing the atria to contract. At the precise moment the atria have completed their contraction, the signal travels via the atrioventricular (AV) bundle to the right ventricle and into the left ventricle, causing the ventricles to contract.

> **KEY FACT**
> An electrical device known as a pacemaker can be implanted to assist or take over initiation of the signal that starts a heartbeat. This may be necessary if the SA node is diseased or damaged.

Heart sounds

Heart sounds may be heard through a stethoscope. Closure of the heart valves produces two main sounds:

- The first sound is a low-pitched 'lubb' which is generated by the closing of the bicuspid and tricuspid valves.
- The second sound is a higher-pitched 'dubb' caused by the closing of the aortic and pulmonary valves.

Pulmonary circulation

The function of pulmonary circulation is to aid respiration. It consists of the circulation of deoxygenated blood from the right ventricle of the heart to the lungs, via the pulmonary arteries, and the circulation of oxygenated blood from the lungs to the heart, via the pulmonary veins. Blood dumps carbon dioxide and becomes oxygenated in the lungs, and is then returned to the left atrium by the pulmonary veins to be passed to the aorta for the general, or systemic, circulation.

Pulmonary circulation is essentially the circulatory system between the heart and the lungs, where a high concentration of blood oxygen is restored and the concentration of carbon dioxide in the blood is lowered.

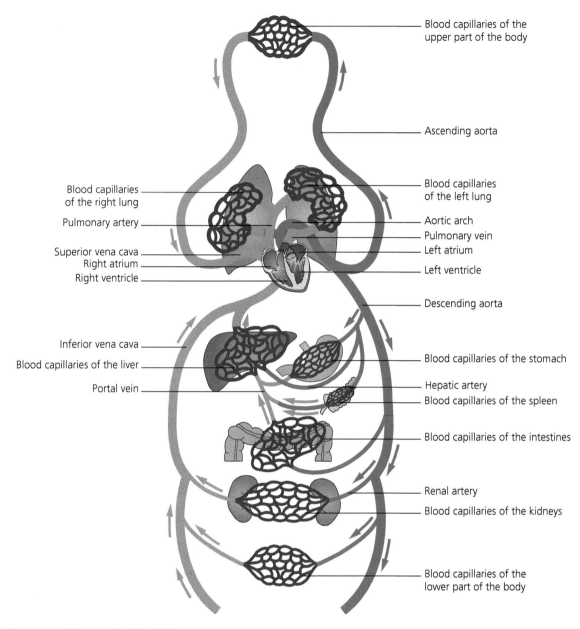

Blood capillaries of the
upper part of the body

Ascending aorta

Blood capillaries
of the right lung

Pulmonary artery

Superior vena cava
Right atrium
Right ventricle

Blood capillaries
of the left lung

Aortic arch
Pulmonary vein
Left atrium
Left ventricle

Descending aorta

Inferior vena cava

Blood capillaries of the liver

Portal vein

Blood capillaries of the stomach

Hepatic artery
Blood capillaries of the spleen

Blood capillaries of the intestines

Renal artery
Blood capillaries of the kidneys

Blood capillaries of the
lower part of the body

▲ The pulmonary and systemic circulation

In practice

The increase in blood flow
during a massage can help
to bring fresh oxygen and
nutrients to the tissues via the
arterial circulation and can aid
the removal of waste products
via the venous circulation. By
boosting blood circulation,
therefore, massage can help to
improve the condition of the
skin and muscle tone.

General or systemic circulation

The systemic circuit is the largest part of the double circulatory
system and it carries oxygenated blood from the left ventricle of the
heart through the branches of the aorta and all around the body.
Deoxygenated blood is returned to the right atrium via the superior and
inferior vena cavae, to be passed to the right ventricle, where it enters
the pulmonary circuit. The function of the systemic circulation is to
bring nutrients and oxygen to all systems of the body and to carry waste
materials away from the tissues for elimination.

Portal circulation

The portal circulation is located within the systemic circuit. It collects blood from the digestive organs (stomach, intestines, gall bladder, pancreas and spleen) and delivers this blood, via the hepatic portal veins, to the liver for processing. The liver has a key function in maintaining proper concentrations of glucose, fat and protein in the blood. The hepatic portal system allows the blood from the digestive organs to take a detour through the liver to process these substances before they enter the systemic circulation.

 Activity

The cardiovascular system

Remind yourself of the following blood vessels and parts of the heart:

- aorta
- pulmonary artery
- bicuspid valve
- right ventricle
- left ventricle
- tricuspid valve
- superior vena cava
- inferior vena cava
- pulmonary veins.

1 Write each one on a separate piece of card or paper.
2 Sort the cards into the correct order to represent the blood flow through the heart.
3 Colour code the cards: use **red** to represent oxygenated blood and **blue** to represent deoxygenated blood.

Main arteries

The aorta (*a-orr-ta*)

The aorta is the main artery of the systemic circuit and it carries oxygenated blood around the body. It is the largest artery in the body, beginning at the top of the left ventricle and then dividing into three main sections, which branch further to supply the whole of the body:

1 The **ascending** part of the aorta has branches which supply the head, neck and the top of the arms.

2 The **descending thoracic** part of the aorta has branches which supply the organs of the thorax.

3 The **descending abdominal** part has branches which supply the legs and organs of the digestive, renal and reproductive systems.

The names of most major arteries are derived from the anatomical structures they serve. For example, the femoral artery is found close to the femur. Arteries are generally deep seated and are found on both sides of the body, identified as either right or left like the sides of the heart.

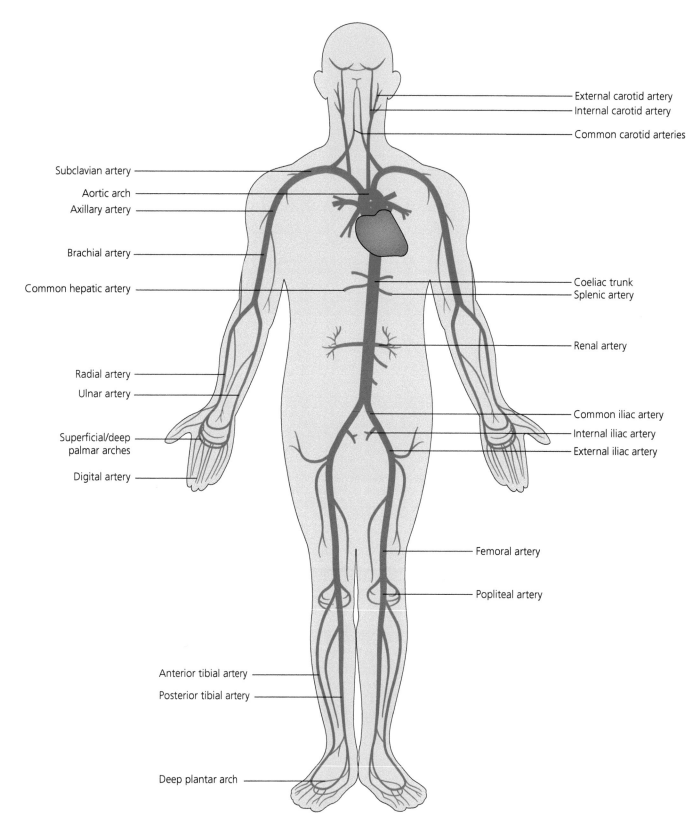

External carotid artery
Internal carotid artery
Common carotid arteries
Subclavian artery
Aortic arch
Axillary artery
Brachial artery
Common hepatic artery
Coeliac trunk
Splenic artery
Renal artery
Radial artery
Ulnar artery
Common iliac artery
Internal iliac artery
External iliac artery
Superficial/deep palmar arches
Digital artery
Femoral artery
Popliteal artery
Anterior tibial artery
Posterior tibial artery
Deep plantar arch

▲ Main arteries

Blood circulation to the head and neck

Blood is supplied to parts within the neck, head and brain through branches of the **subclavian** and **common carotid arteries**. The common cartoid artery extends from the **brachiocephalic (innominate) artery**. It extends on each side of the neck and divides at the level of the larynx into two branches:

- The **internal carotid artery** passes through the temporal bone of the skull to supply oxygenated blood to the brain, eyes, forehead and part of the nose.
- The **external carotid artery** is divided into branches (facial, temporal and occipital arteries) which supply the skin and muscles of the face, and side and back of the head, respectively. This vessel also supplies the superficial parts and structures of the head and neck. These include the salivary glands, scalp, teeth, nose, throat, tongue and thyroid gland.

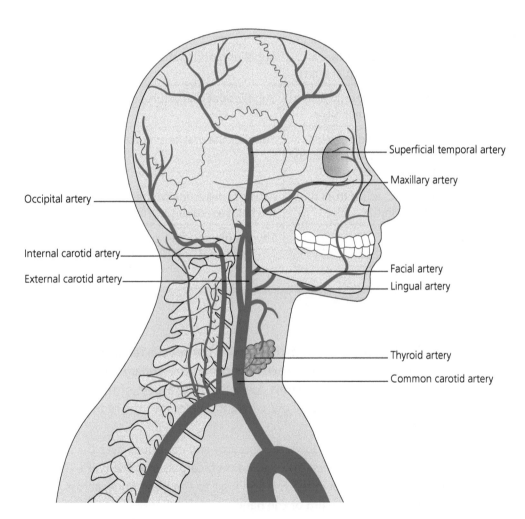

Occipital artery

Internal carotid artery

External carotid artery

Superficial temporal artery

Maxillary artery

Facial artery

Lingual artery

Thyroid artery

Common carotid artery

▲ Blood flow to the head and neck

The **vertebral arteries** are main divisions of the **subclavian arteries**. They arise from the subclavian arteries in the base of the neck near the tip of the lungs and pass upwards through the openings (foramina) of transverse processes of the cervical vertebrae, where they unite to form a single **basilar artery**. The **basilar artery** then terminates by dividing into two posterior **cerebral arteries** that supply the occipital and temporal lobes of the cerebrum.

Main arteries of the face and head

Table 6.4 Arteries of the face and head

Name of artery	Location	Area artery supplies
Common carotid artery (right and left)	Located on each side of the neck	Divide into the external and internal carotid arteries
External carotid artery	Branches off from the common carotid artery	Provides blood supply to the scalp, face and neck
Internal carotid artery	Branches off from the common carotid artery	Supplies blood to the brain
Occipital artery	Branch of the external carotid artery opposite the facial artery Path is below the posterior belly of digastric to the occipital region	Supplies blood to the back of the scalp and sternomastoid muscles, and deep muscles in the back and neck
Facial artery	Branch of the external carotid artery a little above the level of the lingual artery	Supplies blood to the structures of the face
Maxillary artery	Branches from the external carotid artery just deep to the neck of the mandible	Supplies deep structures of the face
Lingual artery	Rises from the external carotid between the superior thyroid artery and facial artery Located easily in the tongue	Supplies the oral floor and tongue
Superficial temporal artery	Major artery of the head, arises from the external carotid artery in the neck	Assists in delivering oxygenated blood from the heart to regions within the neck and head
Thyroid artery	In the neck	Thyroid gland

Blood circulation to the arm and hand

The blood supply to the arm begins with the **subclavian artery** (a branch of the aorta). The subclavian artery becomes the axillary artery and then the **brachial artery**, which runs down the inner aspect of the upper arm to about 1 cm below the elbow, where it divides into the **radial** and **ulnar arteries**.

The **radial artery** runs down the forearm and continues over the carpals to pass between the first and second metacarpals into the palm. The **ulnar artery** runs down the forearm next to the ulnar bone, across the carpals into the palm of the hand.

Together, the **radial** and **ulnar arteries** form two arches in the hand called the **deep** and **superficial arches**. From these arteries others branch to supply blood to the structures of the upper arm, forearm, hand and fingers.

Main arteries of the arm and hand

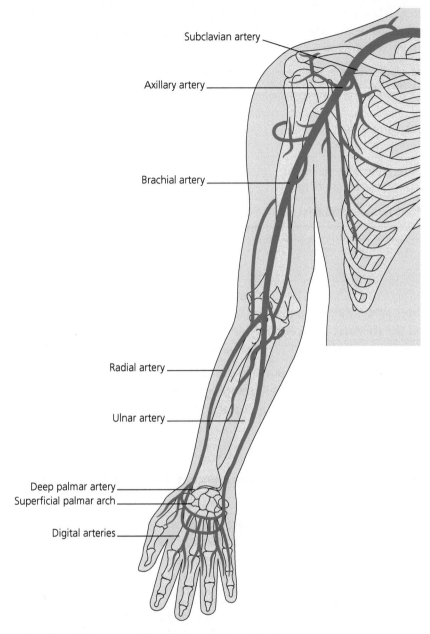

Subclavian artery

Axillary artery

Brachial artery

Radial artery

Ulnar artery

Deep palmar artery
Superficial palmar arch

Digital arteries

▲ Arteries of the arm and hand

Table 6.5 Arteries of the arm and hand

Name of artery	Location	Area artery supplies
Subclavian artery	Below the clavicle	Supplies oxygenated blood to the arm
Axillary artery	Passes through the axilla, just underneath the pectoralis minor muscle	Conveys oxygenated blood to the lateral aspect of the thorax, the axilla and upper limb
Brachial artery	Runs down the inner aspect of the upper arm to about 1 cm below the elbow	Supplies main source of oxygenated blood for the arm

→

Name of artery	Location	Area artery supplies
Ulnar artery	Runs down the forearm next to the ulnar bone, across the carpals into the palm of the hand	Supplies the anterior aspect of the forearm
Radial artery	Runs down the forearm and continues over the carpals to pass between the first and second metacarpals into the palm	Supplies the posterior aspect of the forearm
Superficial palmar arch	Superficial branch of the ulnar artery, lies across the centre of the palm	Supplies blood to palm and fingers
Deep palmar arch	Continuation of the radial artery, lies 1 cm proximal to the superficial palmar arch, across centre of palm	Supplies blood to palm and fingers
Digital arteries	Between the second and third, third and fourth, fourth and fifth fingers	Supplies blood to the second, third, fourth and fifth fingers

Blood circulation to the thoracic wall

The thoracic wall is supplied by branches of the **subclavian artery** and the **thoracic (descending) aorta.** This branch of the aorta spans from the level of T4 to T12, after which it becomes the abdominal aorta.

Branches of the thoracic aorta in descending order include the bronchial arteries, mediastinal arteries, oesophageal arteries, pericardial arteries, superior phrenic arteries, intercostal and subcostal arteries.

Blood circulation to the abdominal wall

The abdominal wall is supplied by branches of the **abdominal aorta**.

Table 6.6 Arteries of the abdominal wall

Name of artery	Location	Area it supplies
Celiac (coeliac) artery	Branches from the abdominal aorta at the twelfth thoracic vertebra	Supplies oxygenated blood to the liver, stomach, abdominal oesophagus, spleen and the superior half of both the duodenum and the pancreas
Hepatic artery	Branch of the celiac trunk that arises from the abdominal aorta at the level of the upper part of the first lumbar vertebra	Supplies oxygenated blood to the liver, stomach, duodenum and pancreas
Renal artery	One of two blood vessels leading off from the abdominal aorta that go to the kidneys	Supplies oxygenated blood to the kidneys
Splenic artery	Branched from the celiac artery (first major branch of the abdominal aorta) and runs above the pancreas to the spleen	Supplies oxygenated blood to the spleen. Has several branches that deliver blood to the stomach and pancreas

In addition to the main visceral branch of abdominal arteries above, other arteries are:

- the superior and inferior mesenteric arteries which supply the small and large intestines

- the suprarenal arteries which supply the adrenal glands
- the gonadal arteries that supply the ovaries and the testes.

The parietal branch of arteries that feed the abdominal wall structures include:

- the inferior phrenic arteries that supply the diaphragm
- the lumbar arteries that supply the spinal cord and muscles and skin of the lumbar region
- the median sacral arteries that supply the sacrum, coccyx and the rectum.

Blood circulation to the leg and foot

The abdominal aorta travels down the length of the trunk to the lower abdomen, where it divides into two arteries (**right** and **left common iliac arteries**) which supply either leg. The common iliac arteries in turn divide into the **internal** and **external iliac arteries**.

The **internal iliac artery** supplies most of the pelvic wall, and the **external iliac artery** becomes the **femoral artery** in the leg.

The **femoral artery** is the artery in the thigh, named after the thigh bone. At the knee the **femoral artery** becomes the **popliteal artery**, which divides into two below the knee. One of these arteries runs down the front of the lower leg and is called the **anterior tibial artery**, while the other runs down the back of the leg and is called the **posterior tibial artery**. This artery divides at the inside of the ankle becoming the **medial plantar artery** on the inside of the foot and the **plantar arch** on the sole of the foot. The **anterior tibial artery** becomes the **dorsal metatarsal artery** on top of the foot.

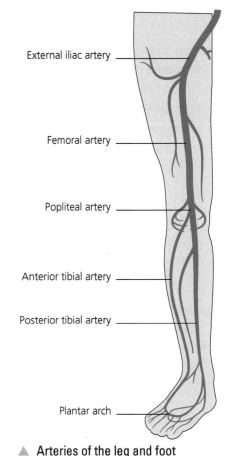

External iliac artery

Femoral artery

Popliteal artery

Anterior tibial artery

Posterior tibial artery

Plantar arch

▲ Arteries of the leg and foot

Main arteries of the leg and foot

Table 6.7 Main arteries of the leg and foot

Name of artery	Location	Area it supplies
Common iliac artery	Passes down along the brim of the pelvis and divides into two large branches	Supplies blood to the pelvic organs, gluteal region and legs
Femoral artery	Main artery in the thigh (continuation of external iliac artery)	Supplies oxygenated blood to the tissues of the leg
Popliteal artery	Behind the knee and the back of the lower leg	Supplies oxygenated blood to the knee joint
Anterior tibial artery	On anterior of lower leg, crossing the anterior aspect of the ankle joint, at which point it becomes the dorsalis pedis artery	Carries oxygenated blood to the anterior compartment of the leg and dorsal surface of the foot, from the popliteal artery
Posterior tibial artery	Branches off from the popliteal artery and runs down the leg, just below the knee	Carries blood to the posterior compartment of the leg and plantar surface of the foot, from the popliteal artery
Plantar arch	Runs from the fifth metatarsal and extends medially to the first metatarsal (of the big toe)	Supplies oxygenated blood to the underside, or sole, of the foot

Main veins

The major veins of the body are the **superior** and **inferior vena cavae**, which convey deoxygenated blood from the other veins to the right atrium of the heart.

- The **superior vena cava** originates at the junction of the two **innominate (briachiocephalic) veins**. It drains blood from the upper parts of the body (head, neck, thorax and arms) above the diaphragm.

- The **inferior vena cava** is formed by the right and left **iliac veins**. It lies posterior to the abdominal cavity and runs along the right side (with the aorta to the left). It receives blood from the lower parts of the body below the diaphragm.

Like arteries, veins are also named for their locations and usually have two branches (right and left). Veins are more superficially placed than arteries.

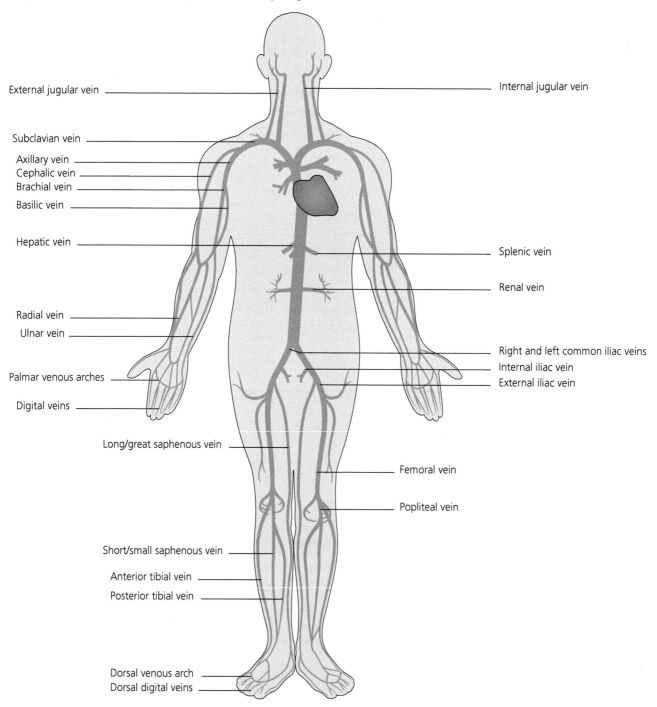

External jugular vein

Subclavian vein

Axillary vein
Cephalic vein
Brachial vein

Basilic vein

Hepatic vein

Radial vein
Ulnar vein

Palmar venous arches

Digital veins

Long/great saphenous vein

Short/small saphenous vein

Anterior tibial vein
Posterior tibial vein

Dorsal venous arch
Dorsal digital veins

Internal jugular vein

Splenic vein

Renal vein

Right and left common iliac veins
Internal iliac vein
External iliac vein

Femoral vein

Popliteal vein

▲ Main veins

Venous drainage of the face and head

The majority of blood draining from the head is passed into three pairs of veins:

1 external jugular veins

2 internal jugular veins

3 vertebral veins.

Within the brain, all veins lead to the internal jugular veins.

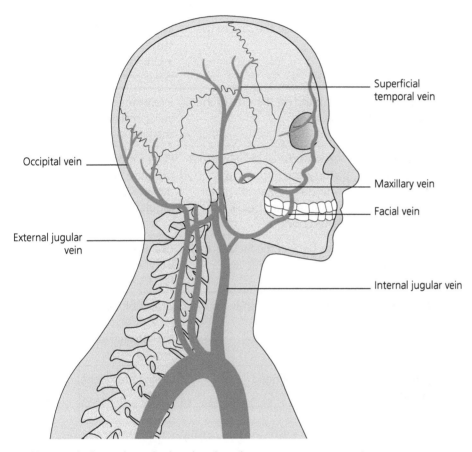

Occipital vein

Superficial temporal vein

Maxillary vein

Facial vein

External jugular vein

Internal jugular vein

▲ Venous drainage from the head and neck

The **external jugular veins** are smaller than the **internal jugular veins** and lie superficial to them. They receive blood from superficial regions of the face, scalp and neck. The **external jugular veins** descend on either side of the neck, passing over the sternomastoid muscles and beneath the platysma. They empty into the right and left **subclavian veins** in the base of the neck.

The **internal jugular veins** form the major venous drainage of the head and neck and are deep veins that run parallel with the **common carotid artery**. They collect deoxygenated blood from the brain and pass downwards through the neck beside the **common carotid arteries** to join the **subclavian veins**.

The **vertebral veins** descend from the transverse openings (or foramina) of the cervical vertebrae and enter the **subclavian veins**. The **vertebral veins** drain deep structures of the neck, such as the vertebrae and muscles.

Main veins of the face and head

Table 6.8 Main veins of the face and head

Name of vein	Location	Area of body it receives venous return from
External jugular vein	Situated on the side of the neck	Drains most of the outer structures of the head, including the scalp and deep portions of the face
Internal jugular vein	Situated on the side of the neck	Drains most of the cerebral veins and outer portions of the face
Common facial vein	Crosses the external carotid artery and enters the internal jugular vein at a variable point below the hyoid bone	Drains most of the blood from the face, draining directly into the internal jugular vein
Anterior facial vein	Lies behind the facial artery	Drains blood from the face before joining the common facial vein
Maxillary vein	Runs alongside the maxillary artery	Drains blood from the face, directs blood flow to the internal and external jugular veins
Superficial temporal vein	Side of the head	Drains the forehead and scalp

Venous drainage of the arm and hands

The venous return of blood from the hand begins with the **palmar arch** and **plexus**, which is a network of capillaries in the palm. The veins that carry deoxygenated blood up the forearm are the **radial vein, ulnar vein** and **median vein**.

The **radial vein** runs parallel to the radius bone of the forearm, the ulnar vein runs parallel to the ulna bone of the forearm and the **median vein** runs up the middle of the forearm. Just above the elbow, the **radial** and **ulnar veins** join to become the **brachial vein** and the **median vein** joins the **basilic vein**, which originates just below the elbow along with the **cephalic vein**.

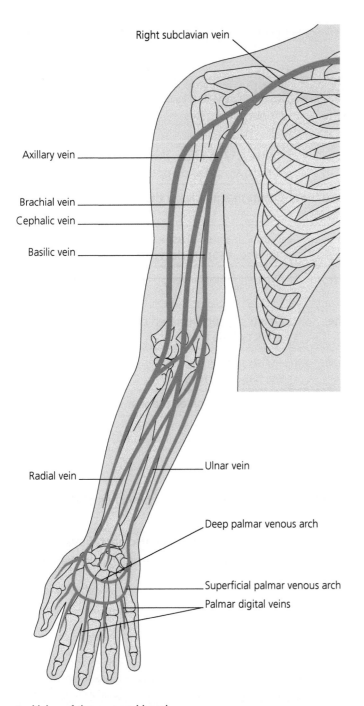

Right subclavian vein

Axillary vein

Brachial vein

Cephalic vein

Basilic vein

Radial vein

Ulnar vein

Deep palmar venous arch

Superficial palmar venous arch

Palmar digital veins

▲ Veins of the arm and hand

As the veins continue over the elbow, they link to form a network that eventually divides with the **basilic vein** joining the **brachial vein**, which then becomes the **axillary vein**. The **cephalic vein** travels up the arm separately and becomes the **subclavian vein** in the upper chest.

Main veins of the arm and hand

Table 6.9 Main veins of the arm and hand

Name of vein	Location	Area of body it receives venous return from
Subclavian vein	Under the clavicle and anterior to the subclavian artery	Upper extremities
Axillary vein	Runs along the medial side of the axillary artery	Conveys blood from the lateral aspect of the thorax, armpit and upper limb
Brachial vein	In the arm between the shoulder and the elbow, runs alongside the brachial artery	Drains the muscles of the upper arm
Basilic vein	Inner side of forearm area	Drains the medial aspect of the upper limbs via numerous superficial veins
Cephalic vein	Runs up the lateral side of the arm from the hand to the shoulder	Drains the dorsal venous network of the hand
Ulnar vein	Runs parallel to the ulna bone of the forearm	Drains oxygen-depleted blood from the medial aspect of the forearm
Radial vein	Runs parallel to the radius bone of the forearm	Assists in draining oxygen-depleted blood from the hand and forearm
Palmar digital veins	Run between the fingers	Carry deoxygenated blood away from the fingers

Venous drainage of thoracic and abdominal walls

The thoracic and abdominal walls are drained by branches of the **brachiocephalic veins.**

Blood from the abdominal organs enters the **hepatic portal system** (page 187) and then from the liver the blood is carried by the **hepatic veins** to the **inferior vena cava.**

Main veins of the thoracic and abdominal walls

Table 6.10 Main veins of the thoracic and abdominal walls

Name of vein	Location	Area of body it receives venous return from
Iliac vein	Lower part of the abdomen, in the pelvic region	Pelvis and lower limbs
Splenic vein	Runs close to the course of the splenic artery (above the pancreas to the spleen)	The spleen, the stomach fundus and part of the pancreas (part of the hepatic portal system)
Renal vein	Posterior of the abdominal wall, connects the kidneys to the inferior vena cava	Kidneys
Hepatic vein	Connects to the liver in the upper right side of the abdominal cavity	Liver
Hepatic portal vein	The upper right quadrant of the abdomen, originating behind the neck of the pancreas	All of the blood draining from the abdominal digestive tract, as well as from the pancreas, gall bladder, and spleen (part of the hepatic portal system)

In addition to the main visceral branch of abdominal veins already mentioned, some other veins are:

- the superior and inferior mesenteric veins, which drain blood from the small and large intestines
- the suprarenal arteries, which supply the adrenal glands
- the gonadal veins, which drain the ovaries and the testes.

The parietal branch of veins, which drain the abdominal wall structures include:

- the inferior phrenic veins, which drain the diaphragm
- the lumbar veins, which drain the spinal cord, and the muscles and skin of the lumbar region
- the median sacral veins, which drain the sacrum, coccyx and rectum.

Venous drainage of the leg and foot

There is a network of veins in the foot that becomes the **dorsal venous arch** on top of the foot. This travels along the inside of the foot to the ankle, where it becomes the **small (short) saphenous vein**. It continues up the back of the whole leg to the thigh, where it is known as the **great (long) saphenous vein**.

Two small veins called the **anterior tibial veins** travel up the front of the lower leg, while two veins, the **posterior tibial veins**, run up the back. These four veins converge just below the knee to become the **popliteal vein** at the back of the knee and then eventually the **femoral vein** in the thigh. The **great saphenous vein** and the **femoral vein** join at the groin and return to the heart via the **inferior vena cava**.

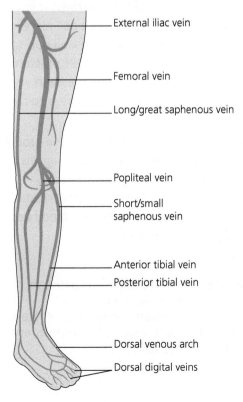

External iliac vein

Femoral vein

Long/great saphenous vein

Popliteal vein

Short/small
saphenous vein

Anterior tibial vein
Posterior tibial vein

Dorsal venous arch
Dorsal digital veins

▲ Veins of the leg and foot

Main veins of the leg and foot

Table 6.11 Main veins of the leg and foot

Main veins of the Leg	Location	Area of body it receives venous return from
Great (long) saphneous vein	Longest vein in the human body, extending from the top of the foot to the upper thigh and groin	Drains blood from the inner part of the foot, the skin and fat of the front and inner aspect of the lower leg, and the skin and fat of the inner part of the thigh
Small (short) saphenous vein	Superficial vein in the posterior of the lower leg	Drains the lateral surface of the leg, and runs up the posterior surface of the leg to drain into the popliteal vein
Femoral vein	Large vein in the groin (continuation of the popliteal vein)	Carries blood back to the heart from the lower extremities
Popliteal vein	Behind the knee and the back of the lower leg	Carries blood from the knee (as well as the thigh and calf muscles) back to the heart
Anterior tibial vein	In the anterior of the lower leg	Originates and receive blood from the dorsal pedis veins, on the back of the foot, and empties into the popliteal vein
Posterior tibial vein	In the posterior of the lower leg	Drains the posterior compartment of the leg and the plantar surface of the foot to the popliteal vein, which it forms when it joins with the anterior tibial vein
Dorsal venous arch	Superficial vein that connects the small saphenous vein and the great saphenous vein	Drains oxygen-depleted blood through the foot

Blood shunting

Along certain circulatory pathways, such as in the intestines, there are strategic points where small arteries have direct connection with veins. When these connections are open, they act as shunts to allow blood in an artery to have direct access to a vein.

These interconnections allow for sudden and major diversions of blood volume, according to the physical needs of the body. This means that treatment should not be given after a heavy meal – increased circulation to the intestines results in a diminished supply to other areas of the body.

Blood pressure

Blood pressure is the amount of pressure exerted by blood on an arterial wall due to the contraction of the left ventricle. Blood pressure may be measured with the use of a sphygmomanometer.

The pressure in the arteries varies during each heartbeat:

- The maximum pressure of the heartbeat is known as the **systolic** (sis-toll-ik) pressure and represents the pressure exerted on the arterial wall during active ventricular contraction. Systolic pressure can, therefore, be measured when the heart muscle contracts and pushes blood out into the body through the arteries.

- The minimum pressure, or **diastolic** (dy-a-stoll-ik) pressure, represents the static pressure against the arterial wall during the pause between contractions. Therefore, the mimimum pressure is when the heart muscle relaxes and blood flows into the heart from the veins.

KEY FACT

Blood pressure is regulated by sympathetic nerves in the arterioles. An increase in stimulation of the sympathetic nervous system, as in exercise, therefore results in a temporary increase in blood pressure.

Factors affecting blood pressure

As blood pressure is the result of the pumping of the heart in the arteries, anything that makes the heart beat faster will raise the blood pressure. Factors that affect blood pressure include:

- excitement
- anger
- stress
- fright
- pain
- exercise
- smoking and drugs.

A normal blood pressure reading is between 100 and 140 mm Hg systolic and between 60 and 90 mm Hg diastolic. Blood pressure is measured in millimetres of mercury and is expressed as a ratio between systolic and diastolic pressures, like this: 120/80 mm Hg.

The pulse

The pulse is a pressure wave that can be felt in the arteries and which corresponds to the beating of the heart. The pumping action of the left ventricle of the heart is so strong that it can be felt as a pulse in some arteries at a considerable distance from the heart. The pulse can be felt at any point where an artery lies near the body's surface. The radial pulse can be found by placing two or three fingers over the radial artery below the thumb.

The average pulse in an adult is between 60 and 80 beats per minute. Factors affecting the pulse rate include:

- exercise
- heat
- strong emotions such as grief, fear, anger or excitement.

The pulse may be palpated (felt) in any place that allows an artery to be compressed near the surface of the body such as at:

- the neck (carotid artery)
- the wrist (radial artery)

- the elbow (brachial artery)
- the groin (femoral artery)
- the back of the knee (popliteal artery)
- the ankle joint (posterior tibial artery)
- the foot (dorsalis pedis artery).

Common pathologies of the cardiovascular system

> **In practice**
>
> Always take a detailed history of a client's symptoms and medical or surgical treatment, so you are aware of any cardiac and/or circulatory problems. If there is a history of cardiovascular illness, seek advice from the client's GP before treating, as this may determine the nature and duration of the proposed treatment.

Anaemia

In anaemia, the haemoglobin level in the blood is below normal. The main symptoms are excessive tiredness, breathlessness on exertion, pallor and poor resistance to infection. There are many causes of anaemia. It may be due to a loss of blood resulting from an accident or operation, chronic bleeding, iron deficiency or due to a blood disease such as leukaemia.

Aneurysm

An aneurysm is an abnormal balloon-like swelling in the wall of an artery. This may be due to degenerative disease (congenital defects, arteriosclerosis) or any condition which causes weakening of the arterial wall, such as trauma, infection or hypertension.

Angina

This is a pain in the left side of the chest and usually radiating to the left arm. It is caused by insufficient blood flow to the heart muscle, usually on exertion or excitement. The pain is often described as constricting or suffocating and can last for a few seconds or minutes. The patient may become pale and sweaty. This condition indicates ischaemic heart disease.

In practice

Stress predisposes people to angina attacks. Massage and other relaxation therapies may help alleviate symptoms by decreasing the activity of the sympathetic nervous system.

Since sudden exposure to extreme heat or cold can bring on an attack, keep the client warm.

Arteriosclerosis

Arteriosclerosis is a circulatory system condition that is characterised by a thickening, narrowing, hardening and loss of elasticity of the walls of the arteries.

In practice

Clients with arteriosclerosis are prone to thrombus formation. Deeper manipulation should not be used as it could encourage a thrombus to dislodge and travel to the lungs, heart or the brain.

Refer clients to their GP before any treatment if they have a history of stroke, heart attack, angina or thrombosis. If treatment is encouraged, use gentle methods and avoid overstimulation.

High blood pressure

High blood pressure is when blood pressure at rest is above normal levels. The World Health Organization defines high blood pressure as consistently exceeding 160 mm Hg systolic and 95 mm Hg diastolic.

High blood pressure is a common complaint and may lead to a stroke or a heart attack, due to the fact that the heart is working harder to force blood through the system. Causes of high blood pressure include:

- smoking
- obesity
- lack of regular exercise
- eating too much salt
- excessive alcohol consumption
- stress.

High blood pressure can be controlled by:

- anti-hypertensive drugs, which help to lower blood pressure
- decreasing salt and fat intake to prevent hardening of the arteries

- keeping weight down
- giving up smoking and cutting down on alcohol consumption
- relaxation and leading a less stressful life.

Low blood pressure

Low blood pressure is when the blood pressure is below normal and is defined by the World Health Organization as a systolic blood pressure of 99 mm Hg or less and a diastolic of less than 59 mm Hg. Low blood pressure may be normal for some people in good health, during rest or fatigue. Very low blood pressure can mean that insufficient blood reaches the vital centres of the brain. Treatment may be by medication, if necessary.

In practice

High and low blood pressure do normally contraindicate treatments but with GP referral and an adaptation of routine, some treatments may be possible. Correct positioning of the couch is essential to maximise comfort of the client with blood pressure problems. Make sure the client is not lying down too long and doesn't get up too fast.

Congenital heart disease

This is a defect in the formation of the heart which usually decreases its efficiency. Defects may take the following forms:

- **ventricular septal defects** – an opening between the right and left ventricle
- **atrial septal defect** – an opening between the right and left atrium
- **coarctation of the aorta** – narrowing of the aorta
- **pulmonary stenosis** – narrowing of the pulmonary artery
- **patent ductus arteriosus** – non-closure of the communication between the pulmonary artery and the aorta that normally exists in the foetus until delivery
- a combination of defects.

The symptoms may vary according to the severity of the defect.

Haemophilia

Haemophilia is a hereditary disorder in which the blood clots very slowly due to deficiency of either of two coagulation factors:

- factor VIII (the antihaemophiliac factor)
- factor IX (the Christmas factor).

The patient may experience prolonged bleeding following an injury, and in severe cases there is spontaneous bleeding into the muscles and joints.

Haemophilia is controlled by a sex-linked gene, which means it is almost exclusively restricted to males. Women can carry the disease and pass it onto their sons without being affected themselves.

Haemorrhoids (piles)

This is an enlargement of the veins in the walls of the anus. They usually form as a result of prolonged constipation.

Heart attack (myocardial infarction)

This is damage to the heart muscles which results from blockage of the coronary arteries. It can cause serious complications including heart failure.

Hepatitis

Hepatitis is an inflammation of the liver caused by viruses, toxic substances or immunological abnormalities. There are several forms:

- **Hepatitis A** – highly contagious and transmitted via the faecal–oral route by ingestion of contaminated food, water or milk. The incubation period is from 15 to 45 days.

- **Hepatitis B** – also known as serum hepatitis, this is more serious than hepatitis A. It lasts longer and can lead to cirrhosis, cancer of the liver, and a carrier state. (A carrier is someone who is infected but is free from disease symptoms.) It has a long incubation period of one and a half to two months. The symptoms may last from weeks to months. The virus is usually transmitted through infected blood, serum or plasma; however, it can spread by oral or sexual contact as it is present in most body secretions.

- **Hepatitis C** – this can cause acute or chronic hepatitis and can also lead to a carrier state and liver cancer. It is transmitted through blood transfusions or exposure to blood products.

High cholesterol

Cholesterol is a fat-like material that is present in the blood and most tissues. A high level of cholsterol in the blood (due to a diet rich in animal fats and refined sugars) is often associated with the degeneration of the walls of the arteries and a predisposition to thrombosis.

Leukaemia

This term refers to any of a group of malignant diseases in which the bone marrow and other blood-forming organs produce an increased number of certain types of white blood cells. Overproduction of these white cells, which are immature or of abnormal form, suppresses the production of normal white cells, red cells and platelets, leading to increased susceptibility to infection. Other manifestations or signs include enlargement of the spleen, liver and the lymph nodes, spontaneous bruising and anaemia.

Pacemaker

This is an artificial electrical device implanted under the skin that stimulates and controls the heart rate by sending electrical stimuli to the heart. It is usually installed to correct an abnormal heart rhythm (heart block) and mostly placed in one side of the upper chest.

> **In practice**
>
> Electrical treatments would be contraindicated in a client with a pacemaker. Seek the GP's advice before offering any other form of treatment.
>
> The site of the pacemaker is likely to be tender and should be avoided, if any suitable treatment is given.

Phlebitis

This condition is an inflammation of the wall of a vein and is most commonly seen in the legs as a complication of varicose veins. A segment of the vein becomes tender and painful, and the surrounding skin may feel hot and appear red.

Thrombosis may develop as a result of phlebitis (thrombophlebitis) with subsequent deep vein thrombosis (DVT). Clots may dislodge and travel to the lungs or other organs with serious consequences.

> **In practice**
>
> The site of phlebitis can be tender and careful handling is essential.
>
> Massage should be avoided so as not to dislodge clots.

Pulmonary embolism

This occurs when a blood clot is carried into the lungs, where it blocks the flow of blood to the pulmonary tissue. It is a very serious condition and can be life threatening, requiring hospitalisation and measures to thin the blood, such as use of warfarin. This condition presents with chest pain, cough and shortness of breath.

Raynaud's syndrome

This is a disorder of the peripheral arterioles, characterised by spasm in the smooth muscle of the fingers and toes. It is generally brought on by cold or emotional upset. The effect is a pallor or discolouration of the skin due to the presence of poorly oxygenated haemoglobin. Affected extremities can become painful and uncomfortable, and this is usually followed by redness and stiffness of the toes and fingers.

> **In practice**
>
> Clients on warfarin (an anticoagulant medication) have an increased risk of bleeding and you should be aware of this with reference to any skin-piercing treatments such as epilation or ear piercing.

Stress

Stress can be defined as any factor which affects physical or emotional health. When the body is under stress, the heart beats faster, increasing the circulation of blood. Excessive or prolonged stress can lead to high blood pressure, coronary thrombosis and heart attack.

Stroke

This occurs when blood flow to the brain is blocked by an embolus (clot) in a cerebral blood vessel. A stroke can result in a sudden attack of weakness on one side of the body, due to the interruption to the flow of blood on the corresponding side of the brain.

A stroke can vary in severity from a passing weakness or tingling in a limb, to a profound paralysis and coma.

Sometimes the term *stroke* is used to describe cerebral haemorrhage when an artery or congenital cyst of blood vessels in the brain bursts, resulting in damage to the brain and causing similar signs to an embolism. Haemorrhage is usually associated with severe headaches and can cause neck stiffness.

Thrombosis

This is a condition in which the blood produces a blood clot. Thrombosis in the wall of an artery obstructs the blood flow to the tissue it supplies. In the brain, this is one of the causes of stroke, and in the heart it results in a heart attack (coronary thrombosis).

Thrombosis may also occur in a vein (DVT). The thrombus (blood clot) may be detached from its site of formation and can be carried in the blood to lodge in another part of the body (see pulmonary embolism).

Varicose veins

Veins become varicose when the valves within them lose their strength. As a result of this, blood flow may reverse or become static (valves normally prevent the backflow of blood). When their function is impaired, veins are unable to prevent the blood from flowing downwards, causing the walls of the affected veins to swell and bulge out, becoming visible through the skin.

Varicose veins may be due to several factors:

- hereditary tendencies
- ageing
- obesity (excess weight), which puts pressure on the walls of the veins
- pregnancy
- sitting or standing still for long periods of time, causing pressure to build up in the vein.

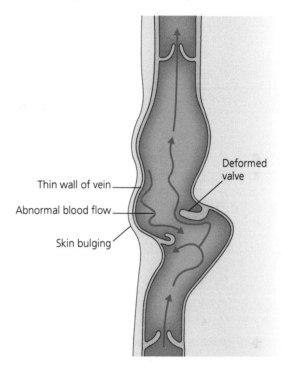

Thin wall of vein

Abnormal blood flow

Skin bulging

Deformed valve

▲ Varicose veins

In practice

As varicose veins can be extremely painful, treatment is contraindicated in the affected area.

Interrelationships with other systems

The cardiovascular system

The cardiovascular system links to the following body systems.

Skin

The circulatory system transports blood rich in nutrients and oxygen to the skin, hair and nails.

Skeletal

Red bone marrow is responsible for the development of blood cells.

Muscular

The heart is a muscular organ and contracts rhythmically and continuously to pump blood around the body.

Lymphatic

The lymphatic system assists the circulatory system in transporting additional waste products away from the tissues in order to maintain blood volume and pressure and to prevent oedema.

Respiratory

The respiratory system oxygenates and deoxygenates blood in the lungs.

Digestive

Nutrients broken down by digestive processeses are transported by blood to the liver to be assimilated by the body.

Nervous

Blood pressure is regulated by sympathetic nerves in the arterioles.

Endocrine

Hormones are carried by blood to their target organs.

Key words

Agranulocyte: one of the two main categories of leucocytes

Antigen: a substance that can trigger an immune response if foreign to the body

Aorta: the main artery of the body, supplying oxygenated blood to the circulatory system

Arteriole: a small branch of an artery leading into capillaries

Artery: a type of blood vessel that carries oxygenated blood away from the heart

Atrium: one of two upper cavities of the heart from which blood is passed to the ventricles

Basophil: a type of granulocyte

Blood: the fluid circulating through the heart, arteries, veins and capillaries of the circulatory system

Blood pressure: the amount of pressure exerted by blood on an arterial wall due to the contraction of the left ventricle

B-lymphocyte: a type of white blood cell that makes antibodies

Capillary: the smallest blood vessel, which unites arterioles and venules

Cardiac cycle: the sequence of events between one heartbeat and the next

Clotting: the process in which blood changes from liquid into a solid state to form a thick lump

Clotting factors: proteins in the blood that control bleeding

Diastolic: the minimum blood pressure when the heart muscle relaxes and blood flows into the heart from the veins

Endocardium: the lining of the heart's cavities

Eosinophil: a type of granulocyte

Erythrocyte: red blood cell, which transports the gases of respiration

Fibrin: an insoluble protein that forms a fibrous mesh during blood clotting

Fibrinogen: a soluble protein present in blood plasma, from which fibrin is produced

Granulocyte: one of the two main categories of leucocytes

Haemoglobin: a red iron–protein complex responsible for transporting oxygen in the blood

Heart: hollow muscular organ which lies in the thorax above the diaphragm and between the lungs, acts as a pump to provide a constant circulation of blood throughout the body

Hypertension (high blood pressure): when the force of blood pushing against the walls of blood vessels is consistently too high

Hypotension (low blood pressure): when the force of blood pushing against the walls of blood vessels is consistently too low

Inferior vena cava: a large vein carrying deoxygenated blood into the heart

Leucocyte: white blood cell, which protects the body against infection and disease

Lumen: a void inside a blood vessel through which blood flows

Lymphocyte: a specialised type of white blood cell

Macrophage: a large white blood cell that ingests foreign particles and infectious micro-organisms by phagocytosis

Monocyte: a large phagocytic white blood cell

Myocardium: a strong layer of cardiac muscle, which makes up the bulk of the heart

Neutrophil: a type of granulocyte

Pericardium: the membrane enclosing the heart, consisting of an outer fibrous layer and an inner double layer of serous membrane

Phagocytosis: the process by which a white blood cell ingests micro-organisms

Plasma: the colourless, liquid part of blood

Portal circulation: circulation of blood to the liver from the small intestine, the right half of the colon and the spleen through the portal vein

Pulmonary artery: an artery carrying blood from the right ventricle of the heart to the lungs for oxygenation

Pulmonary circulation: the circulatory system between the heart and the lungs

Pulmonary vein: a vein carrying oxygenated blood from the lungs to the left atrium of the heart

Pulse: a pressure wave that can be felt in the arteries and which corresponds to the beating of the heart

Septum: a partition separating the two chambers of the heart

Stethoscope: a medical instrument for listening to the action of the heart or breathing

Sphygmomanometer: an instrument for measuring blood pressure

Sinoatrial (SA) node: a specialised piece of heart tissue that generates the electrical impulses to control the heartbeat

Superior vena cava: a large vein carrying deoxygenated blood into the heart

Systemic circulation: part of the cardiovascular system that carries oxygenated blood away from the heart to the body and returns deoxygenated blood back to the heart

Systolic: maximum pressure of the heartbeat, which represents the pressure exerted on the arterial wall during active ventricular contraction

Thrombin: an enzyme in blood plasma which causes the clotting of blood by converting fibrinogen to fibrin

Thrombocyte (platelet): a small fragment of a cell involved in blood clotting

Thromboplastin: a plasma protein that helps with blood coagulation

T-lymphocyte: a type of white blood cell that circulates around the body, scanning for cellular abnormalities and infections

Tunica externa (tunica adventitia): the outermost layer or tunic of a vessel (except capillaries)

Tunica intima: the innermost lining or tunic of a vessel

Tunica media: the middle layer or tunic of a vessel (except capillaries)

Vasoconstriction: constriction of the smooth muscle of a blood vessel, resulting in a decreased vascular diameter

Vasodilation: relaxation of the smooth muscle in the wall of a blood vessel, resulting in an increased vascular diameter

Vein: a type of blood vessel that carries deoxygenated blood towards the heart

Ventricle: one of the two lower chambers of the heart

Venule: a very small blood vessel in the microcirculation that allows blood to return from the capillary beds to drain into the veins

Revision summary

The cardiovascular system

- **Blood** is a type of liquid connective tissue.
- Blood transports substances between the body cells and the external environment to help maintain a stable cellular environment.
- There are four major **blood groups**: A, B, AB and O.
- Blood groups are determined by the presence or absence of two **antigens** – A and B – on the surface of red blood cells.
- The **universal red cell donor** has type **O negative** blood.
- The **universal plasma donor** has type **AB blood**.
- The **percentage composition of blood** is 55% fluid (plasma) and 45% blood cells.
- There are three main types of blood cells – **erythrocytes**, **leucocytes** and **thrombocytes**.
- The function of an **erythrocyte** is transporting oxygen to the cells and carrying carbon dioxide away.

- **Leucocytes** are designed to protect the body against infection.
- **Thrombocytes** are involved in the clotting process.
- **Blood clotting** or coagulation is a biological process that stops bleeding.
- **Haemostasis** is the physiological process by which bleeding ceases. It involves three basic steps: vascular spasm, the formation of a platelet plug, and coagulation, in which clotting factors promote the formation of a fibrin clot.
- There are 12 **clotting factors** in human blood and tissues.
- **Clotting factors** are proteins in the blood that control bleeding.
- Examples of clotting factors include **fibrinogen** (clotting factor I), **prothrombin** (clotting factor II) and **thromboplastin** (clotting factor III).
- There are four main functions of blood – transport, defence, regulation of heat and clotting.
- Blood is carried around the body in vessels called **arteries**, **veins** and **capillaries**.

- **Arteries** carry oxygenated blood away from the heart. They have thick, muscular walls in order to withstand the high pressure of blood.
- **Veins** carry deoxygenated blood towards the heart. They have thinner walls and blood is carried under lower pressure.
- **Capillaries** are the smallest vessels in the circulatory system. They unite **arterioles** and **venules**. Their walls are sufficiently thin to allow dissolved substances in and out of them.
- The **heart** lies in the **thorax** above the diaphragm and between the lungs.
- The heart is composed of three layers of tissue – an outer **pericardium**, a middle **myocardium** and an inner **endocardium**.
- The heart is divided into a right and left side by a partition called a **septum**. Each side is divided into a thin-walled top chamber called an **atrium** and a thick-walled bottom layer called a **ventricle**.
  - The **atria** (top chambers) take in blood from the large veins and pump it to the bottom chambers.
  - The **ventricles** (bottom chambers) pump blood to the body's organs and tissues.
- Blood flows through the heart in four stages.
  - **Stage 1** – deoxygenated blood flows into the right atrium. When the right atrium is full, it empties into the right ventricle.
  - **Stage 2** – when the right ventricle is full, it pushes blood into the pulmonary artery to the lungs, where the blood becomes oxygenated.
  - **Stage 3 (**taking place at the same time as stage 1) – oxygen-rich blood fills the left atrium. When full, the blood passes to the left ventricle.
  - **Stage 4** (taking place at the same time as stage 2) – when the left ventricle is full it forces blood into the aorta and to all parts of the body.
- The **cardiac cycle** is the sequence of events between one heartbeat and the next.
- The duration of the **cardiac cycle** is less than a second.
  - During a cardiac cycle, the **atria** contract simultaneosuly and force blood into the relaxed ventricles.

- The **ventricles** contract strongly and push blood out through the aorta and the **pulmonary artery**.
- As the **ventricles** contract the **atria** relax and fill up with blood.
- Blood is transported as part of a double circuit.
- The **pulmonary circulation** is the circulatory system between the heart and the lungs. It consists of the circulation of deoxygenated blood from the **right ventricle** of the heart to the **lungs** via the **pulmonary arteries** to become oxygenated. Oxygenated blood is then returned to the **left atrium** by the **pulmonary veins**.
- The **systemic circulation** is the largest circulatory system and carries oxygenated blood from the left ventricle of the heart to the aorta and around the body.
- **Blood pressure** is defined as the amount of pressure exterted by blood on an arterial wall due to the contraction of the **left ventricle**.
- The maximum pressure is called the **systolic** pressure and respresents the pressure exerted on the arterial walls during ventricular contraction.
- The lowest pressure is called the **diastolic** pressure and is when the heart muscle relaxes (ventricular relaxation) and blood flows into the heart from the veins.
  - A normal blood pressure reading is between 100 and 140 mm Hg **systolic** and between 60 and 90 mm Hg **diastolic**.
  - **High blood pressure** is when the resting blood pressure is above normal and when consistenly exceeding 160 mm Hg systolic and 95 mm Hg diastolic.
  - **Low blood pressure** is defined as a systolic pressure of 99 mm Hg or less and diastolic of 59 mm Hg.
- The **pulse** is a pressure wave that can be felt in arteries, such as the carotid or brachial arteries, and corresponds to the beating of the heart and the contraction of the left ventricle.
- An average **pulse** is between 60 and 80 beats per minute.

Test your knowledge questions

Multiple choice questions

1 Which of the following accounts for 55% of the composition of blood?
   a hormones
   b haemoglobin
   c fluid or plasma
   d erythrocytes, leucocytes and thrombocytes

2 Which blood cell protects the body against infection?
   a thrombocyte
   b leucocyte
   c erythrocyte
   d platelet

3 Which of the following is **not** a function of blood?
   a transport of oxygen, carbon dioxide, nutrients and hormones
   b protection and defence
   c synthesis of vitamins A, D and E
   d clotting

4 What is the function of an artery?
   a to carry oxygenated blood
   b to carry blood under high pressure
   c to carry blood away from the heart
   d all of the above

5 Which of the following statements is **false**?
   a Veins carry deoxygenated blood.
   b Veins are generally superficial.
   c Veins do not have valves.
   d Veins carry blood towards the heart.

6 What is the function of a capillary?
   a to carry only deoxygenated blood
   b to carry only oxygenated blood
   c to prevent backflow of blood
   d to supply cells and tissues with nutrients

7 What is the name of the blood vessel that carries deoxygenated blood from the heart to the lungs?
   a pulmonary vein
   b aorta
   c pulmonary artery
   d inferior vena cava

8 Which of the following has branches that carry oxygenated blood around the body?
   a left ventricle
   b left pulmonary vein
   c aorta
   d superior vena cava

9 Which of the following best describes the flow of blood through the heart?
   a Blood flows from the capillaries to the veins, to the arteries and then to the aorta.
   b Blood flows into the right atrium, then into the right ventricle, then into the pulmonary artery. The blood returns from the lungs and enters the left atrium, then flows into the left ventricle and into the aorta to all parts of the body.
   c Blood flows into the sinoatrial nodes, then to the right ventricle, aorta and coronary arteries.
   d Blood flows from the brachial artery, to the left ventricle, the pulmonary arteries and onto the right atrium.

10 Where does the blood supply to the arm start?
   a at the brachial artery
   b at the radial artery
   c at the ulnar artery
   d at the subclavian artery

Exam-style questions

11 State two functions of the cardiovascular system. 2 marks

12 List the four main functions of blood. 4 marks

13 Describe one characteristic **and** one function of each of the following types of blood cells:
   a erythrocyte 2 marks
   b leucocyte 2 marks
   c thrombocyte or platelet. 2 marks

14 Name the type of blood cell that is crucial to our immune system. 1 mark

15 Briefly describe each of the following parts of the circulation system:
   a pulmonary circulation 2 marks
   b systemic circulation 2 marks

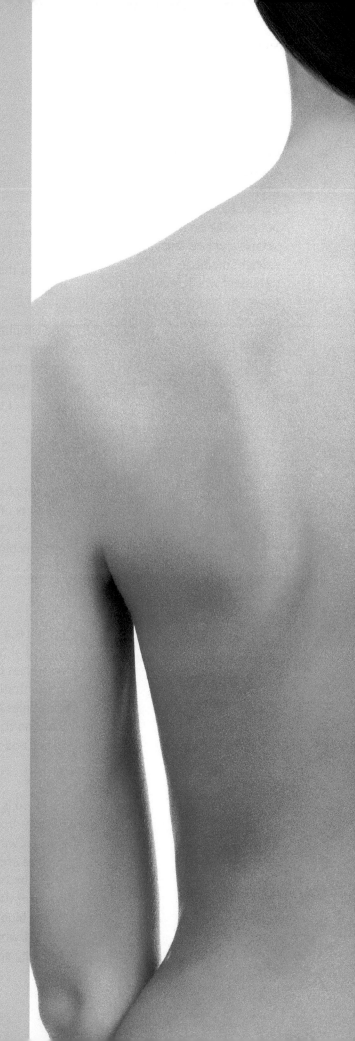

7 The lymphatic system and immunity

Introduction

The lymphatic system is a unidirectional drainage system for the tissues. It helps to provide a circulatory pathway for tissue fluid. This is transported as lymph from the tissue spaces of the body into the venous system, where it becomes part of the blood circulation. Through the filtering action of the lymphatic nodes, along with the actions of specific organs such as the spleen, the lymphatic system also helps to provide immunity against disease.

The human body is equipped with a variety of defence mechanisms that prevent the entry of foreign agents known as pathogens. This defence is called immunity. When working effectively, the immune system protects the body from most infectious micro-organisms. It does this both directly through cells that attack micro-organisms and indirectly by releasing chemicals and protective antibodies.

OBJECTIVES

By the end of this chapter you will understand:

- the functions of the lymphatic system
- the definition of lymph and how it is formed
- the connection between blood and lymph
- the circulatory pathway of lymph
- the lymphatic organs
- the names, positions and drainage of the main lymphatic nodes of the head, neck and the body
- lymphatic organs such as the spleen, tonsils and thymus
- the immune system and response
- common pathologies of the lymphatic system
- the interrelationships between the lymphatic and other body systems.

Functions of the lymphatic system

Drainage of excess fluid from the tissues

The lymphatic system is important for the distribution of fluid and nutrients in the body because it drains excess fluid from the tissues and returns to the blood protein molecules which are unable to pass back through the blood capillary walls because of their size.

Fighting infection

The lymphatic system plays an important part in the body's immune system. The lymphatic nodes help to fight infection by filtering lymph and destroying invading micro-organisms. Lymphocytes are reproduced in the lymph nodes and, following infection, they generate antibodies to protect the body against subsequent infection.

Absorbtion of products of fat digestion

The lymphatic system also plays an important part in absorbing the products of fat digestion from the villi of the small intestine. While the products of carbohydrate and protein digestion pass directly into the bloodstream, fats pass directly into the intestinal lymphatic vessels, known as lacteals.

What is lymph?

Lymph is a transparent, colourless, watery liquid which is derived from intersitial (tissue) fluid and is contained within lymphatic vessels. It resembles blood plasma in composition, except that it has a lower concentration of plasma proteins. This is because some large protein molecules are unable to filter through the cells forming the capillary walls, so they remain in blood plasma. Lymph contains only one type of cell and these are called lymphocytes.

How is lymph formed?

As blood is distributed under pressure to the tissues, some of the plasma escapes from the capillaries and flows around the tissue cells, delivering oxygen, water and nutrients to the cells and picking up cellular waste such as urea and carbon dioxide. Once the plasma is outside the capillary and is bathing the tissue cells, it becomes interstitial (or tissue) fluid. Some of the interstitial fluid passes back into the capillary walls to return to the bloodstream via the veins and some is collected by lymphatic vessels, where it becomes lymph. Lymph is taken through its circulatory pathway and is ultimately returned to the bloodstream.

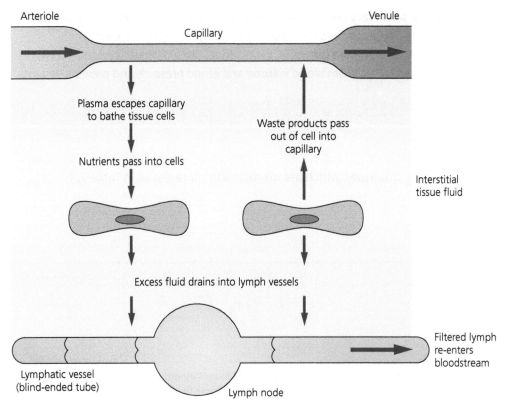

▲ The connection between blood and lymph

The connection between blood and lymph

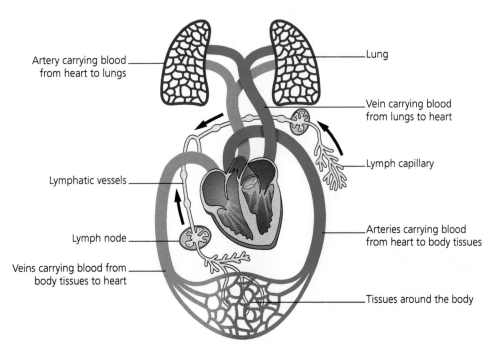

▲ How the lymphatic system works

The lymphatic system is often referred to as a 'secondary circulatory system' as it consists of a network of vessels that assist the blood in returning fluid from the tissues back to the heart. In this way, the lymphatic system is complementary to the circulatory system. After draining the tissues of excess fluid, the lymphatic system returns this fluid to the cardiovascular system. This helps to maintain blood volume and blood pressure and prevent oedema.

The circulatory pathway of lymph

Structures of the lymphatic system

The lymphatic system contains the following structures which are discussed in more detail in Table 7.1.

- lymphatic capillaries
- lymphatic vessels
- lymphatic nodes
- lymphatic collecting ducts.

Overview of the structures of the lymphatic system

Table 7.1 The structures of the lymphatic system

Structure	Description	Function
Lymphatic capillaries	Very small blind-ended tubes, similar in structure to blood capillaries	Drain away excess fluid and waste products from the tissue spaces of the body
Lymphatic vessels	Similar in structure to veins; have one-way valves and thin collapsible walls	Carry the lymph towards the heart
Lymphatic nodes	Oval or bean-shaped structures made of lymphatic tissue and covered by a capsule of connective tissue	Filter lymph, removing micro-organisms, cell debris and other harmful substances
Lymphatic ducts (thoracic and right lymphatic)	The thoracic duct is the largest lymphatic vessel in the body and extends from the second lumbar vertebra up through the thorax to the root of the neck The right lymphatic duct, at the root of the neck, is very short in length	Collect lymph from the whole body and return it to the blood via the subclavian veins

Lymphatic capillaries

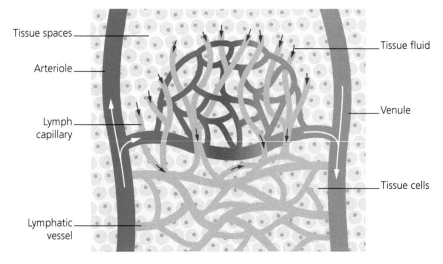

▲ Lymph capillaries in tissue spaces

The lymphatic system is a one-way circulatory pathway. Lymphatic vessels commence as lymphatic capillaries (very small blind-ended tubes) in the tissue spaces of the body. The walls of the lymphatic capillaries are like those of the blood capillaries in that they are a one cell thick, making it possible for tissue fluid to enter. However, they are permeable to substances of larger molecular size than those of the blood capillaries.

The lymphatic capillaries mirror the blood capillaries and form a network in the tissues, draining away excess fluid and waste products from the tissue spaces of the body. Once the tissue fluid enters a lymphatic capillary it becomes lymph and is then collected into larger lymphatic vessels.

KEY FACT

The term *oedema* refers to an excess of fluid within the tissue spaces that causes the tissues to become swollen.

Lymphatic vessels

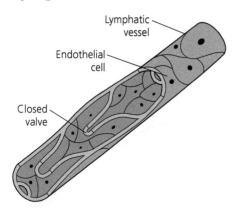

▲ A lymphatic vessel

Lymphatic vessels are similar to veins in that they have thin, collapsible walls and their role is to transport lymph through its circulatory pathway. They have a considerable number of valves, which help to keep the lymph flowing in the right direction by preventing backflow. Superficial lymphatic vessels tend to follow the course of veins by draining the skin, whereas the deeper lymphatic vessels tend to follow the course of arteries, draining the internal structures of the body.

Networks or plexuses of lymphatic channels exist throughout the body. These intertwined channels are found in the following areas:

- **mammary plexus** – lymphatic vessels around the breasts
- **palmar plexus** – lymphatic vessels in the palm of the hand
- **plantar plexus** – lymphatic vessels in the sole of the foot.

The lymphatic vessels carry the lymph towards the heart under steady pressure; about two to four litres of lymph pass into the venous system every day. Once lymph has passed through the lymph vessels, it drains into at least one lymphatic node before returning to the blood circulatory system.

KEY FACT

As the lymphatic system lacks a pump, lymphatic vessels make use of contracting muscles to assist the movement of lymph. Therefore, lymphatic flow is at its greatest during exercise when there is increased contraction of muscle.

Lymphatic nodes

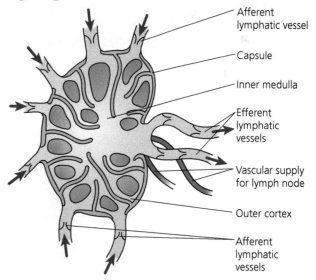

▲ A lymphatic node

Lymphatic nodes occur at intervals along the lymphatic vessels. A lymphatic node is an oval or bean-shaped structure covered by a capsule of connective tissue. It consists of lymphatic tissue and is divided into two regions: an outer cortex and an inner medulla.

There are more than 100 lymphatic nodes in specific locations along the course of the lymphatic vessels. They vary in size from 1 to 25 mm in length and are massed in groups. Some are superficial and lie just under the skin, whereas others are deeply seated and are found near arteries and veins.

Each lymphatic node receives lymph from several afferent lymphatic vessels and blood from small arterioles and capillaries. The valves of the **afferent lymphatic vessels** open towards the node, so that lymph in these vessels can only move towards the node. Lymph flows slowly through the node, moving from the cortex to the medulla, and leaves through an **efferent vessel** which opens away from the node.

The function of a lymphatic node is to act as a filter of lymph – to remove or trap any micro-organisms, cell debris or harmful substances which may cause infection, so that when lymph enters the blood it has been cleared of any foreign matter. When lymph enters a node, it comes into contact with two specialised types of leucocytes:

- **macrophages** – phagocytic in action, these engulf and destroy dead cells, bacteria and foreign material in the lymph
- **lymphocytes** – reproduced within the lymphatic nodes, these neutralise invading bacteria and produce chemicals and antibodies to fight disease.

Once filtered, the lymph leaves the node by one or two efferent vessels which open away from the node. Lymphatic nodes occur in chains so that the efferent vessel of one node becomes the afferent vessel of the next node in the pathway. Lymph drains through at least one lymphatic node before it passes into two main collecting ducts and is returned to the blood.

Cisterna chyli

The cisterna chyli is a large lymphatic vessel situated at the lower end of the thoracic duct. It is part of the lymphatic drainage for the abdomen, as it drains lymph laden with digested fats (chyle) from the small intestine. It also acts like a reservoir as it receives and temporarily stores the chyle before it is collected by the thoracic duct.

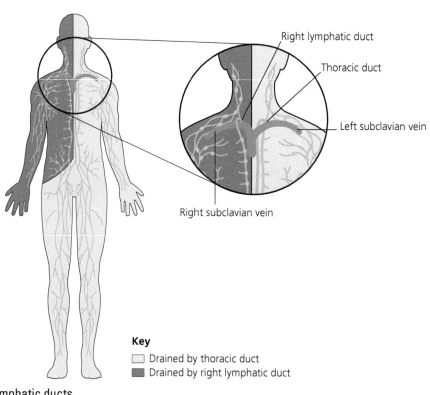

Right lymphatic duct

Thoracic duct

Left subclavian vein

Right subclavian vein

Key

☐ Drained by thoracic duct
■ Drained by right lymphatic duct

▲ Lymphatic ducts

Lymphatic ducts

From each chain of lymphatic nodes, the efferent lymph vessels combine to form lymphatic trunks which empty into two main ducts:

- the thoracic duct
- the right lymphatic duct.

These ducts collect lymph from the whole body and return it to the blood via the subclavian veins.

The thoracic duct is the main collecting duct of the lymphatic system. It is the largest lymphatic vessel in the body and extends from the second lumbar vertebra up through the thorax to the root of the neck. The thoracic duct collects lymph from the left side of the head and neck, left arm, lower limbs and abdomen and drains into the left subclavian vein to return it to the bloodstream.

The right lymphatic duct is very short in length. It lies in the root of the neck and collects lymph from the right side of the head and neck and the right arm and drains into the right subclavian vein to be returned to the bloodstream.

Lymphatic drainage

Movement of lymph throughout the lymphatic system is known as lymphatic drainage and it begins in the lymphatic capillaries. The movement of lymph out of the tissue spaces and into the lymphatic capillaries is assisted by:

- the pressure exerted by the skeletal muscles against the vessels during movement
- changes in internal pressure during respiration
- the compression of lymph vessels from the pull of the skin and fascia during movement.

Lymphatic drainage of the head and neck

The main groups of lymphatic nodes in the head and neck are as follows:

- buccal nodes
- cervical nodes (deep)
- cervical nodes (superficial)
- mastoid nodes (post auricular)
- occipital nodes
- parotid nodes (anterior auricular)
- submandibular nodes
- submental nodes.

Activity

Produce index cards with the name of a structure in the lymphatic system on each one. Include a description and the functions of each system. Divide into groups and put the cards in order of their structures in the circulatory pathway.

KEY FACT

Factors such as muscle tension put pressure on the lymphatic vessels and may block them, interfering with efficient drainage. Taking slow deep breaths can help to stimulate lymphatic flow.

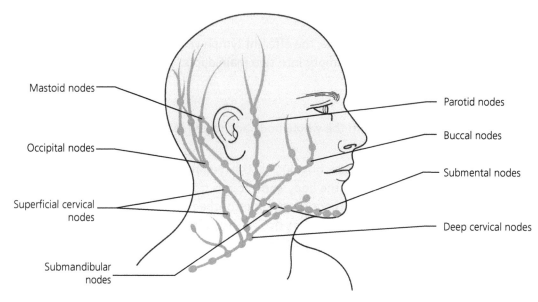

Labels: Mastoid nodes, Occipital nodes, Superficial cervical nodes, Submandibular nodes, Parotid nodes, Buccal nodes, Submental nodes, Deep cervical nodes

▲ Lymphatic nodes of the head and neck

Table 7.2 Lymphatic node groups of the head and neck

Name of lymphatic nodes	Position	Areas from which lymph is drained
Buccal nodes	Located superficially to the buccinator muscle in the cheek	Lower eyelid and anterior of the cheek
Cervical nodes (deep)	Deep within the neck, located along the path of the larger blood vessels (carotid artery and internal jugular vein)	Larynx, oesophagus, posterior of the scalp and neck, superficial part of the chest and arm
Cervical nodes (superficial)	Located at the side of the neck, over the sternomastoid muscle	Lower part of the ear and the cheek region
Mastoid nodes (post auricular)	Behind the ear in the region of the mastoid process	Skin of the ear and the temporal region of the scalp
Occipital nodes (ox-sip-it-tal)	At the base of the skull	Back of scalp and the upper part of the neck
Parotid nodes (anterior auricular)	At the angle of the jaw	Nose, eyelids and ear
Submandibular nodes	Underside of the jaw on either side	Chin, lips, nose, cheeks (submaxillary salivary gland), tongue, mucous membrane that covers the eyeball and under surface of the eyelid
Submental nodes	Middle of the neck under the chin	Central lower lip, the floor of the mouth and the apex of the tongue

KEY FACT

Swollen submandibular nodes usually indicate an active viral or bacterial infection and are commonly associated with infections of the sinuses, eyes and ears.

Lymphatic drainage of the body

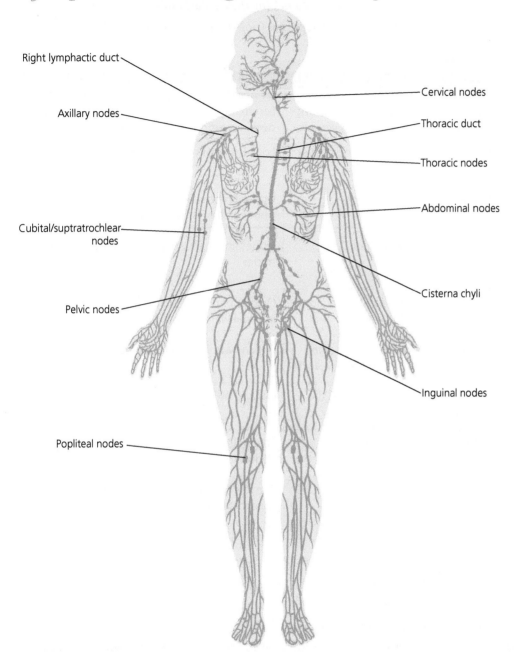

Right lymphactic duct

Axillary nodes

Cubital/suptratrochlear nodes

Pelvic nodes

Popliteal nodes

Cervical nodes

Thoracic duct

Thoracic nodes

Abdominal nodes

Cisterna chyli

Inguinal nodes

▲ Lymphatic nodes of the body

Lymph nodes are mainly clustered at joints, which assist in pumping lymph through the nodes when the joint moves. The superficial lymph nodes are most numerous in the groin, axillae and neck. Most of the deep lymph nodes are found alongside blood vessels of the pelvic, abdominal and thoracic cavities.

The main groups of lymphatic nodes relating to the body are as follows:

- cervical nodes (deep)
- cervical nodes (superficial)
- axillary nodes

- supratrochlear/cubital nodes
- thoracic nodes
- abdominal nodes
- pelvic nodes
- inguinal *(in-gwine-nal)* nodes
- popliteal *(pop-lit-tee-al)* nodes.

Table 7.3 Lymphatic node groups of the body

Name of lymphatic nodes	Position	Area from which lymph is drained
Cervical nodes (deep)	Deep within the neck, located along the path of the larger blood vessels	Larynx, oesophagus, posterior of the scalp and neck, superficial part of the chest and arm
Cervical nodes (superficial)	Located at the side of the neck over the sternomastoid muscle	Lower part of the ear and cheek region
Axillary nodes	In the underarm region	Upper limbs, wall of the thorax, breasts, upper wall of the abdomen
Supratrochlear (soo-pa- trok-lee-er) or cubital nodes	In the elbow region (medial side)	Upper limbs, passing through the axillary nodes
Thoracic nodes	Within the thoracic cavity and along the trachea and bronchi	Organs of the thoracic cavity and from the internal wall of the thorax
Abdominal nodes	Within the abdominal cavity along the branches of the abdominal aorta	Organs within the abdominal cavity
Pelvic nodes	Within the pelvic cavity, along the paths of the iliac blood vessels	Organs within the pelvic cavity
Inguinal (in-gwine-nal) node	In the groin	Lower limbs, external genitalia and lower abdominal wall
Popliteal (pop-lit-tee-al) node	Behind the knee	The lower limbs through deep and superficial nodes

KEY FACT

Thoracic lymph nodes are separated into two types:

- parietal lymph nodes – located in the thoracic wall
- visceral lymph nodes – associated with the internal organs.

Due to their location, abnormalities of the lymph nodes in the thorax, or chest, are not easily detected.

Study tip

It is helpful to think of the lymphatic system as a unidirectional (one-way) drainage system. Although the lymphatic system works alongside the blood circulation, lymph is carried one way, back towards the heart, whereas blood is carried to and from the heart.

When thinking of the circulatory pathway, remember the following key functions:

1 **Drainage** – the lymphatic capillaries drain the tissues spaces of excess fluid.
2 **Transport** – the lymphatic vessels transport lymph back towards the heart.
3 **Filter** – the lymphatic nodes filter the lymph of impurities.
4 **Collect** – the lymphatic ducts collect the lymph before it enters the venous system via the subclavian veins.

Summary of the circulatory pathway of lymph

- Plasma escapes the blood capillaries and bathes the tissue cells.
- Excess fluid flows through a network of lymphatic capillaries.
- Tissue fluid enters lymph vessels, where it becomes lymph.
- Larger lymphatic vessels lead to lymph nodes.
- Lymph passes through at least one lymphatic node, where it is filtered.
- Filtered lymph is collected into lymphatic ducts.
- Collected lymph is drained into the venous system via the subclavian veins.

Lymphatic organs

Lymphatic organs, whose functions are closely related to those of the lymph nodes, are the spleen, tonsils and thymus.

Spleen

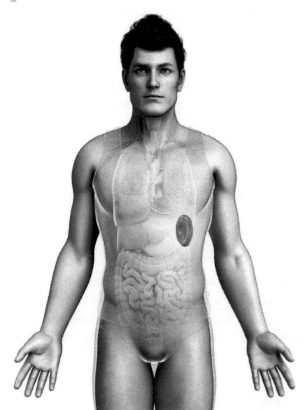

▲ The spleen

The spleen is the largest of the lymphatic organs and is located in the left-hand side of the abdominal cavity between the diaphragm and the stomach. As the spleen is largely a mass of lymphatic tissue, it contains lymph nodes, which produce lymphocytes and macrophages.

The spleen:

- is a major site for filtering out worn-out red blood cells and destroying micro-organisms that are circulating in the blood
- is concerned with protection from disease and the manufacture of antibodies. It functions with the lymphatic system by storing lymphocytes and releasing them as part of the immune response
- serves as a blood reservoir and can release small amounts of blood into the circulation during times of emergency or blood loss.

Tonsils

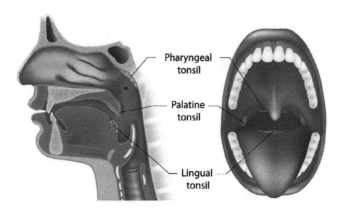

▲ The tonsils

The tonsils are composed of lymphatic tissue, and are located in the oral cavity and the pharynx. There are three different sets of tonsils, all of which provide defence against micro-organisms that enter the mouth and nose:

- the **palatine tonsils** – commonly identified as the tonsils and are located at the back of the throat, one on each side
- the **pharyngeal tonsils** – known as the adenoids, and lie on the wall of the nasal part of the pharynx
- the **lingual tonsils** – found below the tongue.

Thymus

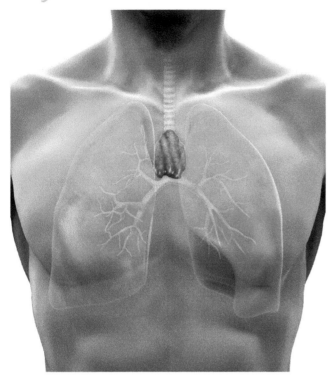

⚠ The thymus

The thymus gland is a triangular-shaped gland composed of lymphatic tissue. It is located in the upper chest above the superior vena cava and below the thyroid, where it lies against the trachea. The function of the thymus is important in newborn babies in promoting the development and maturation of certain lymphocytes and in programming them to become T-cells (specialised types of lymphocytes of the immune system, page 177). The thymus gland begins to atrophy after puberty and becomes only a small remnant of lymphatic tissue in adulthood.

The immune system

The immune system is not a specific structural organ system but more of a functional system. It draws on the structures and processes of each of the organs, tissues and cells of the body and the chemicals produced in them to eliminate any pathogen, foreign substance or toxic material that can be damaging to the body. Immunity can, therefore, be defined as the ability of the body to resist infection and disease by the activiation of specific defence mechanisms.

The human body has a variety of different defence mechanisms. Some are **non-specific** in that they do not differentiate between one threat and another. Others are **specific** as the body mounts its defence against a particular kind of threat.

Non-specific immunity

Non-specific immunity is programmed genetically in the human body from birth. The non-specific defences that are present from birth include:

- mechanical barriers
- chemicals
- inflammation
- phagocytosis
- fever.

Mechanical barriers

These are barriers such as the skin and mucous membranes that line the tubes of the respiratory, digestive, renal and reproductive systems. As long as these barriers remain unbroken, many pathogens are unable to penetrate them.

The respiratory system is lined with mucus-secreting cells to help remove micro-organisms from the respiratory tract. The highly acidic environment in the stomach can help to kill pathogens; saliva also has an antimicrobial effect. Urine helps to deter the growth of micro-organisms in the genito-urinary tract. The pH of the vagina protects against the multiplication and growth of microbes.

Chemicals

Chemicals are released by different cells that play an important role in immunity. There are many different types of chemicals that are involved in immunity including interferons, complements and histamine.

Interferons

These proteins are produced by cells that are infected by viruses. Interferons form antiviral proteins to help protect uninfected cells and inhibit viral growth. There are three types of human interferon:

- **alpha** (from white blood cells)
- **beta** (from fibroblasts)
- **gamma** (from lymphocytes).

Complements

Complements are blood proteins that combine to create substances which stimulate phagocytes to ingest bacteria.

Histamine

This chemical is released by a variety of tissue cells, including mast cells, basophils (a type of white blood cell) and platelets. The release of histamine causes vasodilation to bring more blood to an area of injury or infection. It also increases vascular permeability to allow fluid to enter the damaged area and dilute any toxins.

Inflammation

Inflammation is a sequence of events involving chemical and cellular activation that destroys pathogens and aids in the repair of tissues. It is a tissue response and symptoms include localised redness, swelling, heat and pain. The major actions that occur during an inflammation response include the following.

- Blood vessels dilate, resulting in an increase in blood volume (hyperaemia) to the affected area.
- Capillary permeability increases, causing tissues to become red, swollen, warm and painful.
- White blood cells invade the area and help to control pathogens by phagocytosis.
- In the case of bacterial infections, pus may form.
- Body fluids collect in the inflamed tissues. These fluids contain fibrinogen and other blood factors that promote clotting.
- Fibroblasts may appear and a connective tissue sac may be formed around the injured tissues.
- Phagocytic cells remove dead cells and other debris from the site of inflammation.
- New cells are formed by cellular reproduction to replace dead or injured ones.

Phagocytosis

Neutrophils and monocytes are the most active phagocytic cells of the blood. Neutrophils are able to engulf and ingest smaller particles, while monocytes can phagocytise larger ones. Monocytes give rise to macrophages (large scavenger cells), which become fixed in various tissues and may be attached at the inner walls of the blood and lymphatic vessels.

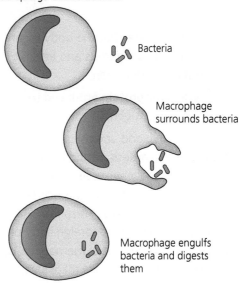

Macrophage detects bacteria

Bacteria

Macrophage surrounds bacteria

Macrophage engulfs bacteria and digests them

▲ Phagocytosis

Fever

An individual is said to have a fever if their body temperature is maintained above 37.28 °C (99 °F). The increase in temperature during a fever tends to inhibit some viruses and bacteria. It also speeds up the body's metabolism and, thereby, increases the activity of defence cells.

Specific immunity

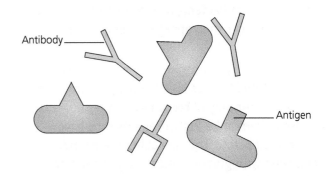

Antibody

Antigen

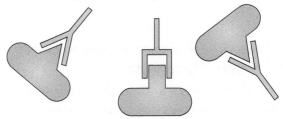

Antibody attaches to antigen and destroys bacteria

▲ The antibody defence system

Immunity involves interaction between two types of molecule:

- an **antigen** – any substance that the body regards as foreign or potentially dangerous and against which elicits antibody production
- an **antibody** – a specific protein produced to destroy or suppress an antigen.

Antibodies circulate in the blood and tissue fluid, killing bacteria or making them harmless. Antibodies can also neutralise poisonous bacterial chemicals called toxins. Specific immunity is the ability to recognise certain antigens and destroy them. It involves responses that are particular to each identified foreign body or unusual substance and calls on special memory cells to help if the foreign body or substance reappears.

The body must be able to identify which foreign bodies and substances cause a threat before any type of response can be initiated.

How antibodies work

Antibodies work in many different ways.

- Some neutralise antigens by combining with them and preventing them from carrying out their usual effects.
- Others may lyse (destroy) the cell on which the antigen is present.
- When some antibodies bind to antigens on the surface of bacteria, they attract other white blood cells like macrophages to engulf the bacteria.

The key cells of specific immunity are a specialised group of white blood cells called lymphocytes. They are capable, not only of recognising foreign agents, but also of 'remembering' the agents they have encountered before. They are able to react more rapidly and with greater force if they encounter the agent again.

The immune response

There are two types of immune response produced by different types of lymphocytes:

- **humoural immunity** – involves the B-cells, which produce free antibodies that circulate in the bloodstream

- **cell-mediated immunity** – effected by helper T-cells, suppressor T-cells and natural killer (NK) cells that recognise and respond to certain antigens to protect the body against their effects.

Lymphocytes develop in the following three ways:

1 T-cells develop in the bone marrow and grow in the thymus gland. They are able to recognise antigens and respond by releasing inflammatory and toxic materials. Specialised T-cells also regulate the immune response, either by amplifying the response (T4 cells) or by suppressing the body's response (T8 cells). Some T-cells develop into memory cells and handle the secondary response on re-exposure to antigens that have already produced a primary response.

2 B-cells grow and develop in the bone marrow. B-cells contain immunoglobulin, an antibody that responds to specific antigens. Some B-cells modify and become non-antigen specific, which means that they have a greater ability to respond to bacterial and viral pathogens. Some B-cells become memory cells and are able to deal with re-exposure to antigens.

3 A type of lymphocyte that does not develop the same structural or functional characteristics as the T-cells or B-cells are the NK cells. They also develop in the bone marrow and when mature can attack and kill tumour cells and virus-infected cells during their initial developmental stage, before the immune system is activated.

Primary and secondary responses of the immune system

The initial response of the body on first exposure to antigens is called the primary response. It normally takes about two weeks after exposure to the antigen for antibody levels to peak. This is due to the fact that B-cells have to become converted to plasma cells that secrete antibodies specifically against the antigen.

If the individual is exposed to the antigen for a second time, the presence of memory cells stimulates rapid production of antibodies (known as the secondary response). Antibody levels are much higher than in the primary response and remain elevated for a very long time. Secondary response can occur even if many years have elapsed since the first exposure to the antigen.

Immunisation

The body may be artificially stimulated into producing antibodies (known as immunisation). This prepares the body in advance to ward off infection. It is carried out by inoculating an individual with a vaccine (a liquid containing antigens powerful enough to stimulate antibody formation without causing disease or harm).

> ## KEY FACT
> Vaccines have been developed against many diseases, including diphtheria, polio, tetanus, whooping cough and measles.

Allergy

Under certain circumstances allergic reactions (abnormal responses) may occur when a foreign substance, or antigen, enters the body. An allergic reaction can only occur if the person has already been exposed to the antigen at least once and has developed an antibody to it.

The type and severity of an allergic reaction depends on the strength and persistence of the antibody evoked by previous exposure to the antigen. These antibodies are located on the cells in the skin or mucous membranes of the respiratory and gastrointestinal tracts.

> ## In practice
> Typical antigens include pollen, dust, feathers, wool, fur, certain foods and drugs. Be aware of products that can trigger allergic reactions.

Allergic reactions may cause symptoms of hayfever, asthma, eczema, urticaria and contact dermatitis. If there is much cellular damage, excessive amounts of histamine may be released causing circulatory failure (anaphylaxis). Anaphylactic shock is an extreme and generalised form of allergic reaction, whereby widespread release of histamine causes swelling (oedema), constriction of the bronchioles, heart failure and circulatory collapse and may even result in death.

Common pathologies of the lymphatic system

Acquired immune deficiency syndrome (AIDS)

This condition results from infection with the human immunodeficiency virus (HIV), which progressively destroys the immunity of the individual. The HIV virus suppressses the body's immune response, allowing opportunist infections to take hold, and results in AIDS.

AIDS patients are more vulnerable to infections than those without the condition. Infections that usually produce mild symptoms in healthy individuals may produce severe symptoms in AIDS patients. Sufferers have an increased risk of developing Kaposi sarcoma, non-Hodgkin lymphoma and cancer of the cervix.

AIDs is caused by contact with infected blood or body fluids. It can be passed on via unprotected sex or through the sharing of unsterilised needles.

Hodgkin's disease

This is a malignant disease of the lymphatic tissues, usually characterised by painless enlargement of one or more groups of lymph nodes in the neck, armpit, groin, chest or abdomen. The spleen, liver, bone marrow and bones may also be involved. Apart from the enlarged nodes, there may also be weight loss, fever, profuse sweating at night and itching.

> ## In practice
> Advice from the client's consultant physician is necessary before undertaking any form of treatment. It is inadvisable to treat if a client is debilitated; otherwise, clients may benefit from a gentle and relaxing treatment.
>
> Note that clients are vulnerable to infection due to reduced immunity and that there is a risk of spreading the disease through lymphatic drainage.

Lupus erythematosus

This is a chronic inflammatory disease of connective tissue affecting the skin and various internal organs. It is an autoimmune disease and can be diagnosed by the presence of abnormal antibodies in the bloodstream. Typical signs are a red, scaly rash on the face, arthritis and progressive damage to the kidneys. Often the heart, lungs and brain are affected by progressive attacks of inflammation, followed by the formation of scar tissue. It can also cause psychiatric illness due to direct brain involvement.

In practice

Care is required when treating patients with lupus erythematosus as skin lesions might be tender, and joint pain and tenderness may be present.

Avoid contact if you are suffering from any infectious illness, as medication for this condition can suppress immunity and clients can be prone to infection.

Oedema

This is an abnormal swelling of body tissues due to an accummulation of tissue fluid. It can be the result of heart failure, liver or kidney disease or chronic varicose veins. The swelling of tissues may be localised, if the oedema is caused by injury or inflammation, or may be more generalised, if it is caused by heart or kidney failure.

Subcutaneous oedema commonly occurs in the legs and ankles due to the influence of gravity, and is a common problem in women before menstruation and in the last trimester of pregnancy.

In practice

Remember that oedema is symptomatic of many disease processes (particularly cardiovascular disease). Therefore, seek advice from the client's GP before offering any form of treatment.

Interrelationships with other systems

The lymphatic system

The lymphatic system links to the following body systems.

Cells and tissues

Lymphatic tissue is a specialised type of tissue found in lymph nodes, spleen, tonsils, the adenoids, walls of the large intestine and glands in the small intestine.

Skin

Lymph vessels are numerous in the dermis of the skin. They form a network allowing the removal of waste from the skin's tissues.

Skeletal

Red bone marrow is responsible for the development of cells found in both blood and lymph.

Muscular

The action of skeletal muscles aids lymphatic drainage.

Circulatory

The lymphatic system aids the circulatory system in that it assists the blood in returning fluid from the tissues back to the heart.

Respiratory

Low pressure in the thorax created by breathing movements aids the movement of lymph.

Digestive

The lymphatic system plays an important part in absorbing the products of fat digestion from the villi of the small intestine.

Key words

Abdominal node: a type of lymphatic node within the abdominal cavity along the branches of the abdominal aorta

Allergic reaction: when the immune system overreacts to a harmless allergen

Antibody: a specific protein produced to destroy or suppress antigens

Antigen: any substance that the body regards as foreign or potentially dangerous and against which it produces an antibody

Axillary nodes: a type of lymphatic node located under the arm

Buccal nodes: a type of lymphatic node located in the cheek

Cisterna chyli – a large lymphatic vessel situated at the lower end of the thoracic duct

Deep cervical nodes: a type of lymphatic node located deep within the neck, along the path of the larger blood vessels

Immunisation: artificial stimulation of the body to produce antibodies

Immunity: the body's ability to resist infection

Inguinal nodes: a type of lymphatic node located in the groin

Lacteals: intestinal lymphatic vessels

Lymph: a transparent, colourless, watery liquid derived from tissue fluid

Lymphatic capillaries: very small blind-ended tubes that arise in the tissue spaces of the body

Lymphatic ducts: one of two lymphatic vessels (right lymphatic and thoracic ducts) that collect and empty lymph into the circulatory system via the subclavian veins

Lymphatic nodes: oval or bean-shaped structures that filter lymph

Lymphatic vessels: thin-walled, valved structures that carry lymph

Mastoid nodes: lymphatic nodes found behind the ear

Non-specific immunity: a type of immunity programmed genetically from birth

Occipital nodes: a type of lymphatic node located at the base of the skull

Oedema: abnormal swelling of the body's tissues due to an accumulation of tissue fluid

Parotid nodes: a type of lymphatic node located at the angle of the jaw

Popliteal nodes: a type of lymphatic node located at the back of the knee

Right lymphatic duct: the lymphatic duct that collects lymph from the right side of the head and neck and the right arm and drains into the right subclavian vein

Specific immunity: the production of antibodies against a particular antigen

Spleen: the largest of the lymphatic organs, concerned with protection from disease and the manufacture of antibodies

Subclavian veins: a type of vein in the arm into which lymph drains

Submandibular nodes: a type of lymphatic node located under the chin

Superficial cervical nodes: a type of lymphatic node located at the side of the neck over the sternomastoid muscle

Suptratrochlear nodes: a type of lymphatic node in the elbow region (medial side)

Thoracic duct: the main collecting duct of the lymphatic system

Thoracic nodes: a type of lymphatic node within the thoracic cavity and along the trachea and bronchi

Thymus: a gland composed of lymphatic tissue located in the upper chest; important in promoting the development and maturation of the lymphatic system in newborn babies

Tissue (interstitial) fluid: intercellular fluid located between the cells of the body tissues

Tonsils: a type of lymphatic tissue located in the oral cavity and the pharynx; there are three pairs (palatine, pharyngeal, lingual)

Revision summary

The lymphatic system

- The lymphatic system is closely associated with the cardiovascular system.

- The lymphatic system assists the circulatory system by draining the tissues of excess fluid and returning the fluid from the tissues back to the heart. This helps to maintain blood volume and blood pressure and to prevent **oedema** (swelling of the tissues).

- The lymphatic system also plays an important role in the body's immune system, as the lymph nodes fight infection and generate antibodies.

- The lymphatic system also absorbs the products of fat digestion through intestinal lymph vessels called the **lacteals**.

- **Lymph** is a clear, colourless, watery fluid derived from tissue fluid and contained within lymph vessels.

- Lymph is similar in composition to blood plasma except that it has a lower concentration of plasma proteins.

- The circulatory pathway of lymph begins with **lymphatic capillaries**, which lie in the tissue spaces between the cells.

- **Tissue (interstitial) fluid** drains into **lymphatic capillaries**, and this fluid becomes **lymph**.

- **Lymphatic capillaries** merge to form larger vessels called **lymphatic vessels**, which convey lymph into and out of structures called **lymph nodes**.

- The main groups of lymph nodes relating to the head and neck include **deep cervical**, **superficial cervical**, **submandibular**, **occipital**, **mastoid** and **parotid nodes**.

- The main group of lymph nodes relating to the body include **superficial cervical**, **deep cervical**, **axillary**, **supratrochlear**, **thoracic**, **abdominal**, **pelvic**, **inguinal** and **popliteal nodes**.

- All lymph passes through at least one node, where it is filtered of cell debris, micro-organisms and harmful substances.

- Once filtered, the lymph is collected into two main ducts – the **thoracic duct** (the largest duct), which collects lymph from the left side of the head and neck, left arm, lower limbs and abdomen, and the **right lymphatic duct**, which collects lymph from the right side of the head and neck and the right arm.

- The collected lymph is then drained into the venous system via the right and left **subclavian veins**.

- Other lymphatic organs include the **spleen**, **tonsils** and **thymus gland**.

- **Immunity** is the ability of the body to resist infection and disease by the activation of defence mechanisms.

- There are two types of **immunity** – specific and non-specific.
  - **Non-specific immunity** is programmed genetically from birth and includes mechanical barriers (skin and mucous membrane), chemicals, inflammation, phagocytosis and fever.
  - **Specific immunity** involves interaction between an **antigen** and an **antibody**.
    - An **antigen** is any substance that the body regards as foreign or potentially dangerous, and against which it produces an antibody.
    - An **antibody** is a specific protein produced to destroy or suppress an antigen.

- There are two types of immune response produced by different types of lymphocytes – **humoural immunity** involving B-cells, which produce free antibodies that circulate in the bloodstream, and **cell-mediated immunity** effected by helper T-cells, suppressor T-cells and NK cells that recognise and respond to certain antigens to protect the body against their effects.

- **Immunisation** is the artificial stimulation of the body to produce antibodies.

- An **allergic reaction** may occur when a foreign substance, or antigen, enters the body.
  - An allergic reaction can only occur if the person has already been exposed to the antigen at least once before and has developed an antibody to it.
  - **Antibodies** are located on the cells in the skin or mucous membranes of the respiratory and gastrointestinal tracts.
  - Typical antigens include pollen, dust, feathers, wool, fur, certain foods and drugs.

Test your knowledge

Multiple choice questions

1 From which of these is lymph derived?
   a plasma proteins
   b tissue fluid
   c blood plasma
   d lymphocytes

2 Lymph is similar in composition to blood except it has lower concentration of:
   a water
   b protein
   c waste
   d hormones.

3 Which of the following is **not** a function of the lymphatic system?
   a production of lymphocytes
   b prevention of oedema
   c production of heat
   d absorption of fat

4 Which of the following cleanse lymph of foreign matter?
   a lymphatic vessel
   b lymphatic node
   c lymphatic capillary
   d lymphatic duct

5 From which type of tissue are lymph nodes made?
   a areolar tissue
   b adipose tissue
   c lymphatic tissue
   d yellow elastic tissue

6 Which of the following statements best describes the structure of a lymph vessel?
   a thick muscular tubes with no valves
   b thin muscular tubes with no valves
   c thick collapsible walls with valves
   d thin collapsible walls with valves

7 Which of these promote lymph flow?
   a pressure exerted by skeletal muscles during movement
   b compression of lymph vessels from the pull of the skin and fascia during movement
   c changes in internal pressure during respiration
   d all of the above

8 Which of the following drains lymph from the lower limbs?
   a cervical nodes
   b axillary nodes
   c popliteal nodes
   d supratrochlear nodes

9 Which of the following nodes drain lymph from the back of the scalp and the upper part of the neck?
   a occipital
   b parotid
   c deep cervical
   d superficial cervical

10 Where are axillary nodes situated?
   a neck
   b groin
   c underarm
   d elbow

Exam-style questions

11 State two functions of the lymphatic system.
   2 marks

12 Name the type of cell that produces antibodies.
   1 mark

13 Name the lymph nodes that drain lymph from the lower limbs, external genitalia and lower abdominal wall.
   1 mark

14 Name the lymph nodes located in the elbow region.
   1 mark

15 a Name the type of vessel that enters a lymph node.
   1 mark
   b Name the type of vessel that exits a lymph node.
   1 mark

16 State two structural features of lymphatic vessels.
   2 marks

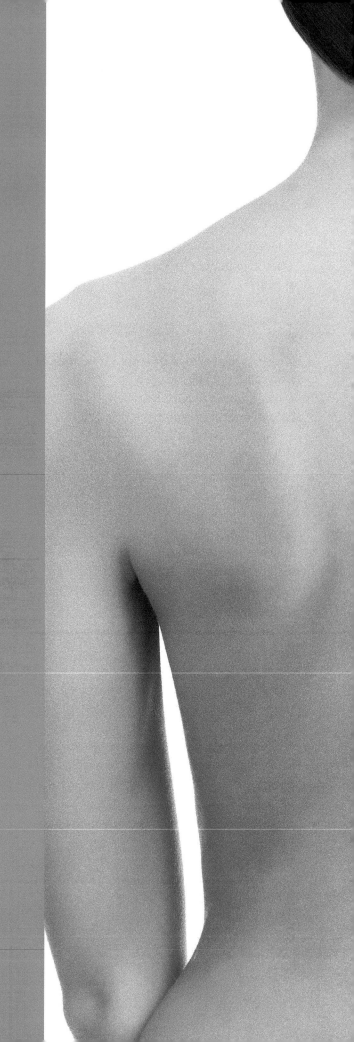

8 The respiratory system

Introduction

The respiratory system consists of the nose, nasopharynx, pharynx, larynx, trachea, bronchi and lungs, which together provide the passageway for air, in and out of the body. Oxygen is needed for survival by every cell of the body. Respiration is the process by which the living cells of the body receive a constant supply of oxygen and by which carbon dioxide is removed. Our respiratory system serves us in many ways, exchanging oxygen and carbon dioxide, detecting smell, producing speech and regulating pH.

OBJECTIVES

By the end of this chapter you will understand:

- the functions of the respiratory system
- the structure and functions of the main parts of the respiratory system
- the process of the interchange of gases in the lungs
- the mechanisms of external and internal respiration
- the theory of olfaction
- the importance of correct breathing
- pathologies of the respiratory system
- the interrelationships between the respiratory and other body systems.

Functions of the respiratory system

The primary role of the respiratory system is to provide oxygen to the body's cells while also removing carbon dioxide; other functions include providing a sense of smell, speech and helping to maintain homeostasis.

- **Exchange of gases** – oxygen and carbon dioxide exchange is the primary function of the respiratory system and is necessary in order to sustain life.
- **Olfaction** – specialised nerve endings embedded in the nasal cavity send impulses for the sense of smell to the brain.
- **Speech** – the vocal cords in the larynx aid in producing speech.
- **Homeostasis** – the respiratory system has homeostastic functions in that it maintains the oxygen and carbon dioxide levels in the blood and it also dissipates heat.

In practice

It is important for therapists to have a good knowledge of the respiratory system in order to understand how breathing may be affected during a treatment. A massage, for instance, deepens respiration and improves lung capacity by relaxing any tightness in the respiratory muscles.

Understanding the mechanism of breathing helps therapists when teaching clients deep breathing exercises as part of a stress management or relaxation programme.

The structures of the respiratory system

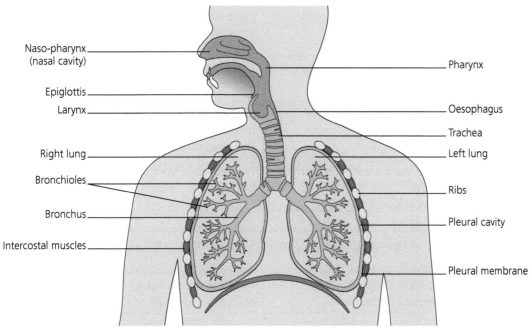

Naso-pharynx (nasal cavity) — Epiglottis — Larynx — Right lung — Bronchioles — Bronchus — Intercostal muscles — Pharynx — Oesophagus — Trachea — Left lung — Ribs — Pleural cavity — Pleural membrane

▲ Structures of the respiratory system

Overview of the structures of the respiratory system

The respiratory system contains the following structures, as discussed in Table 8.1.

- nose
- nasopharynx
- pharynx
- larynx
- trachea
- bronchi
- lungs.

Study tip

When studying structures of the respiratory system it helps to think of a tree with branches.

- Air is breathed in through the nose and travels down the pharynx and larynx.
- It travels from the larynx to the trachea, which is the trunk of the tree.
- The trachea then divides into two branches (right and left bronchi), which then subdivide into bronchioles (imagine these as smaller branches and twigs) inside the lungs.
- At the end of the bronchioles are the tiny air sacs. The alveoli are where the lungs and the bloodstream exchange carbon dioxide and oxygen. Alveoli are like the leaves of the tree.

Table 8.1 The structures of the respiratory system

Structure	Description	Function
Nose	Lined with cilia and mucous membrane	Inhales air Moistens, warms and filters the air Senses smell
Nasopharynx	Upper part of the nasal cavity behind the nose, lined with mucous membrane	Continues to filter, warm and moisten the incoming air
Pharynx (far-rink-s)	Large, muscular tube lined with mucous membrane Lies behind the mouth and between the nasal cavity and the larynx	Acts as a passageway for air, food and drink Resonating chamber for sound
Larynx (lar-rink-s)	Short passage connecting the pharynx to the trachea	Provides a passageway for air between the pharynx and the trachea Produces sound
Trachea (trak-kee-a)	Tube anterior to the oesophagus, extends from the larynx to the upper chest Composed of smooth muscle and up to 20 c-shaped rings of cartilage	Transports air from the larynx into the bronchi
Bronchi	Two short tubes (similar in structure to the trachea) which lead to each lung	Carry air into the lungs
Lungs	Cone-shaped spongy organs situated in the thoracic cavity on either side of the heart	Facilitate the exchange of the gases oxygen and carbon dioxide

The nose

The nose is divided into the right and left cavities. It is lined with tiny hairs called cilia, which begin to filter the incoming air, and mucous membrane, which secretes a sticky fluid called mucus to prevent dust and bacteria from entering the lungs. The nose moistens, warms and filters the air and is an organ which senses smell.

The nasopharynx

The nasopharynx is the upper part of the nasal cavity behind the nose and is lined with mucous membrane. The Eustachian tubes from the middle ears open into the nasopharynx so that air pressure inside the ear can be adjusted to prevent damage to the eardrum. At the back of the nasopharynx there is lymphoid tissue, including the adenoids. Due to the close proximity of the Eustachian tubes to the throat, throat infections can easily spread to the ear.

> **KEY FACT**
>
> The sinuses are air-filled spaces located within the maxillary, frontal, ethmoid and sphenoid bones of the skull. These spaces open into the nasal cavity and are lined with mucous membrane, which is continuous with the lining of the nasal cavity. Consequently, mucous secretions can drain from the sinuses into the nasal cavity. If this drainage is blocked by membranes that are inflamed and swollen because of nasal infections or allergic reactions, the pressure of the accumulating fluids may cause a painful sinus headache.

The nasopharynx continues to filter, warm and moisten the incoming air.

The pharynx

The pharynx or throat is a large muscular tube lined with mucous membrane that lies behind the mouth and between the nasal cavity and the larynx. The tonsils are found at the back of the pharynx. The pharynx serves as an air and food passage but cannot be used for both purposes at the same time, otherwise choking results. Air is warmed and moistened further as it passes through the pharynx.

The larynx

The larynx (voice box) is a short passage connecting the pharynx to the trachea. The larynx is a box-like cavity with rigid walls. It contains the vocal cords and stiff pieces of cartilage, such as the Adam's apple, which prevent collapse and obstruction of the airway. The vocal cords are bands of elastic ligaments that are attached to the rigid cartilage of the larynx by skeletal muscle. When air passes over the vocal cords they vibrate and produce sound. The opening into the larynx from the pharynx is called the glottis. During the process of swallowing, the glottis is covered by a flap of tissue called the epiglottis which prevents food from 'going down the wrong way'. The larynx provides a passageway for air between the pharynx and the trachea.

The trachea

The trachea, or windpipe, is a tube anterior to the oesophagus that extends from the larynx to the upper chest. It is composed of smooth muscle and up to 20 c-shaped rings of cartilage, which serve a dual purpose. The incomplete section of the ring allows the oesophagus to expand into the trachea when a food bolus is swallowed and the rings help to keep the trachea permanently open. The trachea passes down into the thorax and connects the larynx with the bronchi, which pass into the lungs.

The bronchi

The bronchi are two short tubes similar in structure to the trachea that carry air into each lung. They are lined with mucous membrane and ciliated cells and, like the trachea, contain cartilage to hold them open.

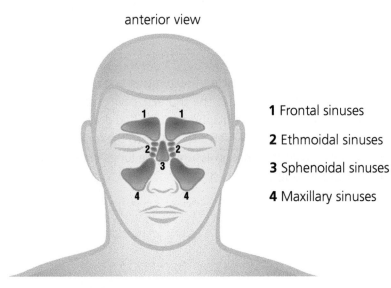

anterior view   lateral view

1 Frontal sinuses

2 Ethmoidal sinuses

3 Sphenoidal sinuses

4 Maxillary sinuses

▲ The paranasal sinuses

The mucus traps solid particles and cilia move them upwards, preventing dirt from entering the delicate lung tissue. The bronchi subdivide into bronchioles in the lungs. These subdivide yet again and finally end in very small air-filled sacs called alveoli.

The lungs

The lungs are paired, cone-shaped spongy organs situated in the thoracic cavity on either side of the heart. The left lung has two lobes and the right lung has three lobes. The right lung is thicker and broader than the left and is also slightly shorter than the left, as the diaphragm is higher on the right side to accommodate the liver which lies below it. Internally, the lungs consist of millions of tiny air sacs called alveoli which are arranged in lobules and resemble bunches of grapes. The function of the lungs is to facilitate the exchange of the gases oxygen and carbon dioxide. In order to carry out this function efficiently, the lungs have several important features:

- a very large surface area (about $100\,m^2$), provided by approximately 300 million alveoli
- a thin, permeable membrane surrounding the walls of the alveoli
- a thin film of water lining the alveoli, which is essential for dissolving oxygen from the incoming air
- thin-walled blood capillaries forming a network around the alveoli, which absorb oxygen from the air breathed into the lungs and release carbon dioxide into the air breathed out of the lungs.

The structures enclosed within the lungs are bound together by elastic and connective tissue. On the outside, the lungs have two layers of a serous membrane called pleura, an outer parietal layer that lines the thoracic cavity and an inner visceral layer that is attached to the surface of the lungs. Between the visceral and parietal pleurae is the pleural cavity, which contains a lubricating fluid secreted by the membranes that reduces friction between the lungs and the chest wall.

The diaphragm (di-a-fram)

The diaphragm is the chief muscle of respiration and is a dome-shaped muscular partition that separates the thoracic cavity from the abdominal cavity. During contraction, the diaphragm is pulled down, decreasing the pressure in the chest cavity, which sucks air into the lungs. Relaxation of the diaphragm causes it to rise, pressure in the lungs increases and air is pushed out of the lungs.

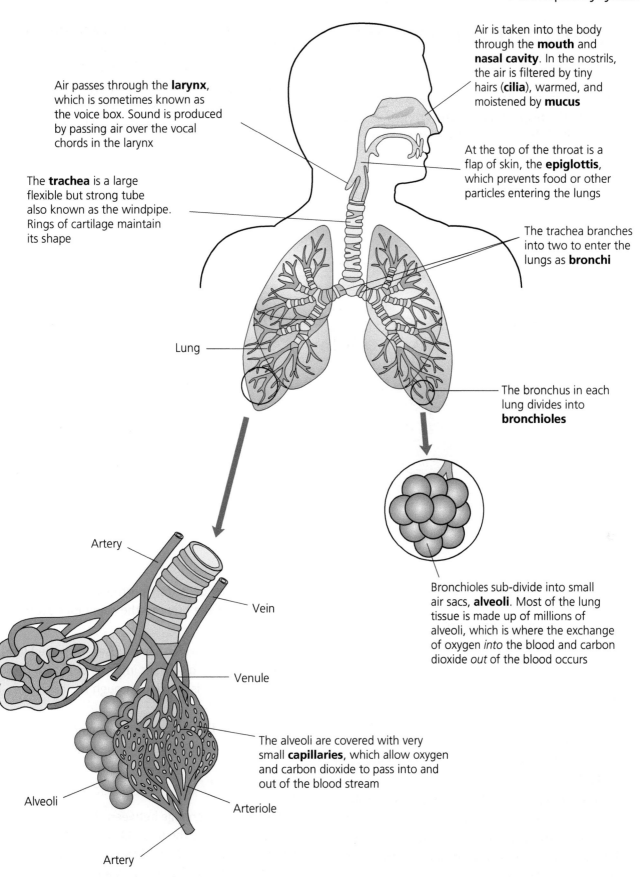

Air passes through the **larynx**, which is sometimes known as the voice box. Sound is produced by passing air over the vocal chords in the larynx

The **trachea** is a large flexible but strong tube also known as the windpipe. Rings of cartilage maintain its shape

Air is taken into the body through the **mouth** and **nasal cavity**. In the nostrils, the air is filtered by tiny hairs (**cilia**), warmed, and moistened by **mucus**

At the top of the throat is a flap of skin, the **epiglottis**, which prevents food or other particles entering the lungs

The trachea branches into two to enter the lungs as **bronchi**

Lung

The bronchus in each lung divides into **bronchioles**

Artery

Vein

Venule

Bronchioles sub-divide into small air sacs, **alveoli**. Most of the lung tissue is made up of millions of alveoli, which is where the exchange of oxygen *into* the blood and carbon dioxide *out* of the blood occurs

The alveoli are covered with very small **capillaries**, which allow oxygen and carbon dioxide to pass into and out of the blood stream

Alveoli

Arteriole

Artery

▲ The respiratory system

The interchange of gases in the lungs

Oxygen and carbon dioxide exchange is the primary function of the respiratory system. Oxygen is needed by every cell of the body and is delivered by way of the circulatory system. Carbon dioxide, a waste product of cell metabolism, is produced by every cell in the body and is removed by the circulatory system.

The interchange of gases in the lungs involves the absorption of oxygen from the air into the blood and release of carbon dioxide from the blood into the alveolar air, from where it is expelled from the body.

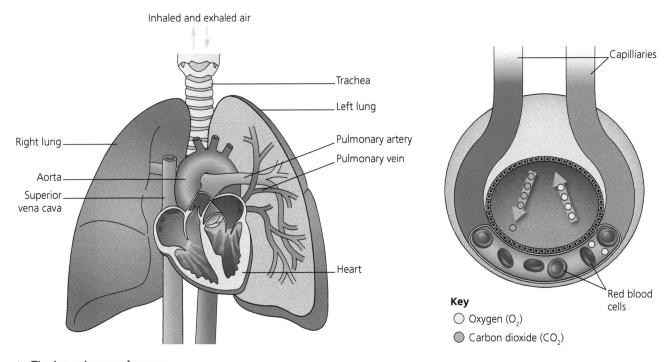

▲ The interchange of gases

Inhaled and exhaled air

Trachea

Left lung

Right lung

Pulmonary artery

Pulmonary vein

Aorta

Superior vena cava

Heart

Capilliaries

Red blood cells

Key

○ Oxygen (O_2)

● Carbon dioxide (CO_2)

External respiration

This refers to gas exchange in the lungs between the blood and air in the alveoli that came from the external environment. The respiration process is as follows:

1 During inhalation, air is taken in through the nose and mouth. It flows along the trachea and bronchial tubes to the alveoli of the lungs, where oxygen diffuses through the thin film of moisture lining the alveoli.

2 Oxygen diffuses from the air inside the alveoli, across the alveolar walls and into the blood capillaries. The oxygen binds to the haemoglobin inside erythrocytes and is transported to the cells throughout the body.

3 Carbon dioxide is transported by the blood from the cells of the body to the capillaries attached to the alveoli.

4 The carbon dioxide then diffuses from the blood across the alveolar walls into the air inside the alveoli, which is exhaled through the nose and mouth.

5 Oxygen and carbon dioxide exchange across the walls of the alveoli at the same time.

Internal/tissue respiration

This is the gas exchange between the blood and the tissues throughout the body. Oxygen diffuses from the blood into the cells and carbon dioxide diffuses from the cells into the bloodstream.

The mechanism of respiration

Rib movements in breathing

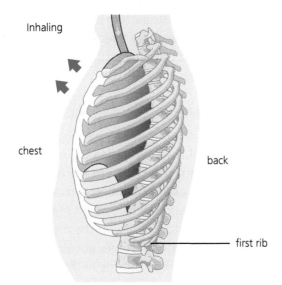

Inhaling

chest

back

first rib

Inhaling. The diaphragm and intercostal muscles contract, pulling the ribs upward. This increases the volume of the chest cavity, drawing air into the lungs.

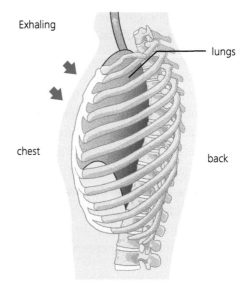

Exhaling

lungs

chest

back

Exhaling. The contracted muscles relax, the ribs fall slightly and decrease the volume of the chest. Air is forced out of the lungs.

How the diaphragm works

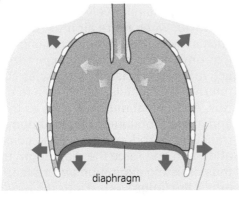

Inhaling

diaphragm

Inhaling. As the rib cage expands (*arrows, above*), the diaphragm contracts and flattens downwards, enlarging the chest cavity.

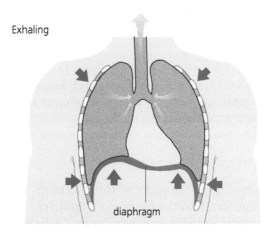

Exhaling

diaphragm

Exhaling. The diaphragm relaxes and is pressed up by the abdominal organs, returning to its dome shape. The chest narrows, driving air out of the lungs.

▲ The mechanism of respiration

The mechanism of respiration is the means by which air is drawn in and out of the lungs (breathing). It is an active process in which the muscles of respiration contract to increase the volume of the thoracic cavity.

The major muscle of respiration is the diaphragm. Air is moved in and out of the lungs by the combined action of the diaphragm and the intercostal muscles.

Inspiration

During inspiration, the dome-shaped diaphragm contracts and flattens, increasing the volume of the thoracic cavity. The diaphragm is responsible for bringing approximately 75% of the volume of air into the lungs.

The external intercostal muscles are also involved in respiration and on contraction they increase the depth of the thoracic cavity by pulling the ribs upwards and outwards. The external intercostal muscles are responsible for bringing approximately 25% of the volume of air into the lungs.

The combined contraction of the diaphragm and the external intercostals increases the thoracic cavity, which decreases the pressure inside the thorax so that air from outside of the body enters the lungs.

Expiration

During normal respiration, the process of expiration is passive (without conscious effort) and is brought about by the relaxation of the diaphragm and the external intercostal muscles, along with the elastic recoil of the lungs. This increases the internal pressure inside the thorax so that air is pushed out of the lungs.

KEY FACT

Other accessory muscles which assist in inspiration include the sternomastoid, serratus anterior, pectoralis minor, pectoralis major and the scalene muscles in the neck.

 Activity

Breathe in slowly for a count of three. During inspiration, focus on the changing volume of your rib cage. Hold for a count of one and then breathe out slowly for a count of three. Focus on what happens to the rib cage on expiration.

Breathing rate

The normal breathing rate is 12 to 15 breaths per minute, although this may increase during exercise and stress, and decrease during sleep. Breathing takes place rhythmically, with inspiration lasting for about two seconds and expiration for approximately three seconds.

Regulation of breathing

Breathing, like the beating of the heart, occurs continuously and rhythmically without conscious thought. The basic pattern of breathing can be modified by voluntary intervention but the underlying mechanism is essentially automatic; it continues when we are asleep and even when we are unconscious.

Nervous control

Breathing is controlled by a group of neurones in the parts of the brain called the medulla oblongata and the pons, also known as the respiratory centre. Nerve cells, called chemoreceptors, which are found in the aorta and the carotid arteries send impulses to the respiratory centre in the medulla oblongata of the brain with messages about the levels of oxygen and carbon dioxide in the blood. When the levels of carbon dioxide and oxygen need adjusting, a nerve impulse is sent to the respiratory muscles and breathing rate and depth changes accordingly. The medulla oblongata controls the rate and depth of respiration and the pons moderates the rhythm of the switch from inspiration to expiration.

KEY FACT

Breathing is a relatively passive process. However, when more air must be exhaled, such as when coughing or playing a wind instrument, the process of expiration becomes active. This is assisted by muscles such as the internal intercostals, which help to depress the ribs. Abdominal muscles, such as the external and internal obliques, rectus abdominus and transversus abdominus, help to compress the abdomen and force the diaphragm upwards, thus assisting expiration and squeezing more air out of the lungs.

Modified respiratory movements

Speech/talking

Speech is achieved when air flows from the lungs, causing the vocal cords to vibrate. The sound is then reverberated in the vocal spaces created by the tongue, soft palate, lips and jaw. During speech, forced inspiration and forced expiration are needed to create sound. In forced inspiration, the rib cage is elevated and the thoracic cavity enlarged by the accessory muscles. During forced expiration, the trunk and abdomen muscles pull the rib cage down and the thoracic cavity volume is reduced, forcing air out of the lungs.

Singing

During singing, there is a need to inhale quickly and deeply, then exhale slowly and steadily. Therefore, singing requires a higher rate of breathing than speaking does, as well as elongation of the breath cycle.

Singers need to learn how to extend the normal breath cycle by remaining in the inspiratory position for as long as possible, maintaining a raised sternum (but not raised shoulders or clavicle), no (or minimal) chest displacement, allowing the muscles of the lateral abdominal wall to stay close to the position of inhalation and delaying rib cage collapse.

Coughing

Coughing is caused by a stimulus in the air passages, particularly the larynx. A cough involves a deep inspiration, then closure of the glottis, followed by a violent expiratory effort. This is accompanied by two, three, or more sudden openings and closures of the glottis, so that rapidly repeated blasts of air pass through the upper air passages and out of the (generally open) mouth.

> **KEY FACT**
> The glottis is the part of the larynx that includes the vocal cords and the slit-like opening between them. It affects voice modulation through expansion and contraction.

Crying

Crying (sobbing) is a series of short, sudden expirations and rapid convulsive inspirations.

Hiccup

A hiccup is an unexpected inspiratory spasm, usually of the diaphragm, with the entrance of the air being checked by the sudden closure of the glottis.

Laughing

Laughing consists of a full inspiration, followed by a long series of very short and rapid expiratory efforts together with the facial expressions typically associated with laughter.

> **KEY FACT**
> Laughter helps to reduce stress hormones and triggers the release of the feel-good hormones, endorphins, leading to a sense of wellbeing and happiness.

Sighing

Sighing is a long, slow inspiration, quickly followed by a long slow expiration.

Sneezing

Sneezing is purely a reflex act as it is impossible to produce a sneeze voluntarily. It is caused by a stimulus to the nose or eyes which triggers impulses to the respiratory centre from the nasal and other branches of the fifth nerve.

A sneeze consists of a deep inspiration and closure of the glottis, followed by a single explosive expiration and sudden opening of the glottis and nostrils.

Yawning

Yawning involves a very long, deep inspiration which completely fills the chest, accompanied by wide opening of the mouth.

Olfaction

Olfaction is the sense that detects different smells. The sense of smell can evoke emotional responses due to its close link with the endocrine system. The process of olfaction involves the nervous system as smells received as stimuli at the nose are transmitted by nerve impulses to be perceived by the brain.

The structures of the olfactory system

The features of the olfactory system are:

- **Nose** – this is the organ of olfaction (smell).
- **Mucous membrane** – this lines the nose, moistens the air passing over it and helps to dissolve the odorous gas particles passing through the nasal cavity. The mucous membrane has a very rich blood supply and warmth from the blood flowing

through the tiny capillaries raises the temperature of the air as it passes through the nose.

- **Cilia** – these are the tiny, mucous-covered hairs inside the nose. They are highly sensitive protrusions of the olfactory cells and are covered in receptors.

- **Olfactory cells** – these lie embedded in the mucous membrane in the upper part of the nasal cavity. These sensory nerve cells are specially adapted for detecting smell. Each olfactory cell has a long nerve fibre called an axon leading out of the main body of the cell. This transmits any information received at the receptor to the brain.

- **Olfactory bulb** – this is the area of the brain, situated in the cerebral cortex, which perceives smell.

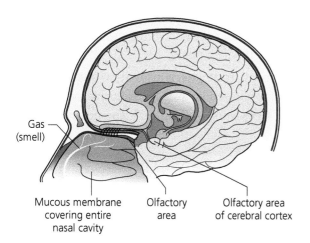

Gas (smell)

Mucous membrane covering entire nasal cavity

Olfactory area

Olfactory area of cerebral cortex

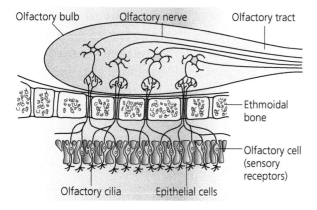

Olfactory bulb   Olfactory nerve   Olfactory tract

Ethmoidal bone

Olfactory cell (sensory receptors)

Olfactory cilia   Epithelial cells

▲ Olfaction

The theory of olfaction

Olfaction is categorised as a 'special sense' as odour perception uses specialised structures that

transmitted information directly to the brain. The process of olfaction involves reception, transmission and preception. Consider, for example, how we detect the smell of an essential oil.

Reception

1. The volatile particles of an essential oil evaporate on contact with air.

2. The volatile molecules disperse through the air and some enter the nose.

3. The odiferous particles of the essential oil dissolve in the mucus that lines the inner nasal cavity.

Transmission

1. The dissolved aromatic molecules join to the receptors in the cilia which protrude from the olfactory receptor cells.

2. The olfactory receptor cells have a long nerve fibre called an axon, and an electrochemical message of the aroma is transmitted along the axons of receptor cells to join the olfactory nerves.

3. The fibres of the olfactory nerves pass through the cribriform plate of the ethmoid bone in the roof of the nose to reach the olfactory bulb, from where the odorant signal is relayed to other parts of the brain.

Perception

1. Once the message reaches the olfactory bulb, the olfactory impulses pass into the olfactory tract and then directly to the cerebral cortex, where the smell is perceived.

2. The temporal lobe of the brain contains the primary olfactory area, which is directly connected to the limbic area – this is concerned with emotions, memory and sex drive.

3. The olfactory bulb also connects closely with the hypothalamus, the nerve centre which governs the endocrine system.

KEY FACT

Most nerves in the body transmit nerve impulses to the spinal cord and then on to the brain. However, in the case of the olfactory cells, the nerve fibres connect directly with the olfactory bulb of the brain and, therefore, have a powerful and immediate effect on the emotions.

The importance of correct breathing

During exercise, the demand for oxygen in muscle cells goes up, so the rate and depth of breathing increases. The breathing rate can more than double during vigorous exercise. Correct breathing is very important as it ensures that all the body's cells receive an adequate amount of oxygen and dispose of enough carbon dioxide to enable them to function efficiently. It is important to note that breathing affects both our physiological and psychological state. Deep breathing exercises can help to increase the vital capacity and function of the lungs.

Common pathologies of the respiratory system

Asthma

This condition presents as attacks in which the patient experiences shortness of breath and difficulty in breathing due to spasm or swelling of the bronchial tubes. This is caused by hypersensitivity to allergens such as pollens of various plants, pet hair, dust mites and various proteins in foodstuffs such as shellfish, eggs and milk. Asthma may be made worse by exercise, anxiety, stress or smoking. It can run in families and may also be associated with hayfever and eczema.

> ### In practice
> Always obtain a detailed history during the consultation stage, specifically the triggers that bring on an asthma attack. If the client has a history of allergies, ensure they are not allergic to any preparations you are proposing to use.
>
> Position the clients according to their individual comfort, usually in a semi-reclined position. The client should have their medications to hand, in case of an attack.

Bronchitis

This is a chronic or acute inflammation of the bronchial tubes. Chronic bronchitis is common in smokers and may lead to emphysema (which is caused by damage to the lung's alveolar structure). Acute bronchitis can result from recent cold or flu.

Cancer of the lung

This may be caused by chronic inhalation of pollutants, such as cigarette smoke and asbestos fibres. Usually, there are no symptoms in the initial stages and it is often detected only in the advanced stages. Late symptoms include chronic cough, hoarseness, difficulty in breathing, chest pain, blood in sputum, weight loss and weakness.

Emphysema

This is a chronic obstructive pulmonary disease (COPD) in which the alveoli of the lungs break down, reducing the surface area for the exchange of oxygen and carbon dioxide. Severe emphysema causes breathlessness which is made worse by infection. It is commonly associated with chronic bronchitis, smoking and advancing age.

Hayfever

This is an allergic reaction involving the mucous passages of the upper respiratory tract and the conjunctiva of the eyes, caused by pollen or other allergens. It causes blocked or streaming nose, sneezing and itchy, watery eyes.

Pleurisy

This is an inflammation of the pleura of the lung. It presents on breathing deeply as an intense stabbing pain over the chest. The patient has difficulty breathing, so respiration is shallow and rapid, and fever is present. Pleurisy may develop as a complication of pneumonia, tuberculosis or trauma to the chest.

Pneumonia

Pneumonia is a bacterial infection of the lung in which the alveoli become filled with inflammatory cells and the lung becomes solid. Symptoms include fever, malaise and headache, together with a cough and chest pain.

Rhinitis

This condition is the inflammation of the mucous membrane of the nose, causing a blocked or runny nose. It may be caused by a virus infection, such as a cold, or by an allergic reaction.

Sinusitis

This condition involves inflammation of the paranasal sinuses. It is usually caused by a virus, such as a cold, or can result from bacterial infection or allergy. Nasal congestion blocks the opening of the sinus into the nasal cavity, causing build-up of pressure in the sinus. The condition presents with nasal congestion followed by a mucous discharge from the nose. The pain is located in a specific area, depending on the sinuses affected. If the frontal sinuses are affected, the major symptom is a headache over one or both eyes. If the maxillary sinuses are affected, one or both cheeks hurt and the patient may experience a feeling like toothache in the upper jaw.

Stress

Stress can be defined as any factor which affects physical or emotional health. Examples of excessive respiratory stress include exacerbation of asthma and frequent colds.

Tuberculosis (TB)

This infectious disease is caused by the bacteria (bacillus) *Mycobacterium tuberculosis*. Transmission of tuberculosis is usually via droplet infection, and hence the most common site of initial infection is the lungs. Infection can also result from drinking unpasteurised milk from infected cows. TB is characterised by the formation of nodules in the body tissues. Symptoms include coughing, sneezing, night sweats, fever, weight loss and blood in the sputum. Enlarged lymph nodes can also be an indication of TB. Immunisation against TB is possible, using the BCG (Bacillus Calmette-Guérin) vaccine.

Interrelationships with other systems

The respiratory system

The respiratory system links to the following body systems.

Cells and tissues

Squamous and ciliated are types of simple epithelium that line the respiratory system.

Skin

Oxygen that is absorbed through the respiratory process is carried to the skin via capillaries to facilitate cell renewal.

Skeletal

The bones of the thorax (sternum, ribs and 12 thoracic vertebrae) provide vital protection for the organs of respiration (heart and lungs).

Muscular

The mechanism of respiration uses the combined action of the diaphragm and the intercostal muscles.

Circulatory

Blood transports oxygen that is breathed into the lungs around the body to the cells and transports carbon dioxide from the cells to the lungs to be exhaled.

Nervous

Breathing is an involuntary response that results from the stimulation of the respiratory centre in the medulla and the pons of the brain.

Endocrine

The hormone adrenaline, produced by the adrenal glands, is released into the bloodstream to change the rate of the breathing when the body is under stress.

Digestive

The mouth and the pharynx link the respiratory and digestive systems.

Key words

Adenoids: lymphoid tissue at the back of the nasopharynx

Alveoli: tiny air sacs of the lungs which allow for rapid gaseous exchange

Bronchi: two short tubes which lead to and carry air into each lung

Bronchioles: small branches into which a bronchus divides

Chemoreceptors: nerve cells found in the aorta and the carotid arteries that send impulses to the respiratory centre in the medulla oblongata of the brain about the levels of oxygen and carbon dioxide in the blood

Cilia: tiny hairs that protect the nasal passageways and other parts of the respiratory tract by filtering out dust

Diaphragm: the chief muscle of respiration; a dome-shaped muscular partition that separates the thoracic cavity from the abdominal cavity

Diffusion: the movement of a fluid from an area of higher concentration to an area of lower concentration (gas exchange in the lungs)

Epiglottis: a flap of tissue covering the glottis that prevents food from 'going down the wrong way'

Expiration: the act of expelling air out of the lungs

External respiration: the process by which external air is drawn into the body in order to provide the lungs with oxygen, and used carbon dioxide is expelled from the body

Glottis: part of the larynx consisting of the vocal cords and the slit-like opening between them

Inspiration: the act of drawing air into the lungs

Intercostal muscles: muscle groups situated between the ribs and that contribute to the respiration

Internal respiration: gas exchange between the blood and the tissues throughout the body

Larynx: a tube-shaped organ in the neck that contains the vocal cords

Lungs: paired, cone-shaped, spongy organs situated in the thoracic cavity on either side of the heart that facilitate gas exchange

Mucous membrane: a sheet of tissue (or epithelium) lining all body channels that communicate with the air

Nasopharynx: the upper part of the pharynx, connecting with the nasal cavity above the soft palate

Nose: the organ of smell

Olfaction: sense of smell

Olfactory bulb: area of the brain that perceives olfaction

Olfactory cells: sensory nerve cells that are adapted for sensing smell

Pharynx: a membrane-lined cavity behind the nose and mouth, connecting them to the oesophagus

Pleura: a pair of serous membranes lining the thorax and enveloping the lungs

Sinuses: air-filled spaces located within the maxillary, frontal, ethmoid and sphenoid bones of the skull

Trachea: the windpipe

Revision summary

The respiratory system

- The respiratory organs include the **nose**, **nasopharynx**, **pharynx**, **larynx**, **trachea**, **bronchi**, **bronchioles** and **lungs**.

- The respiratory organs act with the cardiovascular system to supply oxygen and remove carbon dioxide from the blood.

- The **nose** is lined with **cilia** and **mucous membrane**, and is adapted for warming, moistening and filtering air, and detecting smells.

- Smell is perceived by specialised **olfactory cells** which connect directly with the olfactory bulb in the brain.

- The **pharynx** or throat connects the nasal cavity to the larynx.

- As well as providing an air passage between the nasal cavity and **larynx**, the **pharynx** also serves as a food passage for the digestive system.

- The **larynx** is a short passage that connects the pharynx with the trachea and contains the vocal cords.

- The **trachea** (windpipe) is made up mainly of cartilage and passes down into the thorax to connect the larynx with the bronchi, which pass into the lungs.

- The **lungs** are situated in the thoracic cavity on either side of the heart.

- Internally, the lungs consist of tiny air sacs called **alveoli** which provide a very large surface area for the exchange of the gases oxygen and carbon dioxide.

- The interchange of gases occurs as a result of simple **diffusion**.
  - During **inhalation**, oxygen is taken in through the **nose** and **mouth**, along the **trachea** and **bronchi** to the **lungs**, where it diffuses through a thin film of moisture that lines the **alveoli**.
  - Oxygen then diffuses across the permeable membrane surrounding the **alveoli** to be taken up by the red blood cells. Oxygen-rich blood is carried to the heart and pumped to the cells of the body.
  - Carbon dioxide, collected from respiring cells, diffuses from the capillary walls into the **alveoli**, passes through the **bronchi** and **trachea**, and is exhaled through the **nose** and **mouth**.

- Air is moved in and out of the lungs by the combined action of the **diaphragm** and the **intercostal muscles**.
  - During **inspiration**, the combined contraction of the **diaphragm** and the **external intercostals** increases the volume of the **thoracic cavity**, which decreases the pressure inside the thorax so that air enters the lungs.
  - The process of **expiration** is passive and is brought about by the relaxation of the **diaphragm** and the **external intercostals**, and the elastic recoil of the lungs.

- **Modified respiratory movements** are needed for talking (speech), singing, sighing, laughing, coughing, crying, hiccups, sneezing and yawning.

Test your knowledge questions

Multiple choice questions

1 Which of the following is a function of the respiratory system?
   a produces speech
   b detects smell
   c exchanges oxygen and carbon dioxide
   d all of the above

2 Which of the following statements is **false**?
   a The pharynx serves as a food and air passage.
   b The larynx contains the vocal cords.
   c The pharynx provides a passageway between the larynx and the bronchi.
   d The bronchi subdivide into bronchioles in the lungs.

3 Which of the following opens into the nasopharynx so that air pressure inside the ear can be adjusted?
   a glottis
   b Eustachian tube
   c adenoids
   d sinuses

4 What is another name for the throat?
   a larynx
   b pharynx
   c epiglottis
   d trachea

5 The *windpipe* is the common name for which of these structures?
   a pharynx
   b epiglottis
   c larynx
   d trachea

6 What is the name of the tiny air sacs in the lungs which provide a large surface area for diffusion?
   a surfactants
   b alveoli
   c pleura
   d bronchioles

7 The trachea is made up of mainly:
   a spongy tissue
   b mucous membrane
   c cartilage
   d cilia.

8 Which of the following respiratory structures produces sound?
   a pharynx
   b larynx
   c nasopharynx
   d trachea

9 What is a typical normal breathing rate?
   a 12 to 15 breaths per minute
   b 10 to 12 breaths per minute
   c 15 to 20 breaths per minute
   d 20 to 25 breaths per minute

10 Which of the following brings about inspiration?
   a the combined relaxation of the diaphragm and the internal intercostal muscles
   b the combined contraction of the diaphragm and the external intercostal muscles
   c the combined relaxation of the diaphragm and the external intercostal muscles
   d the combined contraction of the diaphragm and the internal intercostal muscles

Exam-style questions

11 Name the sticky fluid that prevents dust and bacteria from entering the lungs.   1 mark

12 State two factors that may cause the breathing rate to increase.   2 marks

13 State two muscles that become active when more air needs to be expelled from the lungs.   2 marks

14 Define these terms:
   a *external respiration*   2 marks
   b *internal respiration*.   2 marks

15 Name the two parts of the brain that control breathing.   2 marks

16 Describe the mechanism of respiration.   4 marks

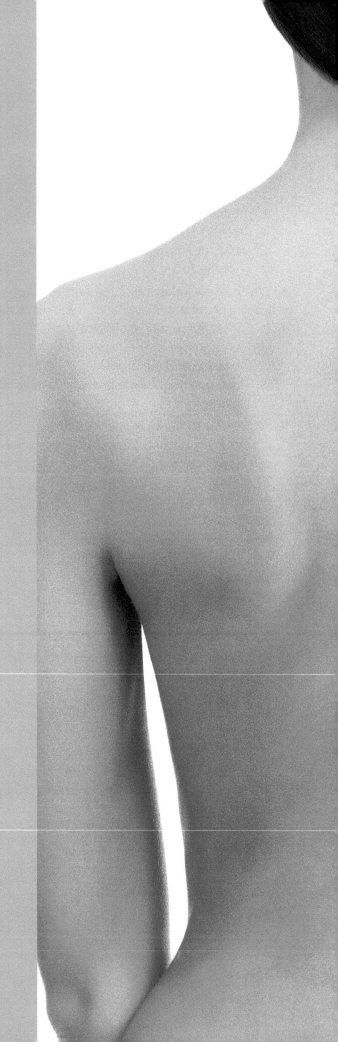

9 The nervous system

Introduction

The anatomical structures of the nervous system include the brain, spinal cord and nerves, which together form the main communication system for the body. The nervous system is the body's control centre and is, therefore, responsible for receiving and interpreting information from inside and outside the body.

The nervous system receives, interprets and integrates all stimuli to effect an appropriate response. It is also responsible for all mental processes and emotional responses, and works intimately with the endocrine system to help regulate body processes.

OBJECTIVES

By the end of this chapter you will understand:

- the functions of the nervous system
- the organisation of the nervous system
- the characteristics of nervous tissue
- the structure and function of different types of neurones
- the transmission of nerve impulses
- an outline of the principal parts of the nervous system
- the sense organs
- common pathologies of the nervous system
- the interrelationships between the nervous and other body systems.

> **In practice**
>
> Therapists need to have a comprehensive knowledge of the nervous system in order to understand the effects of treatments. Some treatments stimulate nerves, others have the ability to relax the client.
>
> Having knowledge of the nervous system can also help therapists to understand the effects of stress on the body.
>
> Therapists communicate with their clients through their respective nervous systems.

Functions of the nervous system

Neurology is the science of nerves and the nervous system, and especially of the diseases that affect them.

The nervous system has three main functions:

1 It **detects changes** both within the body (the internal environment) and outside the body (the external environment).

2 It **analyses sensory information**, stores some data and makes decisions about how to respond. This is called **integration**.

3 It may **respond** to stimuli by initiating muscular contractions or glandular secretions.

Organisation of the nervous system

The nervous system has two main parts which possess unique structural and functional characteristics:

1 The **central nervous system (CNS)** – this is the main control system and it consists of the brain and the spinal cord.

2 The **peripheral** (pur-rif-fur-ral) **nervous system (PNS)** – this system can be subdivided into:
   - the **somatic** nervous system
   - the **autonomic** (aw-toe-nom-ik) nervous system.

Peripheral nervous system (PNS)

Somatic nervous system

This contains 31 pairs of spinal nerves and 12 pairs of cranial nerves, and governs the impulses from the CNS to the skeletal muscles.

Automatic nervous system

This supplies impulses to smooth muscles, cardiac muscle, skin, special senses, proprioceptors (sensory nerve endings located in muscles and tendons that transmit information to coordinate muscular activity), organs and glands. The autonomic nervous system consists of:

- the **sympathetic** system
- the **parasympathetic** system.

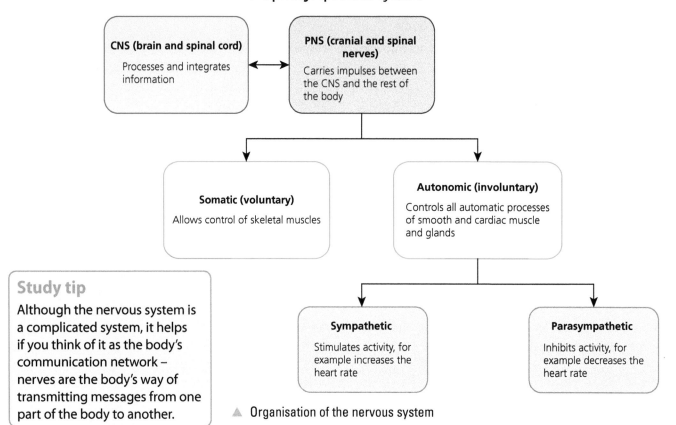

CNS (brain and spinal cord)
Processes and integrates information

PNS (cranial and spinal nerves)
Carries impulses between the CNS and the rest of the body

Somatic (voluntary)
Allows control of skeletal muscles

Autonomic (involuntary)
Controls all automatic processes of smooth and cardiac muscle and glands

Sympathetic
Stimulates activity, for example increases the heart rate

Parasympathetic
Inhibits activity, for example decreases the heart rate

▲ Organisation of the nervous system

Nervous tissue

There are two types of nervous tissue – **neuroglia** and **neurones**.

Neuroglia or glial cells are a special type of connective tissue of the CNS that is designed to support, nourish and protect the neurones. Glial cells are smaller and more numerous than neurones. They are unable to transmit impulses and never lose their ability to divide by mitosis.

The functional unit of the nervous system is the neurone, a specialised nerve cell that is designed to receive stimuli and conduct impulses. The nervous system contains billions of interconnecting, impulse-conducting neurones. Neurones have two major properties:

1 **excitability** – the ability to respond to a stimulus and convert it to an electrical impulse
2 **conductibility** – the ability to transmit the impulses to other neurones, muscles and glands.

Neurones also occur in groups called ganglia outside the CNS and as single cells, known as a ganglion, in the walls of organs.

Parts of a neurone

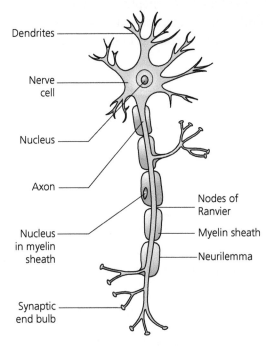

Dendrites

Nerve cell

Nucleus

Axon

Nucleus in myelin sheath

Synaptic end bulb

Nodes of Ranvier

Myelin sheath

Neurilemma

▲ The structure of a nerve cell

Although neurones vary in their shape and size, they all have three basic parts:

1 **Cell body** – this has a central nucleus that is surrounded by cytoplasm and contains standard organelles such as mitochondria and a Golgi body.

2 **Dendrites** – these are highly branched extensions of the nerve cell. These neural extensions receive and transmit stimuli towards the cell body.

3 **Axon** – this is long, single nerve fibre extending from the cell body. The function of the axon is to transmit impulses away from the cell body.

Other parts of a neurone's structure are discussed below.

Myelin (*my-a-lyn*) sheath

This is a fatty, insulating sheath that covers the axon. Its function is to insulate the nerve and accelerate the conduction of nerve impulses along the length of the axon. The myelin sheath is produced by Schwann cells (large flat cells containing a nucleus and cytoplasm) which wrap themselves around the axon in a spiral fashion, layer after layer.

KEY FACT

Thick myelinated nerve fibres allow nervous signals to be transmitted very quickly, for example pain fibres. In contrast, hot and cold receptor fibres are non-myelinated and their signals are transmitted more slowly.

Neurilemma

This is a delicate membrane that surrounds the axon and consists of a layer of one or more Schwann cells. The neurilemma plays an important role in the regeneration of PNS nerve fibres.

Nodes of Ranvier

The nodes of Ranvier are gaps in the myelin sheath situated at intervals of 2–3mm along the length of the axon. During neural activity, impulses jump from one node to the next, resulting in a faster rate of conduction.

Synapse

This is the tiny gap between one neurone and the next, across which nerve impulses have to pass. When a nerve impulse reaches a synapse, a chemical called a neurotransmitter is released. This diffuses across the gap and triggers an electrical impulse in the next neurone.

Synaptic cleft

This is the name of the space between neurones at a synapse, also called a synaptic gap.

Axon terminals

These are also called synaptic ends, bulbs, or feet.

The ends of the axons have bulb-like structures containing sacs (synaptic vesicles) that store neurotransmitters. These are chemicals that facilitate the transmission of impulses between neurones across synapses.

Neurotransmitters

The two most common neurotransmitters released by neurones of the autonomic nervous system are:

● **Acetylcholine** – a neurotransmitter which causes muscles to contract, activates pain responses and regulates endocrine and rapid eye movement (REM) sleep functions.

- **Norepinephrine** (noradrenaline) – a neurotransmitter designed to mobilise the brain and body for action. Norepinephrine release is lowest during sleep, rises during wakefulness, and reaches a higher level during situations of stress or danger, such as in the so-called fight-or-flight response.

Types of neurones

There are three types of neurones, as shown in Table 9.1.

Table 9.1 Types of neurones

Sensory/ afferent neurones	Receive stimuli from sensory organs and receptors, and transmit the impulse to the spinal cord and brain Sensations transmitted by the sensory neurones include heat, cold, pain, taste, smell, sight and hearing
Motor/efferent neurones	Conduct impulses away from the brain and the spinal cord to muscles and glands in order to stimulate them into carrying out actions
Association (mixed) neurones	Link sensory and motor neurones, helping to form the complex pathways that enable the brain to interpret incoming sensory messages, decide on what should be done and send out instructions along motor pathways to keep the body functioning properly

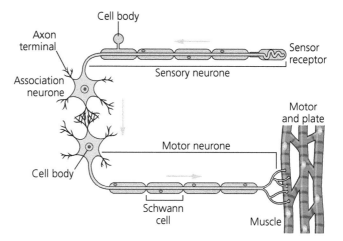

▲ A simple nerve pathway

The transmission of nerve impulses

Neurones are responsible for neurotransmission; the conduction of electrochemical impulses throughout the nervous system. Neurone activity is triggered by:

- **mechanical** stimuli – touch and pressure
- **thermal** stimuli – heat and cold
- **chemical** stimuli – from external chemicals or from chemicals released by the body, such as histamine.

Nerve impulses are transmitted electrochemically, by the movement of charged chemicals called ions into and out of the cell. At rest, positive potassium ions are located inside the nerve cell body and positive sodium ions are located outside the cell membrane. Other ions within the cell cause the inside of the cell to be negatively charged in comparison to the outside. The cell is said to be polarised due to the charge difference across the membrane.

When a nerve is stimulated, the membrane becomes permeable to sodium and the positively charged ions flow in, depolarising the cell. Eventually, the inside of the cell becomes positive and the outside of the membrane is negative. Then, potassium ions flow out of the cell to reverse the depolarisation until the outside of the membrane is slightly more positively charged than at rest. After this, the balance of potassium and sodium ions inside and outside the cell returns to normal.

This wave of polarisation and depolarisation conducts the nerve impulse along the neurone from dendrite to axon.

The neurones are not physically joined together. The junction where nerve impulses are transmitted from one neurone to another (or from a neurone to a muscle cell or gland) is called a synapse.

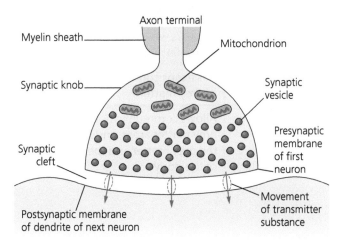

Myelin sheath
Axon terminal
Mitochondrion
Synaptic knob
Synaptic vesicle
Synaptic cleft
Presynaptic membrane of first neuron
Postsynaptic membrane of dendrite of next neuron
Movement of transmitter substance

▲ The conduction of a nerve impulse across a synapse

Nerve impulses are relayed from one neurone to another by a chemical transmitter substance. The transmitter is released by the first neurone and it diffuses across the synapse to stimulate the next neurone. Synapses can pass nerve impulses in one direction only and are important in co-ordinating the actions of neurones.

A particular kind of synapse occurs at the junction between a nerve and a muscle – this is known as a motor point and is the point at which the nerve supply enters the muscle.

How a motor impulse causes contraction of skeletal muscle

1 A motor impulse is initiated in the brain.
2 The motor stimulus travels along the length of the axon to the effector muscle.
3 The motor impulse reaches the motor point of the muscle fibres.
4 The nerve impulse is passed across the neuromuscular junction by a chemical transmitter substance that diffuses across the synapse.
5 The muscle fibres contract and effect the desired movement.

The principal parts of the nervous system

The central nervous system

The CNS consists of the brain and spinal cord and is covered by a special type of connective tissue membrane called the **meninges**. The meninges have three layers:

1 **dura mater** – this is the outer, protective, fibrous connective tissue sheath that covers the brain and spinal cord
2 **pia mater** – this is the innermost layer which is attached to the surface of organs and is richly supplied with blood vessels to nourish the underlying tissues
3 **arachnoid mater** – this provides a space for the blood vessels and circulation of cerebrospinal fluid.

Cerebrospinal fluid

This is a clear fluid derived from the blood and secreted into the inner cavities of the brain. It carries some nutrients to the nerve tissue and takes waste away, but its main function is to protect the CNS by acting as a shock absorber for the delicate nervous tissue.

The brain

The brain is an extremely complex mass of nervous tissue that lies within the skull. It is the main communication centre of the nervous system and its function is to analyse the nerve stimuli received and co-ordinate the correct responses. The main parts of the brain include the cerebrum, thalamus, cerebellum and brainstem.

Like the spinal cord, the brain is mainly made of grey matter and white matter arranged in distinct layers.

Grey matter and white matter

The differences between grey and white matter are outlined in Table 9.2.

Table 9.2 Grey matter and white matter

Grey matter	White matter
Primarily associated with processing and cognition	Acts as a relay to co-ordinate communication between different brain regions
In the cortex of the brain, the place where all the higher mental processing takes place	In the brain, connects the various parts of the cortex so that information can be transported for further processing and integration
Occupies 40% of the brain	Fills 60% of the brain
Composed of nerve cell bodies	Composed of myelinated nerve fibre tracts
Has no myelin sheath	Myelinated
Has a grey colour because of the grey nuclei in the cells	Appears white due to the presence of myelin
Does not have extended axons	Has axons connecting different parts of grey matter
Processing is concluded in the grey matter	Allows communication to and from grey matter areas, and between the grey matter and the other parts of the body

Study tip

When trying to understand the relationship between grey matter and white matter, it can be helpful to think of a computer network: the grey matter equates to the actual computers, whereas the white matter is represented by the network cables that connect the computers together.

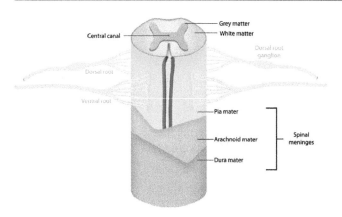

▲ Grey matter and white matter

The principal parts of the brain

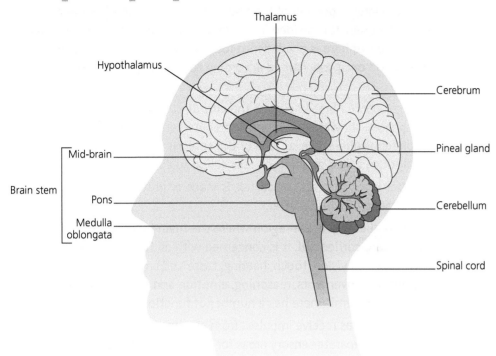

Thalamus

Hypothalamus

Cerebrum

Mid-brain

Brain stem

Pons

Medulla oblongata

Pineal gland

Cerebellum

Spinal cord

▲ Principal parts of the brain

Table 9.3 Overview of the principal parts of the brain

Part of brain	Location	Function
Cerebrum	Largest part of the brain Makes up the front and top part of the brain	Intelligence Emotions
Thalamus	Lies to either side of the forebrain	Relays sensory impulses to the cerebral cortex
Hypothalamus	Small structure lying beneath the thalamus	Governs many important homeostatic functions (hunger, thirst, temperature regulation, anger, aggression, hormones, sexual behaviour, sleep patterns and consciousness)
Pineal gland	Pea-sized mass of nervous tissue attached by a stalk to the central part of the brain Attached to the upper portion of the thalamus	Secretes melatonin Regulates circadian rhythms
Cerebellum	Cauliflower-shaped structure located at the posterior of the cranium, below the cerebrum	Co-ordinates skeletal muscles, posture and balance
Brainstem	Enlarged continuation of the spinal cord	Connects the brain with the spinal cord Contains control centres for heart, lungs and intestines

Cerebrum

This is the largest portion of the brain and makes up the front and top part of the brain. It is divided into two large cerebral hemispheres. Each cerebral hemisphere is divided into four lobes – **frontal**, **temporal**, **parietal** and **occipital**, named according to the skull bones that lie over them.

A mass of nerve fibres known as the **corpus callosum** bridges the hemispheres, allowing communication between corresponding centres in each hemisphere. The surface of the cerebrum is made up of convolutions called gyri and creases called sulci.

The outer layer, the cerebral cortex, is made of grey matter and the inner layer is made of white matter.

The cerebral cortex is the region where the main functions of the cerebrum are carried out. It is concerned with all forms of conscious activity such as vision, touch, hearing, taste and smell, as well as control of voluntary movements, reasoning, emotion and memory. The cortex of each cerebral hemisphere has a number of functional areas:

- **Sensory areas** receive impulses from sensory organs all over the body. There are separate sensory areas for vision, hearing, touch, taste and smell.

- **Motor areas** have motor connections (through motor nerve fibres) with voluntary muscles all over the body.

- **Association areas** where links are made between information from the sensory areas and remembered information from past experiences. Conscious thought then analyses these links and decisions are made, which often result in conscious motor activity controlled by motor areas.

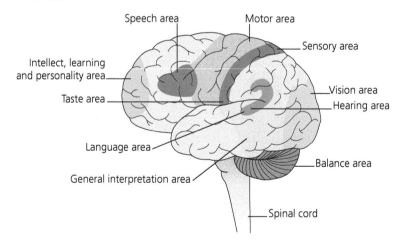

▲ Functional areas of the brain

KEY FACT

The brain requires a continuous supply of glucose and oxygen as it is unable to store glycogen, unlike the liver and muscles.

Thalamus

Lying deep in the cerebral hemispheres in each side of the forebrain are one of two egg-shaped masses of grey matter called the thalami. The thalami are relay and interpretation stations for the sensory messages (except olfaction) that enter the brain before they are transmitted to the cortex.

Hypothalamus

This small structure lies beneath the thalamus and governs many important homeostatic functions. It regulates the autonomic nervous and endocrine systems by governing the pituitary gland. It controls hunger, thirst, temperature regulation, anger, aggression, hormones, sexual behaviour, sleep patterns and consciousness.

Pineal gland

This is a pea-sized mass of nerve tissue attached by a stalk to the central part of the brain. It is located deep between the cerebral hemispheres, where it is attached to the upper portion of the thalamus. The pineal gland secretes a hormone called melatonin, which is produced from serotonin. The pineal gland is involved in the regulation of circadian rhythms. These are patterns of repeated activity that are associated with the environmental cycles of day and night, such as sleep–wake rhythms. The pineal gland is also thought to influence mood.

Cerebellum

The cerebellum is a cauliflower-shaped structure located at the posterior of the cranium, below the cerebrum. It is the brain's second largest region. Like the cerebrum, it has two hemispheres with an outer cortex of grey matter and an inner core of white matter. The cerebellum is concerned with muscle tone, the co-ordination of skeletal muscles and balance.

Brainstem

The brainstem contains three main structures:

1 **Midbrain** – this contains the main nerve pathways connecting the cerebrum and the lower nervous system. It also mediates certain visual and auditory reflexes that co-ordinate head and eye movements with vision and hearing.

2 **Pons** – this is below the midbrain and relays messages from the cerebral cortex to the spinal cord and helps regulate breathing.

3 **Medulla oblongata** – this vital part of the brain is an enlarged continuation of the spinal cord and connects the brain with the spinal cord. Control centres within the medulla oblongata include those for the heart, lungs and intestines. The medulla also controls gastric secretions and reflexes such as sweating, sneezing, swallowing and vomiting.

Blood–brain barrier

The blood–brain barrier is a selective, semipermeable (allowing some substances to pass through) wall of blood capillaries with a thick basement membrane. It prevents, or slows down, the passage of some drugs and other chemicals, and keeps disease-causing organisms such as viruses from travelling into the central nervous system via the bloodstream.

Spinal cord

The spinal cord provides the nervous tissue link between the brain and other organs of the body and is the centre for reflex actions which provide fast responses to external or internal stimuli.

The spinal cord is an extension of the brainstem, which extends from an opening at the base of the skull down to the second lumbar vertebra.

It forms a two-way information pathway between the brain and the rest of the body via the spinal nerves. It is protected by three layers of tissue called the meninges and by cerebrospinal fluid. Its function is to relay impulses to and from the brain. Sensory tracts conduct impulses to the brain and motor tracts conduct impulses from the brain.

Grey matter is located in the centre of the spinal cord. In cross section it is shaped like a butterfly, and consists of cell bodies of interneurons and motor neurons, as well as neuroglia cells and unmyelinated axons. The projections that give the grey matter its characteristic H-shape (the butterfly wings) in cross section are called horns.

White matter surrounds the grey matter in the spinal cord, acting as a relay and co-ordinating communication between different brain regions.

KEY FACT

Damage to the grey matter may lead to tingling and muscle weakness.

Reflex action

A reflex action is a rapid and automatic response to a stimulus without any conscious input from the brain.

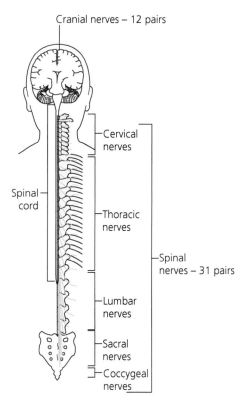

▲ The spinal cord

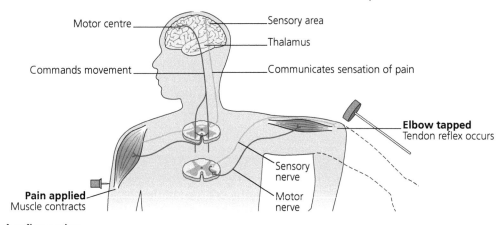

▲ A reflex action

Reflexes are essentially designed to protect the body. A reflex action, sometimes called a reflex arc, is a neural relay cycle for a quick motor response to a harmful sensory stimulus. It is mediated through the spinal cord by a sensory (afferent) neurone and a motor (efferent) neuron.

A reflex arc allows a quicker response than a situation in which a sensory impulse would have to go all the way to the brain to be analysed and the correct response selected, followed by transmission of a motor impulse all the way back from the brain.

A typical example of a reflex action is illustrated by a hand that touches a hot object and is

immediately removed. The stimulus triggers a sensory impulse which travels along the dorsal root to the spinal cord and a motor impulse that travels back again.

Two synaptic transmissions occur at the same time. One synapse continues the impulse along a sensory neurone to the brain, the other immediately relays the impulse to an interneuron which transmits it to a motor neurone.

The motor neurone delivers the impulse to a muscle (or gland) producing an immediate response, in this example withdrawing the hand from the hot object.

The peripheral nervous system

The PNS contains all the nerves outside of the CNS. It consists of cable-like nerves that link the CNS to the rest of the body. The PNS can be subdivided into:

- the somatic nervous system
- the autonomic nervous system.

The somatic nervous system contains:

- 31 pairs of spinal nerves (nerves originating from the spinal cord)
- 12 pairs of cranial nerves (nerves originating from the brain).

Somatic nervous system

Pairs of spinal nerves (31)

These nerves pass out of the spinal cord and each has two thin branches that link it with the autonomic nervous system. Spinal nerves receive sensory impulses from the body and transmit motor signals to specific regions of the body, thereby providing two-way communication between the CNS and the body.

Each of the spinal nerves are numbered and named according to the level of the spinal column from which they emerge. There are:

- 8 cervical nerves
- 12 thoracic nerves
- 5 lumbar nerves
- 5 sacral nerves
- 1 coccygeal spinal nerve.

Each spinal nerve is divided into several branches, forming a network of nerves or plexuses which supply different parts of the body (Table 9.4).

Table 9.4 Types of nerve plexuses

Nerve plexus	Location	Area(s) of the body it supplies
Cervical	Neck	Skin and muscles of the head, neck and upper region of the shoulders
Brachial	Top of the shoulder	Skin and muscles of the arm, shoulder and upper chest
Lumbar	Between the waist and hip	Front and sides of the abdominal wall and part of the thigh
Sacral	Base of the abdomen	Skin and muscles and organs of the pelvis
Coccygeal (cox-e-gee-al)	Base of the spine	Skin in the area of the coccyx and the muscles of the pelvic floor

Pairs of cranial nerves (12)

These nerves connect directly to the brain. Between them they provide a nerve supply to sensory organs, muscles and skin of the head and neck. Some of the nerves are mixed, containing both motor and sensory neurones, while others are either sensory or motor.

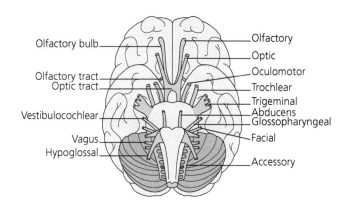

▲ The cranial nerves

Table 9.5 Types of cranial nerves

Cranial nerve	Type of nerve	Description
Olfactory	Sensory	Nerve of olfaction
Optic	Sensory	Nerve of vision
Oculomotor	Mixed nerve	Innervates both internal and external muscles of the eye and a muscle of the upper eyelid
Trochlear	Motor nerve	Smallest of the cranial nerves Innervates the superior oblique muscle of the eyeball, which helps you look upwards
Abducens	Mixed nerve	Innervates only the lateral rectus muscle of the eye, which helps you look to the side
Facial	Mixed nerve	Conducts impulses to and from several areas in the face and neck The sensory branches are associated with the taste receptors on the tongue and the motor fibres transmit impulses to the muscles of facial expression
Vestibulocochlear	Sensory nerve	Transmits impulses generated by auditory stimuli and stimuli related to equilibrium, balance and movement
Glossopharyngeal	Mixed nerve	Supplies motor fibres to part of the pharynx and to the parotid salivary glands, and sensory fibres to the posterior third of the tongue and the soft palate
Vagus	Mixed	Has branches to numerous organs in the thorax and abdomen, as well as to the neck Supplies motor nerve fibres to the muscles for swallowing, and to the heart and organs of the chest cavity Sensory fibres carry impulses from the organs of the abdominal cavity and the sensation of taste from the mouth

Cranial nerve	Type of nerve	Description
Accessory	Motor	Innervates muscles in the neck and upper back, such as the trapezius and the sternomastoid, as well as muscles of the palate, pharynx and larynx
Hypoglossal	Motor	Innervates the muscles of the tongue
Trigeminal	Mixed	Containing motor and sensory nerves that conduct impulses to and from several areas in the face and neck Controls the muscles of mastication (the masseter, temporalis and pterygoids) Has three main branches: ● the ophthalmic branch carries sensations from the eye, nasal cavity and skin of the forehead, upper eyelid, eyebrow and part of the nose ● the maxillary branch carries sensations from the lower eyelid, upper lip, gums, teeth, cheek, nose, palate and part of the pharynx ● the mandibular branch carries sensations from the lower gums, teeth, lips, palate and part of the tongue

The autonomic nervous system

This is the part of the nervous system that controls the automatic actions of smooth and cardiac muscle and the activities, of glands. It is divided into the **sympathetic** and **parasympathetic** divisions, which have complementary (balancing) responses.

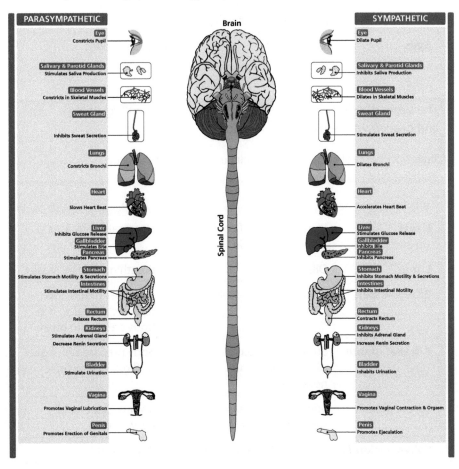

▲ The autonomic nervous system

Effects of the sympathetic and parasympathetic nervous systems

Table 9.6

Part of body	Effects of sympathetic stimulation	Effects of parasympathetic stimulation
Heart	Increases heart rate	Slows down heart rate
Lungs	Dilates bronchi to increase respiration	Slows down breathing rate
Blood vessels	Dilates blood vessels	Constricts blood vessels
Adrenal glands	Stimulates release of adrenaline	
Sweat glands	Stimulates/increases the secretion of sweat	
Digestive	Reduces peristalsis	Increases peristalsis
Liver	Increases conversion of glycogen to glucose by liver	Increases conversion of glucose to glycogen
Bladder	Relaxes bladder	Contracts bladder
Skin	Constricts arterioles so less blood flows near skin surface (skin looks pale)	
Eyes	Dilates pupils	Constricts pupils

> **Study tip**
>
> When comparing the sympathetic and parasympathetic nervous systems, think of the sympathetic nervous system as the 'day nerve' that makes us active and allows us to work, and the parasympathetic system as the 'night nerve', which permits us to rest, sleep and renew our strength for the next working day.

The sympathetic system

The sympathetic system prepares the body for expending energy and dealing with emergency situations.

Table 9.7 Effects of sympathetic stimulation on parts of the body

Part of body	Effects of sympathetic stimulation	Body response
Heart	Increases rate of contraction of cardiac muscle	Heart rate increases
Lungs	Dilates bronchi	Breathing rate increases
Blood vessels	Dilates blood vessels	Increases blood flow to muscles and so the body's ability to move
Adrenal glands	Stimulates release of adrenaline and noradrenaline	Body prepared for fight-or-flight response
Sweat glands	Stimulates/increases the secretion of sweat	Sweaty palms and nervousness
Salivary glands	Decreased secretion of saliva	Dry mouth
Digestive	Reduces peristalsis	May feel constipated

➡

Part of body	Effects of sympathetic stimulation	Body response
Liver	Increases conversion of glycogen to glucose by liver	Provides extra glucose for tissues (may get 'sugar high')
Bladder	Relaxes bladder and closes sphincter muscles	Body can go long periods without urinating
Skin	Constricts arterioles	Less blood flows near skin surface (skin looks pale)
Eyes	Dilates pupils	Improves vision

The parasympathetic system

This balances the action of the sympathetic division by working to conserve energy and create the conditions needed for rest and sleep. It slows down the body processes, except digestion and the functions of the genito-urinary system. In general, the actions of the parasympathetic system contrast with those of the sympathetic system and the two systems work in opposition to regulate the internal workings of the body.

KEY FACT

The sympathetic stimulation of the autonomic nervous system is increased by the release of the hormone adrenaline from the adrenal medulla. This is an example of the nervous and endocrine system working synergistically (together).

Table 9.8 Effects of parasympathetic stimulation on parts of the body

Part of body	Effects of parasympathetic stimulation	Body response
Heart	Slows down the rate of contraction of the cardiac muscle	Heart rate slows down, blood pressure reduces
Lungs	Constricts bronchi	Breathing rate slows down and becomes deeper
Blood vessels	Constricts blood vessels	Increased ability to sit still
Adrenal glands	No effect	
Sweat glands	No effect	
Salivary glands	Increases secretion of saliva	Stimulates digestion
Digestive	Increases peristalsis	May digest food better
Liver	No effect	
Bladder	Contracts bladder and relaxes sphincter muscles	More frequent urination
Skin	No effect	
Eyes	Constricts pupils	

 **Activity**

Imagine you are feeling stressed – perhaps due to an exam. Identify the systems of the body and describe how they are affected by stress.

KEY FACT

The sympathetic and parasympathetic nervous systems are finely balanced to ensure the optimum functioning of organs of the body.

Sense organs

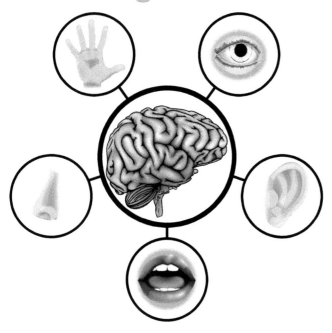

▲ The sense organs

The special senses are:

- touch – mediated through the skin
- olfaction – using the nose
- taste – using the tongue
- sight – through the eyes
- hearing – using the ears.

The special senses are all linked to the nervous system. For instance, olfactory receptors in the nose connect directly with the olfactory bulb in the brain; nerve receptors in the eye send messages to the visual cortex in the brain to enable images to be interpreted.

Skin (touch)

There are numerous sensory nerve endings in the skin that are sensitive to touch, pain and changes in temperature (see Chapter 2).

Nose (olfaction)

The specialised chemoreceptors of the olfactory nerve cells in the nose pick up information about an incoming odour and pass it to the olfactory bulb in the brain to be analysed.

Tongue (taste)

Chemosensitive receptors are concentrated on the papillae (projections of the tongue). Within the papillae are tiny taste buds, which are round in structure. These are formed from bundles of cell bodies and nerve endings of the seventh, ninth and tenth cranial nerves. The taste hairs are stimulated by food and drink that is placed in the mouth, sending messages in the form of electrical impulses to the taste area in the cerebrum for interpretation.

Eyes (sight)

The human eyes are the organs of vision and are responsible for around 80% of all the information our brain receives. The human eye is the organ which gives us the sense of sight, allowing us to learn more about the surrounding world than we can with any of the other four senses.

Functions of the eye

The functions of the eyes are to:

1 provide vision
2 provide depth perception
3 help with balance
4 produce tears.

Structure of the eye

The eye consists of three layers:

1 an outer fibrous layer
2 a middle vascular layer
3 an inner nervous layer.

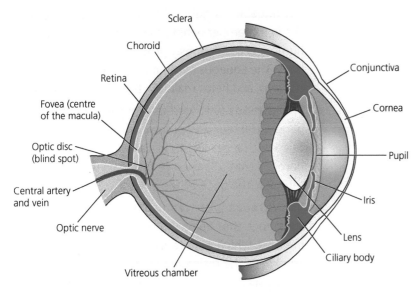

Sclera
Choroid
Retina
Fovea (centre
of the macula)
Optic disc
(blind spot)
Central artery
and vein
Optic nerve
Vitreous chamber
Conjunctiva
Cornea
Pupil
Iris
Lens
Ciliary body

▲ The structure of the eye

The outer fibrous layer

This consists of the:

- **conjunctiva** (con-junk-ti-vaa) – the mucous membrane that covers the front of the eye and lines the inside of the eyelids
- **cornea** (corr-nee-a) – the outer, transparent structure at the front of the eye that covers the iris, pupil and anterior chamber; it is the eye's primary light-focusing structure
- **sclera** (sk-ler-raa) – the sclera is the white of the eye; it is made up of a tough outer coat that protects the entire eyeball, giving it shape and rigidity
- **trabecular meshwork** – this is spongy tissue located near the cornea, through which aqueous humour flows out of the eye.

The middle vascular layer

The middle vascular layer consists of the following parts:

- **Iris** – the coloured ring of tissue behind the cornea that regulates the amount of light entering the eye by adjusting the size of the pupil. The iris is pigmented, and this determines the colour of the eyes.
- **Pupil** – the adjustable opening at the centre of the iris through which light enters the eye. The pupil regulates the amount of light entering the eyeball.
- **Choroid** (cor-roy-d) – the layer of the eye behind the retina containing blood vessels that nourish the retina by providing nutrients. It also absorbs scattered light.
- **Ciliary body** – part of the eye, above the lens, that contains processes that secrete aqueous humour. It also contains muscles that alter the shape of the lens for near or far vision. The ciliary body is closely associated with the lens and the trabecular meshwork.

KEY FACT
The main purpose of the cornea is to help focus light as it enters the eye.

In practice
It is important that contact lenses are removed for eye treatments. Since lenses rest on the cornea, they could cause irritation during treatment. There is also a risk of damage to the lenses.

KEY FACT

Aqueous humour is a watery fluid that nourishes the lens and cornea, and helps to produce intraocular pressure (the pressure inside the eyeball that helps to maintain its shape).

- **The lens** – the transparent structure suspended behind the iris that helps to focus light on the retina; it primarily provides a fine-tuning adjustment to the primary focusing structure of the eye, which is the cornea.

The inner nervous layer of the eyes

The inner nervous layer of the eyes consists of the following parts:

- The **retina** is the light-sensitive layer of tissue that lines the back of the eye and is located near the optic nerve. The purpose of the retina is to:
  - receive light that the lens has focused
  - convert the light into neural signals
  - send these signals to the brain for visual recognition.
- The retina has two types of cells that initiate these nerve impulses; the **rods** and **cones**, which are classed as photoreceptors (light-sensitive nerve cells). Rods and cones convert the light from our retinas into electrical impulses, which are sent by the optic nerve to the brain, where an image is produced.
  - Cones are the cells that are responsible for daylight vision, allowing us to see images in colour and in detail. There are three kinds, each responding to a different wavelength of light: red, green and blue. In darkness, the cones do not function at all.
  - Rods are responsible for night vision, peripheral (side) vision and detecting motion. They are more sensitive to light; therefore, they allow us to see in low light situations, but they do not allow us to see colour.
- The **macula** is the portion of the eye at the centre of the retina that processes clear straight-ahead vision. It's how we see form, colour, and detail in our direct line of sight.
- The **fovea** is the pit or depression at the centre of the macula that provides the greatest visual acuity.

- The **aqueous** (a-qui-us) **humour** is a liquid which sits in a chamber behind the cornea. Aqueous humour is continuously produced by the ciliary body. The aqueous humour nourishes the cornea and the lens, and helps maintain the shape of the eye.
- The **vitreous** (vitt-tree-us) **cavity** is located behind the lens and in front of the retina. It is filled with a gel-like fluid, called the vitreous humour that fills the eye from the lens to the retina at the back of the eye.
- The **vitreous** (vitt-tree-us) **humour** is a jelly-like substance that helps to produce the intraocular pressure which maintains the shape of the eyeball and keeps the retina in place.
- The **optic nerve** is the bundle of nerve fibres at the back of the eye that carries visual messages from the retina to the brain.

KEY FACT

As people age, the vitreous humour liquefies and shrinks, and collagen and proteins become stringy. These stringy entities float around, casting shadows on the retina. These are called 'floaters' and they may appear as specks, strings or other shapes, visible just out of the corner of the eye.

How the eyes work

> **Study tip**
>
> The human eye functions a bit like a digital camera, as shown in these steps.

1 Light enters the eye through the **cornea**, the clear front surface of the eye, which acts like a camera lens.

2 The **iris** works like the diaphragm of a camera, controlling how much light reaches the back of the eye. It does this by automatically adjusting the size of the pupil which, in this scenario, functions like a camera's aperture.

3 The eye's crystalline lens sits just behind the **pupil** and acts like an autofocus camera lens, focusing on close and approaching objects.

4 Focused by the **cornea** and the crystalline lens, the light makes its way to the retina. This is the light-sensitive lining in the back of the eye. Think of the

retina as the electronic image sensor of a digital camera. The retina's job is to convert images into electronic signals and send them to the **optic nerve**.

5 The optic nerve then transmits these signals to the **visual cortex** of the brain, which creates our sense of sight. This is like viewing camera images on a screen.

Ears (hearing)

The ear is a multifaceted organ that connects the CNS to the external head and neck. The ear structure as a whole is divided into three parts:

- outer ear
- middle ear
- inner ear.

All three parts of the ear's structure have different, but important, features that work collectively to facilitate hearing and balance.

All three parts of the ear are important for detecting sound. They work together to move sound waves from the outer part through the middle part and into the inner ear. The ears collect sounds, process them and send electrical signals about sound to your brain.

Functions of the ear

The two functions of the ears are:

- hearing
- balance.

Hearing

- Sound waves are collected by the outer ear and travel into the ear canal until they reach the eardrum.
- The eardrum passes the vibrations through the middle ear bones (ossicles) into the inner ear. Inside part of the inner ear (the cochlea), there are thousands of tiny hair cells.
- The hair cells change the vibrations into electrical signals that are sent to the brain through the hearing nerve.
- The brain experiences this as hearing and may process knowledge about what sound has been detected.

Balance

As well as providing hearing, our ears help keep us balanced.

Inside the inner ear, there are three small, hollow loops called the semicircular canals. These loops are filled with liquid and have thousands of microscopic hairs.

When you move your head, the liquid in the semicircular canals moves, too. In turn, the liquid moves the tiny hairs, which send nerve messages to your brain about the position of your head. In less than a second, your brain sends messages to the right muscles so that you keep your balance.

> **Study tip**
>
> Sometimes the liquid in the semicircular canals keeps moving after the body has stopped moving. Consider the effects of a roller coaster ride. When you stop spinning and step off the ride, the fluid in your semicircular canals is still moving. The hairs inside the canals sense movement, even though you are actually standing still. This can produce a feeling of dizziness, because your brain is getting conflicting messages from your eyes and ears and, therefore, is confused about the position of your head. Once the fluid in the semicircular canals stops moving, your brain gets a message that supports what you are seeing, and you regain your balance.

Structure of the ear

The outer ear

The outer ear is made up of:

- the auricle or pinna
- the ear canal (external auditory canal or meatus)
- the tympanic membrane (eardrum).

The function of the outer part of the ear is to collect sound. Sound waves travel through the auricle and the auditory canal, a short tube that ends at the eardrum.

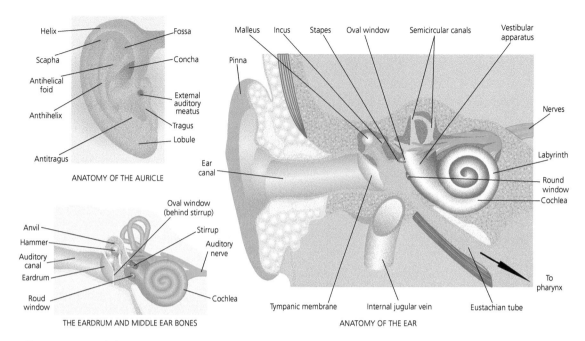

ANATOMY OF THE AURICLE

THE EARDRUM AND MIDDLE EAR BONES

ANATOMY OF THE EAR

▲ The structure of the ear

The **auricle** or **pinna** is made up of fibroelastic cartilage and skin. It has three different parts:

1 the **tragus** – the prominence over and in front of the external auditory canal or auditory meatus

2 the **helix** – a fold surrounding the auricle

3 the **lobe** (or lobule) – the fleshy part at the base of the ear.

The auricle is the only visible part of the ear, the rest of which is housed within the temporal bone of the skull. It is the first part of the ear to come into contact with sound. The auricle of the external ear helps collect sound waves travelling through the air and directs them into the external auditory canal.

The **external auditory canal** (also called the ear canal or external auditory meatus is a passageway that leads from the outside of the head to the tympanic membrane (eardrum) of each ear. After entering the canal, the sound waves pass to the end of the tube and cause pressure changes on the eardrum. The eardrum moves back and forth in response and thus reproduces the vibrations of the sound wave source.

The **tympanic** (tim-pan-ik) **membrane** or eardrum is a thin layer of cone-shaped tissue that separates the auditory canal from the middle ear. It facilitates hearing by transmitting sound vibrations from the air to the bones in the middle ear.

Cerumen (ear wax) is a yellow waxy secretion that is produced inside the ear canal. Cerumen has an important function in that it assists in cleaning and lubricating the ear canal, as well as protecting against bacteria, fungi and the penetration of insects.

KEY FACT

Excessive or impacted cerumen can obstruct the external auditory canal and press against the eardrum, causing conductive hearing loss.

The middle ear

The middle ear is a small air-filled cavity found between the outer ear and the inner ear. Its function is to transmit sound from the outer ear to the inner ear. It is partitioned from the outer ear by the eardrum and from the inner ear by a bony partition containing two windows: the oval window and the round window.

The middle ear contains three tiny auditory bones called **ossicles** (oss-ik-culls):

● the **hammer** or **malleus** (mall-ee-us)

● the **anvil** or **incus**

● the **stirrup** or **stapes** (stay-pees).

The bones of the middle ear are surrounded by a small cavity called the tympanic (tim-pan-ik) cavity.

The handle of the hammer is attached to the inner surface of the eardrum and, when the eardrum vibrates, it causes the hammer to move. The hammer hits the anvil, which in turn hits the stapes, which is attached to the oval window, a membrane covered opening. The oval window then transmits the sound wave into the inner ear.

The **Eustachian** (yoo-stay-shun) **tube**, also known as the auditory tube, is part of the middle ear and links to the nasopharynx. It controls the pressure within the middle ear, making it equal with the air pressure outside the body.

The inner ear

The inner ear (labyrinth) is entirely enclosed within the temporal bone. The parts of the inner ear include:

- The **cochlea** (cock-lee-er) – the portion of the inner ear that results in the final detection of sounds within the CNS, transforming sounds into signals that get sent to the brain where the signals are interpreted as hearing.

- The **semicircular canals** – three tiny, fluid-filled tubes that help with balance. When the head moves around, the liquid inside the semicircular canals sloshes about and moves the tiny hairs that line each canal.

- The **vestibule** – a bony chamber, located between the cochlea and the semicircular canals, which contains membranous structures that serve both hearing and balance. The vestibule's main function is to detect changes in gravity and linear acceleration.

KEY FACT

The semicircular canals are lined with tiny hairs called cilia. The fluid contained in the membranous labyrinth of the inner ear is called endolymph.

When the head moves, it causes the endolymph to move the cilia. The actions of the cilia are relayed to the brain, which senses this as motion.

How the ear works

The process of hearing sounds happens like this:

1 Sound waves enter the ear canal and cause the eardrum to vibrate.

2 Vibrations pass through three connected bones (malleus, incus, stapes) in the middle ear.

3 These vibrations set fluid in motion in the inner ear.

4 Moving fluid bends thousands of delicate hair-like cells, which convert the vibrations into nerve impulses.

5 Nerve impulses are carried to the brain by the auditory nerve.

6 These impulses are processed in the auditory cortex located in the temporal lobes of the cerebral cortex in brain, which are situated above the ears, and we experience hearing.

The eyes and ears are such important sense organs, constantly collecting information for the brain to interpret, that their malfunction can have a significant effect on health and wellbeing.

In practice

For client comfort and wellbeing the eyes should be closed during some treatments. Treatments around the eyes could cause stinging and irritation if the eyes are not protected.

Care should be taken if a client wears a hearing aid to prevent product entering the ears, where it could cause the equipment to malfunction.

Activity

Research the following disorders that affect the eyes and ears. Discuss the effects these disorders may have on sight and hearing, and on the overall wellbeing of the person (consider, for example, the complete loss of vision and hearing).

Disorders of the eye

- Blepharitis
- Cataract
- Conjunctivitis
- Glaucoma

Disorders of the ear

- Glue ear
- Labyrinthitis
- Meniere's disease
- Tinnitus
- Vertigo

Common pathologies of the nervous system

Anxiety

Anxiety involves experiencing feelings of fear and worry. As an illness, it can vary from mild to severe stress and can include panic attacks and severe phobias that can be socially, psychologically and physically disabling.

It presents with feelings of dread and is associated with palpitations, rapid breathing, sweaty hands, tremor (shakiness), dry mouth, indigestion, sensations of butterflies in the stomach, occasional diarrhoea and generalised aches and pains in the muscles. Some of these signs are similar to features of mild-to-moderate agitated depression.

The causes of anxiety include genetic and behavioural predisposition, a traumatic experience or physical illness such as hyperthyroidism.

Bell's palsy

This is a disorder of the seventh cranial nerve (facial nerve) that results in paralysis on one side of the face. The disorder usually comes on suddenly and is commonly caused by inflammation around the facial nerve as it travels from the brain to the exterior. It may be caused by pressure on the nerve due to a tumour, injury to the nerve, infection of the meninges or inner ear, or dental surgery. Diabetes, pregnancy and hypertension are other causes.

The condition may present with a drooping of the mouth on the affected side, due to flaccid paralysis of the facial muscles. There may also be difficulty in puckering the lips due to paralysis of the orbicularis oris muscle. Other symptoms include:

- diminished or complete loss of sense of taste, if the nerve has been affected proximal to the branch which carries taste sensations
- difficulty tightly closing the eye and creasing the forehead
- inability to puff out the cheeks and problems with food getting caught between the teeth and cheeks, if the buccinator muscle is affected
- excessive tearing from the affected eye
- pain near the angle of the jaw and behind the ear.

Between 80–90% of individuals recover spontaneously and completely in around one to eight weeks. Corticosteroids may be used to reduce the inflammation of the nerve.

Cerebral palsy

This condition is caused by damage to the CNS of the baby during pregnancy, delivery or soon after birth. The damage to the brain could be due to bleeding, lack of oxygen or other injuries. The signs and symptoms of this condition depend on the area of the brain that is affected, but may include some or all of the following:

- speech impairment and difficulty swallowing
- learning difficulties
- spastic muscle tone, making co-ordinated movements difficult
- hyperexcitable muscles; small movements, touch and emotional stress can increase muscle spasticity
- abnormal posture and gait due to muscle spasticity
- abnormal involuntary movements of the limbs that may be exaggerated on voluntarily performing a task
- general muscle weakness
- seizures
- hearing and visual impairment.

Depression

Symptoms include lowered mood, changes in appetite and sleep pattern, lack of concentration, interest and enjoyment, constipation and loss of libido. There may also be suicidal thinking, death wish or active suicide attempts.

Depression can be the result of chemical imbalance, usually related to serotonin and noradrenalin. The causes of depression can be endogenous (no apparent external cause but thought to be linked to genetic predisposition), physical illness or disability, or emotional trauma, such as bereavement or loss of relationship.

People who are depressed may present in a wide variety of sometimes opposing ways. Behaviours may include:

- being tearful, or restless, or irritable
- slowness in speech, thinking and remembering
- comfort eating or loss of appetite
- difficulty maintaining eye contact and social withdrawal
- difficulty sleeping or sleeping a lot
- excessive use of drugs or alcohol
- self-harm.

The severity of depression can vary from mild to psychotic, manifested by hallucinations, delusions, paranoia and other thought disorders.

Epilepsy

This is a neurological disorder which makes the individual susceptible to recurrent and temporary seizures. Epilepsy is a complex condition and classifications of types of epilepsy are not definitive. Types of epilepsy are:

- **Generalised** – this may take the form of major tonic–clonic seizures (formerly known as *grand mal*). At the onset, the patient falls to the ground unconscious with their muscles in a state of spasm (tonic phase). This is followed by convulsive movements (clonic phase). The patient may bite their tongue and urinary incontinence may occur. Movements gradually cease and the patient may rouse in a state of confusion, complaining of a headache or may fall asleep.
- **Partial** – this may be idiopathic (of unknown cause) or a symptom of structural damage to the brain. In one type of partial idiopathic epilepsy, often affecting children, seizures may take the form of absences (formerly known as *petit mal*), in which there are brief spells of unconsciousness lasting for a few seconds. The eyes stare blankly and there may be fluttering movements of the lids and momentary twitching of the fingers and mouth. This form of epilepsy seldom appears before the age of three or after adolescence. It often subsides spontaneously in adult life but may be followed by the onset of generalised or partial epilepsy.
- **Focal** – this is partial epilepsy due to brain damage (either local or from a stroke). The nature of the seizure depends on the location of the damage in the brain. In a Jacksonian motor seizure the convulsive movements may spread from the thumb to the hand, arm and face.
- **Psychomotor** – this type of epilepsy is caused by a dysfunction of the cortex of the temporal lobe of the brain. Symptoms may include hallucinations of smell, taste, sight and hearing. Throughout an attack the patient is in a state of clouded awareness and afterwards may have no recollection of the event.

In practice

Always refer to the client's GP regarding their type of epilepsy and nature of seizures. If they are on controlled medication, the chances of a seizure are minimal; however, caution is advised due to the complexity of this condition. Avoid any form of electrotherapy.

Headache

This is a pain affecting the head, excluding facial pain. It can result from diseases of the ear, nose and throat, such as sinusitis. Eye problems, which can often be corrected by glasses, may also cause headache. Types of headaches include:

- **Simple headache** – this may occur at times of stress, during menstruation, the day after heavy alcohol consumption and as a symptom of a cold or flu. These are transient and normally settle spontaneously, or after simple analgesia.
- **Chronic headaches** – these are daily headaches, often caused by tension. The pain can be severe and disabling. Pain can affect the front of the head, be located behind the eyes or may be felt in the whole head. The client sometimes describes the pain as being like a tight band around the head.
- **Cervical spine headache (cervicalgia)** – this is normally in the back and sides of the head and can present with neck pain.
- **Migraine headache** – this is a specific form of headache, usually unilateral (affecting one side of the head), associated with nausea or vomiting, and may be accompanied by visual disturbances such as scintillating light waves or zigzag lines.
- **Intracranial** (inside brain) **disease headaches** – these are headaches caused by diseases such as brain tumours. They can present with nausea and vomiting, and may cause other neurological signs and symptoms.

Herpes zoster (shingles)

This is an infection by the chicken pox virus of the sensory nerves. Lesions resemble herpes simplex, with erythema and blisters along the lines of the nerves.

The most commonly affected areas are the back and upper chest wall. This condition is very painful due to acute inflammation of one or more of the peripheral nerves. Severe pain may persist at the site of shingles for months or even years after apparent healing of the skin.

Meningitis

This is an inflammation of the meninges due to infection by viruses or bacteria. Meningitis presents with an intense headache, fever, loss of appetite, intolerance to light and sound, and rigidity of muscles, especially those in the neck. In severe cases there may be convulsions, vomiting and delirium, leading to death. The different types of meningitis are:

- **Bacterial meningitis** – bacterial infection of the meninges is treated with large doses of antibiotics.
- **Meningococcal meningitis** – this specific form of bacterial meningitis is caused by a bacterium called *Neisseria meningitidis* and is characterised by a non-blanching haemorrhagic rash that may occur anywhere on the body. The symptoms appear suddenly and the infection may very quickly culminate in life-threatening septicaemia.
- **Viral meningitis** – this does not respond to drugs but normally has a relatively benign prognosis.

Migraine

This is a specific form of headache, usually unilateral (affecting one side of the head), associated with nausea or vomiting, and visual disturbances such as scintillating light waves or zigzag lines. The client may experience a visual aura (visual disturbance) before an attack actually happens. This is usually called a classical migraine. There are other types of migraine:

- **ophthalmoplegic migraine** – this causes painful, red and watery eyes
- **neuropathic migraine** – causes one-sided paralysis and weakness of the face and body
- **abdominal migraine** – this can affect children with recurring attacks of abdominal pain, sometimes accompanied by nausea and vomiting.

Depending on severity, migraines can be treated with simple analgesia (pain medication) or more specialised anti-migraine medication.

> ### In practice
> Stress and tension can increase the frequency of migraines. Women are more likely to have migraines during premenstrual periods, when they are taking the contraceptive pill, during the menopause or when starting hormone replacement therapy (HRT).
>
> Clients should avoid treatments during acute migraine attacks.

Motor neurone disease (MND)

This is a progressive degenerative disease of the motor neurones of the nervous system. It tends to occur in middle age and causes muscle weakness and wasting.

Multiple sclerosis (MS)

This is a disease of the CNS in which the myelin (fatty) sheath covering the nerve fibres is destroyed and various functions become impaired, including movement and sensations. Multiple sclerosis is characterised by relapses and remissions. It can present with blindness or reduced vision, and can lead to severe disability within a short period. It can also cause incontinence, loss of balance, tremor and speech problems. Depression and mania can occur.

Myalgic encephalomyelitis

Myalgic encephalomyelitis (ME, or chronic fatigue syndrome) is characterised by extreme disabling fatigue that has lasted for at least six months and is made worse by physical or mental exertion and is not resolved by bed rest. The symptom of fatigue is often accompanied by some of the following: muscle pain or weakness, poor co-ordination, joint pain, slight fever, sore throat, painful lymph nodes in the neck and armpits, depression, inability to concentrate and general malaise.

People in any age group can contract ME, but recently a higher incidence has been reported in children and adolescents.

Neuralgia

Neuralgia presents as attacks of pain along the entire course or branch of a peripheral sensory nerve. A common example is trigeminal neuralgia, which affects the trigeminal nerve in the face.

Neuritis

This is inflammation or disease of a single nerve or several nerves. There are different causes, such as infection, injury or poison. Neuritis causes pain along the length of the nerve and/or loss of the use of structures supplied by the nerve.

Parkinson's disease

This disease is caused by damage to the grey matter of the brain, known as basal ganglia. It causes involuntary tremors of limbs with stiffness and a shuffling gait. The face lacks expression and movements are slow. Clients may suffer from depression, confusion and anxiety.

Sciatica

This is lower back pain which can also affect the buttock and thigh. On occasions, it radiates to the leg and foot. In severe cases, it can cause numbness and weakness of the lower limb. It can result from prolapse of the discs between the spinal vertebrae, tumour or blood clot (thrombosis). Diabetes or heavy alcohol intake can also produce symptoms of sciatica. This condition tends to recur and may require strong analgesia or surgery in severe cases.

Stress

Stress can be defined as any factor that affects physical or emotional wellbeing. Signs of stress in the nervous system include anxiety, depression, irritability, headache, back pain and excessive tiredness.

Interrelationships with other systems

The nervous system

The nervous system links to the following body systems.

Cells and tissues

Nervous tissue is a specialised type of tissue which can detect stimuli and transmit them as electrical nerve impulses.

Skin

The skin is a highly sensitive organ with many sensory nerve endings which respond to touch, temperature and pressure.

Skeletal

The skeleton provides protection for the spinal cord and the brain.

Muscular

The brain sends impulses to muscles via motor nerves in order to effect movement.

Circulatory

Blood transports vital oxygen to the nerve cells. The medulla oblongata in the brain is the control centre for the heart. The sympathetic nervous system prepares the body for activity by increasing the heart rate. The parasympathetic nervous system activates the resting heart rate.

Respiratory

Oxygen that is inhaled into the body is carried to nerve cells to enable them to function properly. Without oxygen, nerve cells become damaged and die, causing irreversible damage. The sympathetic nervous system prepares the body for activity by increasing the respiration rate. The parasympathetic nervous system activates the resting respiratory rate.

Endocrine

The endocrine system works closely with the nervous system in order to maintain homeostasis in the body.

Digestive

The nervous system influences the actions of the digestive system. The effects of the sympathetic nervous system include increased conversion of glycogen to glucose by the liver and decreased secretion of saliva. The effects of the parasympathetic nervous system include increased gastrointestinal activity and stimulated salivation.

Key words

Anvil (incus): one of the three small auditory bones located in the middle ear

Aqueous humour: watery fluid that nourishes the lens and cornea of the eye, and helps produce intraocular pressure

Auditory nerve: a bundle of nerve fibres that carries information about sound between the cochlea and the brain

Autonomic nervous system: part of the nervous system responsible for control of the bodily functions that are not consciously directed, such as breathing, heartbeat and digestive processes

Axon: a long, single nerve fibre extending from the cell body, whose function is to transmit impulses away from the cell body

Brain: the central organ of the nervous system

Brainstem: the portion of the brain that is continuous with the spinal cord and comprises the medulla oblongata, pons, midbrain, and parts of the hypothalamus

Central nervous system (CNS): part of the nervous system consisting of the brain and spinal cord

Cerebrospinal fluid: the fluid that flows in and around the brain and spinal cord, to help cushion them from injury and provide nutrients

Cerebellum: the portion of the brain that is in the back of the head, between the cerebrum and the brainstem

Cerebral cortex: the outer layer of the cerebrum that is concerned with all forms of conscious activity, such as vision, touch, hearing, taste and smell, as well as control of voluntary movements, reasoning, emotion and memory

Cerebrum: the largest and anterior part of the brain, divided into two hemispheres

Cerumen: a yellowish, waxy substance secreted into the ear canals

Choroid: a layer of the eye behind the retina that provides nutrients to the retina

Ciliary body: a circular structure located above the lens that secretes aqueous humour and also contains (ciliary) muscles that alter the shape of the lens for near or far vision

Cochlea: part of the inner ear responsible for transforming sounds into signals that get sent to the brain, where they are experienced as hearing

Conjunctiva: the mucous membrane that covers the front of the eye and lines the inside of the eyelid

Cornea: the outer, transparent structure at the front of the eye that help focus light as it enters the eye

Cranial nerves: the nerves of the brain which emerge from or enter the skull (the cranium); there are 12 pairs (accessory, abducens, facial, glossopharyngeal, hypoglossal, olfactory, optic, oculomotor, trochlear, trigeminal, vestibulocochlear and vagus)

Dendrites: short branched extensions of a nerve cell, along which impulses received from other cells at synapses are transmitted to the cell body

Endolymph: watery fluid contained within the labyrinth of the inner ear

Eustachian tube (auditory tube): a tube that links the nasopharynx to the middle ear, and controls the pressure within the middle ear

External auditory canal: also called the ear canal; a passageway that leads from the outside of the head to the tympanic membrane (eardrum) of each ear

Fovea: a depression at the centre of the macula that provides the greatest visual acuity

Ganglia/ganglion: a mass of nerve tissue existing outside the CNS

Grey matter: darker tissue of the brain and spinal cord, consisting mainly of nerve cell bodies and branching dendrites

Hammer (malleus): one of the three small auditory bones located in the middle ear

Hypothalamus: a region of the forebrain below the thalamus which co-ordinates both the autonomic nervous system and the activity of the pituitary

Incus (anvil): one of the three small auditory bones located in the middle ear

Iris: the coloured ring of tissue behind the cornea that regulates the amount of light entering the eye by adjusting the size of the pupil

Labyrinth: the rigid, bony outer wall of the inner ear in the temporal bone, consisting of three parts: the vestibule, semicircular canals, and cochlea

Lens: the transparent structure suspended behind the iris that helps to focus light on the retina

Macula: the functional centre of the retina that provides the best colour vision

Malleus (hammer): one of the three small auditory bones located in the middle ear

Medulla oblongata: the continuation of the spinal cord within the skull, forming the lowest part of the brainstem and containing control centres for the heart and lungs

Meninges: three membranes (the dura mater, arachnoid, and pia mater) that line the skull and vertebral canal and enclose the brain and spinal cord

Mixed neurone: a nerve containing both sensory and motor fibres

Motor neurone: a nerve cell which passes impulses from the brain or spinal cord to a muscle or gland

Myelin sheath: a fatty insulating sheath that covers the axon of a neurone

Neurilemma: the fine delicate membrane that surrounds the axon of a neurone

Neuroglia: a type of connective tissue of the CNS that is designed to support, nourish and protect the neurones

Neurone: nerve cell

Neurotransmitter: a chemical that is released from a nerve cell to transmit an impulse across a synapse to another nerve, muscle, tissue or organ

Optic nerve: a bundle of nerve fibres at the back of the eye that transmits visual information from the retina to the brain

Parasympathetic nervous system: part of the autonomic nervous system that controls the body at rest

Peripheral nervous system (PNS): the part of the nervous system outside of the brain and spinal cord

Photoreceptors: light-sensitive nerve cells (rods and cones)

Pineal gland: a small endocrine gland located on the back portion of the third cerebral ventricle of the brain; secretes melatonin

Pinna: the only visible part of the ear; collects sound waves

Plexuses: a network of nerves

Pons: part of the brainstem that links the medulla oblongata and the thalamus

Pupil: an adjustable opening at the centre of the iris through which light enters the eye

Reflex action: an automatic response to a stimulus that does need any conscious thought

Retina: a thin layer of tissue that lines the back of the eye on the inside; its role is to receive light that the lens has focused, convert the light into neural signals, and send these signals on to the brain for visual recognition

Sclera: the white of the eye

Semicircular canals: three tiny, fluid-filled tubes in the inner ear that help with sense of balance

Sensory neurone: a neurone that transmits nerve impulses from a sense organ towards the CNS

Somatic nervous system: part of the PNS associated with the voluntary control of body movements via skeletal muscles

Spinal cord: an extension of the brainstem which extends from an opening at the base of the skull down to the second lumbar vertebra

Spinal nerves: mixed nerves, which carry motor, sensory, and autonomic signals between the spinal cord and the body; there are 31 pairs, one on each side of the vertebral column (8 cervical, 12 thoracic, 5 lumbar, 5 sacral and 1 coccygeal)

Stapes (stirrup): one of the three small auditory bones located in the middle ear

Sympathetic nervous system: part of the autonomic nervous system that prepares the body for intense physical activity, and mediates the fight-or-flight response

Synapse: a junction between two nerve cells

Thalamus: a structure in the middle of the brain, located between the cerebral cortex and the midbrain, that works to co-ordinate several important processes, including consciousness, sleep, and sensory interpretation

Tympanic membrane (eardrum): a membrane that transfers sound vibrations from the outer ear to the middle ear

Vestibule: a bony chamber, located between the cochlea and the semicircular canals

Vitreous membrane: a transparent jelly-like fluid filling the eyeball behind the lens

White matter: the paler tissue of the brain and spinal cord, consisting mainly of nerve fibres with their myelin sheaths

Revision summary

The nervous system

- The nervous system helps regulate homeostasis and integrate all body activities by sensing changes, interpreting them and reacting to them.
- The **central nervous system** (CNS) consists of the brain and the spinal cord.
- The **peripheral nervous system** (PNS) consists of the **somatic** nervous system, consisting of the cranial and spinal nerves, and the **autonomic** (involuntary) nervous system.
- There are two types of nervous tissue:
  - **Neurone** – this is a functional unit of the nervous system. The neurone is designed to receive stimuli and conduct impulses.
  - **Neuroglia** – this is a specialised type of connective tissue that supports, nourishes and protects neurones.
- Neurones have two major properties – excitability and conductibility.
- Most **nerve cells**, or neurones, consist of a cell body, many dendrites and usually a single axon.
- There are three main types of neurones – sensory, motor and mixed.
  - **Sensory neurones** conduct impulses from receptors to the CNS.
  - **Motor neurones** conduct impulses to effectors (muscles and glands).
  - **Mixed neurones** conduct impulses to other neurones.
- The junction where nerve impulses are transmitted from one neurone to another is called a **synapse**.

- **Impulses** are relayed from one neurone to another by a chemical transmitter substance which is released by the neurone to carry impulses across the synapse to stimulate the next neurone.
- The **central nervous system** (brain and spinal cord) is covered by a protective type of connective tissue in three layers called the **meninges**.
- The parts of the brain include the **cerebrum, thalamus, hypothalamus, pituitary gland, pineal gland, cerebellum** and **brainstem**.
  - The **cerebrum** is the largest part of the brain and is concerned with all forms of conscious activity. It has sensory areas which control vision, touch, hearing, taste and smell, as well as motor areas which control voluntary movements and association areas which control reasoning, memory and emotions.
  - The **thalamus** is a relay and interpretation centre for all sensory impulses, except olfaction.
  - The **hypothalamus** controls hunger, thirst, temperature regulation, anger, aggression, hormones, sexual behaviour, sleep patterns and consciousness.
  - The **pituitary gland** is a lobed structure attached by a stalk to the hypothalamus.
  - The **pineal gland** is involved in the regulation of circadian rhythms and is thought to influence mood.
  - The **cerebellum** is concerned with the co-ordination of skeletal muscles, muscle tone and balance.
  - The **brainstem** contains the midbrain, pons and medulla oblongata.
    - The **midbrain** contains certain visual and auditory reflexes that co-ordinate head and eye movements with vision and hearing.
    - The **pons** relays messages from the cerebral cortex to the spinal cord and helps regulate breathing.
    - The **medulla oblongata** contains control centres for the heart, lungs and intestines.
- The **spinal cord** is an extension of the brainstem and its function is to relay impulses to and from the brain.

- A **reflex action** is a rapid and automatic response to a stimulus without any conscious action of the brain.
- The **PNS** contains all the nerves outside of the CNS and can be subdivided into the **somatic nervous system** and the **autonomic nervous system**.
  - The **somatic nervous system** contains **31 pairs of spinal nerves** (nerves originating from the spinal cord) and **12 pairs of cranial nerves** (nerves originating from the brain).
  - The 31 pairs of spinal nerves are **8 cervical, 12 thoracic, 5 lumbar, 5 sacral** and **1 coccygeal**.
  - Each **spinal nerve** is divided into several branches, forming a network of nerves or **plexuses** which supply different parts of the body.
  - The 12 pairs of cranial nerves connect directly to the brain. They are **olfactory, optic, oculomotor, trochlear, trigeminal, abducens, facial, vestibulocochlear, glossopharyngeal, vagus, accessory** and **hypoglossal**.
- **The autonomic nervous system** is the part of the nervous system that controls the automatic actions of smooth and cardiac muscle and the activities of glands. It is divided into the **sympathetic** and **parasympathetic** divisions.
  - The function of the **sympathetic** system is to prepare the body for expending energy and dealing with emergency situations.
  - The **parasympathetic** system balances the action of the sympathetic division by working to conserve energy and create the conditions needed for rest and sleep. It slows down the body processes, except digestion and the functions of the genito-urinary system.
- The **sense organs** include the **nose** (olfaction), **tongue** (taste), **eyes** (sight), **ears** (hearing) and **skin** (touch).
- The human eye is the organ of vision.
  - Its main function is to convert light into electrical nerve impulses, which then travel to the visual cortex in the brain where they are interpreted.
  - The eye is divided into three layers: an **outer fibrous layer, a middle vascular layer** and an **inner nervous layer**.

- The eye produces two types of gel-like fluids: **aqueous humour** and **vitreous humour**.
- The human eye functions much like a digital camera.
- Focused by the cornea and the crystalline lens, the light makes its way to the retina, which is the light-sensitive lining in the back of the eye.
- The retina acts like the electronic image sensor of a digital camera, converting images into electrical signals and sending them to the optic nerve.
- The optic nerve then transmits these signals to the visual cortex of the brain, which creates our sense of sight.
- The human ear is the organ of hearing.
  - The ear is structured in three parts (outer, middle and inner ear), that work collectively to co-ordinate hearing and balance.
  - The outer ear is made up of fibroelastic cartilage and skin, and has three different parts; the **tragus**, **helix** and **lobule**.
  - The **auricle** or **pinna** is the only visible part of the ear; it helps collect sound waves travelling through the air and directs them into the **external auditory canal** or **meatus**.
  - The **tympanic membrane**, otherwise known as the eardrum, facilitates hearing by transmitting sound vibrations from the air in the auditory canal to the bones in the middle ear.
  - **Earwax**, also known as **cerumen**, is a yellowish, waxy substance secreted in the ear canals. It assists in cleaning and lubrication of the ear canal.
  - The middle ear is a small air-filled cavity found between the outer ear and the inner ear. Its function is to transmit sound from the outer ear to the inner ear.
  - The middle ear contains three tiny auditory bones (ossicles): the **hammer** (malleus), the **anvil** (incus) and the **stirrup** (stapes) and is connected to the throat via the Eustachian tube.
  - The Eustachian tube links the nasopharynx to the middle ear. It controls the pressure within the middle ear, making it equal with the air pressure outside the body.
  - The handle of the hammer is attached to the inner surface of the eardrum and when the eardrum vibrates it causes the hammer to move. The hammer hits the anvil, which in turn hits the stapes, which then transmits the sound wave into the inner ear.
  - The **inner ear (labyrinth)** is entirely enclosed within the temporal bone.
  - The parts of the inner ear include:
    - the **cochlea** – responsible for transforming sounds into signals that get sent to the brain for hearing
    - the **semicircular canals** – three tiny, fluid-filled tubes in the inner ear that help with balance
    - the **vestibule** – a bony chamber, located between the cochlea and the semicircular canals, whose function is to detect changes in gravity and linear acceleration.
- The sequence of hearing a sound is:
  - Sound waves enter the ear canal and cause the eardrum to vibrate.
  - Vibrations pass through three connected bones (**malleus**, **incus** and **stapes**) in the middle ear; this sets fluid in motion in the inner ear.
  - Moving fluid bends thousands of cilia which convert the vibrations into nerve impulses, which are carried to the brain by the **auditory nerve**.
  - In part of the brain's cerebral cortex, the **auditory cortex**, these impulses are interpreted as hearing.

Test your knowledge questions

Multiple choice questions

1 What are the two major divisions of the nervous system?
   a the central nervous system and peripheral nervous system
   b the central nervous system and autonomic nervous system
   c the brain and the spinal cord
   d the peripheral nervous system and the brain

2 What are the three basic parts of a neurone?
   a cell body, sensory and afferent nerves
   b cell body, nucleus and axon
   c cell body, axon and dendrites
   d cell body, motor and efferent nerves

3 Where do sensory nerves send messages?
   a from the brain and spinal cord
   b to and from the brain and spinal cord
   c to the brain and spinal cord
   d none of the above

4 What is the name of the part of the brain that houses the thalamus and hypothalamus?
   a cerebrum
   b brainstem
   c cerebellum
   d medulla oblongata

5 Which part of the brain is concerned with all forms of conscious activity?
   a cerebrum
   b thalamus
   c hypothalamus
   d medulla oblongata

6 What are the connective tissue membranes that envelop the central nervous system called?
   a cerebrospinal membranes
   b meninges
   c myelin sheaths
   d synapses

7 Which part of the brain contains vital control centres for the heart, lungs and intestines?
   a hypothalamus
   b mid brain
   c medulla oblongata
   d cerebellum

8 What is the name of the junction where nerve impulses are transmitted from one neurone to another?
   a neurotransmitter
   b synapse
   c dendrite
   d axon

9 The point where the nerve supply enters the muscle is the:
   a motor impulse
   b motor transmitter
   c motor point
   d muscle fibre.

10 What type of neurone stimulates muscles to produce movement?
   a sensory
   b motor
   c afferent
   d efferent

Exam-style questions

11 Name the two parts of the central nervous system (CNS).   2 marks

12 What is the somatic nervous system and what does it consist of?   2 marks

13 What is meant by the autonomic nervous system, and what parts does it consist of? 2 marks

14 Describe the functions of these parts of a neurone:
   a myelin sheath
   b axon.   2 marks

15 Name the two most common neurotransmitters released by neurons of the autonomic nervous system.   2 marks

16 State two differences between sensory and motor nerves.   4 marks

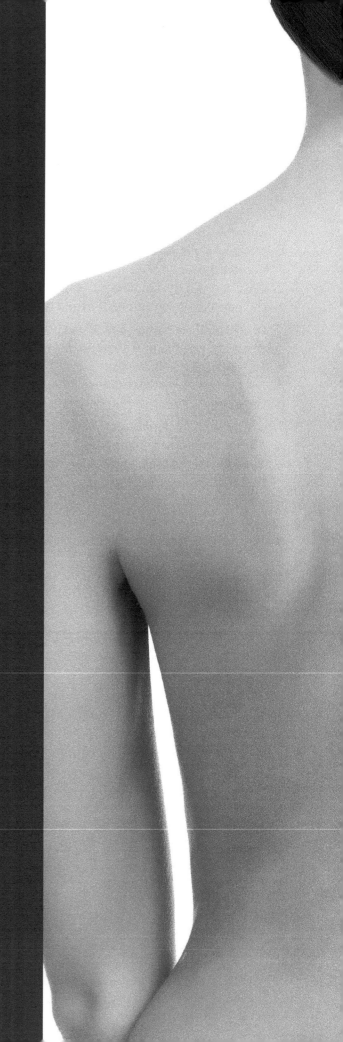

10 The endocrine system

Introduction

The endocrine system comprises a series of internal secretions called hormones and the glands that secrete them. Hormones are chemical messengers that help to regulate body processes to provide a constant internal environment. The endocrine system works closely with the nervous system. Nerves enable the body to respond rapidly to stimuli, whereas the endocrine system causes slower and longer-lasting effects.

OBJECTIVES

By the end of this chapter you will understand:

- the functions of the endocrine system
- the definition of a hormone
- the location of the main endocrine glands of the body
- the principal hormone secretions from the main endocrine glands and their effects on the body
- the natural glandular changes that occur in the body during puberty, menstruation, pregnancy and menopause
- common pathologies of the endocrine system
- the interrelationships between the endocrine and other body systems.

Functions of the endocrine system

The functions of the endocrine system are to:

- produce and secrete hormones, which regulate body activities such as growth, development and metabolism
- maintain the body during times of stress
- contribute to the reproductive process.

Hormones

A hormone is a chemical messenger or regulator that is secreted by an endocrine gland and which travels via the bloodstream to influence the activity of a destination organ. Some hormones have a slow action over a period of years, such as growth hormone from the anterior pituitary, while others have a quick action, such as adrenaline from the adrenal medulla.

The endocrine glands are ductless glands and the hormones they secrete pass directly into the bloodstream to influence the activity of another organ or gland. The main endocrine glands are as follows:

- pituitary gland
- thyroid gland
- parathyroid glands
- adrenal glands
- islets of Langerhans
- ovaries in the female
- testes in the male.

Overview of the endocrine glands

Table 10.1 Location of endocrine glands

Endocrine gland	Location
Thymus gland	Behind the sternum, between the lungs
Pituitary (pit-tu-it-tur-ree) gland	Attached by a stalk to the hypothalamus of the brain
Thyroid gland	In the neck on either side of the trachea
Parathyroid glands	Four small glands situated on the posterior of the thyroid gland
Adrenal glands	Two triangular-shaped glands which lie on top of each kidney
Pancreas (islets of Langerhans)	Situated behind the stomach between the duodenum and the spleen
Ovaries	Situated in the lower abdomen below the kidneys
Testes	Situated in the groin in a sac called the scrotum

In practice

It is important for therapists to have a comprehensive knowledge of the endocrine system in order to understand the actions of hormones and their roles in the healthy functioning of the body. Over- or under-secretion of particular hormones results in disorders and diseases. For instance, hypersecretion of the hormone testosterone in women can lead to hair growth in the male sexual pattern.

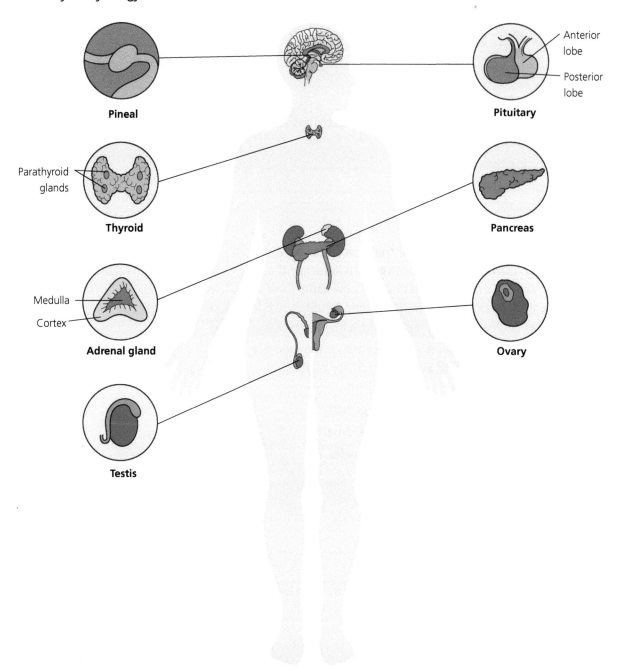

▲ The endocrine glands

The thymus gland

The thymus gland, despite containing glandular tissue and producing several hormones, is much more closely associated with the immune system than with the endocrine system. It serves a vital role in the training and development of disease-fighting T-cells or T-lymphocytes, an extremely important type of white blood cell.

Thymosin is the hormone of the thymus, and it stimulates the development of T-cells, which defend the body from potentially deadly pathogens such as bacteria, viruses and fungi.

The thymus gland, located behind the sternum and between the lungs, is only active until puberty. After puberty, the thymus starts to shrink and is replaced by fat.

The immune system fights disease and infection via T-cells that migrate to the different lymph nodes throughout the body. The T-cells start migrating to the lymph nodes once they reach full maturity in the thymus.

Lymphomas such as Hodgkin and non-Hodgkin occur when lymphocytes (small white blood cells) develop into cancers.

Though the thymus gland is only active until puberty, its double-duty function as an endocrine *and* lymphatic gland means that it plays a significant role in long-term health.

The pituitary gland

This is a lobed structure attached by a stalk to the hypothalamus of the brain. For many years the pituitary gland was referred to as the 'master' endocrine gland because it secretes several hormones that control other endocrine glands. However, the pituitary itself has a master – the hypothalamus.

The hypothalamus

The hypothalamus is a small region of the brain that is the major integrating link between the nervous and endocrine systems. Hormones of the pituitary are controlled by **releasing** or **inhibiting** hormones produced by the hypothalamus.

The hypothalamus produces releasing or inhibiting hormones as a result of stimulation in the brain. This has a cascading effect on the pituitary, which in turn produces its own hormones that stimulate other glands. For example, thyrotrophin (a releasing hormone) is produced by the hypothalamus, which signals the pituitary gland to secrete thyroid-stimulating hormone, which controls the growth and activity of the thyroid gland.

The pituitary gland consists of two main parts:

1 an anterior lobe
2 a posterior lobe.

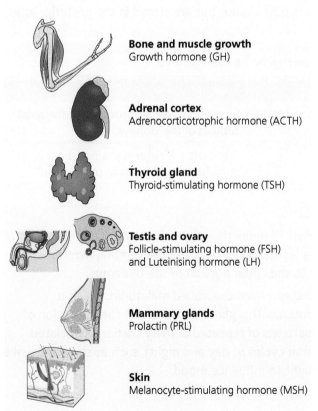

Bone and muscle growth
Growth hormone (GH)

Adrenal cortex
Adrenocorticotrophic hormone (ACTH)

Thyroid gland
Thyroid-stimulating hormone (TSH)

Testis and ovary
Follicle-stimulating hormone (FSH) and Luteinising hormone (LH)

Mammary glands
Prolactin (PRL)

Skin
Melanocyte-stimulating hormone (MSH)

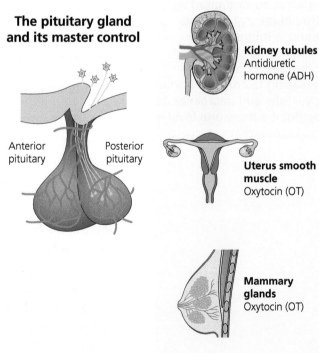

The pituitary gland and its master control

Anterior pituitary   Posterior pituitary

Kidney tubules
Antidiuretic hormone (ADH)

Uterus smooth muscle
Oxytocin (OT)

Mammary glands
Oxytocin (OT)

▲ The pituitary and its master control

Anterior lobe

The principal hormones secreted by the anterior lobe of the pituitary are shown in Table 10.2.

Table 10.2 Hormones secreted by the anterior lobe

Hormone	Effects
Thyroid-stimulating hormone (TSH)	Controls the growth and activity of the thyroid gland
Adrenocorticotrophic (add-dree-no-corr-tick-co tro-fik) hormone (ACTH)	Stimulates and controls the growth and hormonal output of the adrenal cortex
Gonadotrophic (go-nad-oo-tro-fik) hormones **The gonadotrophic hormones include:** **a) Follicle-stimulating hormone (FHS)** **b) Luteinising hormone (LH)**	Control the development and growth of the ovaries and testes In women FSH stimulates the development of the Graafian follicle in the ovary, which secretes the hormone oestrogen In men FSH stimulates the testes to produce sperm In women LH helps to prepare the uterus for the fertilised ovum In men LH acts on the testes to produce testosterone
Prolactin	Stimulates the secretion of milk from the breasts following birth
Melanocyte-stimulating hormone (MSH)	Stimulates the production of melanin in the basal cell layer of the skin

KEY FACT

Endocrine glands in the body have a feedback mechanism which is co-ordinated by the pituitary gland. This gland is influenced by the hypothalamus, which increases its output of releasing factors if a hormone level falls and decreases its output if a hormone level in the bloodstream starts to rise.

Posterior lobe

The posterior lobe of the pituitary secretes two hormones which are manufactured in the hypothalamus but are stored in the posterior lobe (Table 10.3).

Table 10.3 hormones secreted by the posterior lobe

Hormone	Function
Antidiuretic hormone (ADH)	Increases water reabsorption in the renal tubules of the kidneys
Oxytocin	Stimulates the uterus during labour and stimulates the breasts to produce milk

Pineal gland

This is a pea-sized mass of nerve tissue attached by a stalk to the central part of the brain. It is located deep between the cerebral hemispheres, where it is attached to the upper portion of the thalamus.

The pineal gland secretes a hormone called melatonin, which it synthesises from serotonin. This gland is involved in the regulation of circadian rhythms (patterns of repeated activity that are associated with the environmental cycles of day and night), such as sleep and wake cycles. It is also thought to influence mood.

Thyroid gland

The thyroid gland is found in the neck, situated on either side of the trachea and is controlled by the anterior lobe of the pituitary.

The principal secretions of the thyroid gland are shown in Table 10.4.

Table 10.4 Hormones secreted by the thyroid gland

Hormone	Effects
Triiodothyronine (T3)	The active form of thyroxine, the thyroid hormone, is triiodothyronine (T3) T3 performs essential roles in the functioning of the heart, the metabolic rate, brain development, digestive functions, the upkeep of the bones and muscle control
Thyroxine (T4)	Thyroxine (T4) is a hormone the thyroid gland secretes into the bloodstream Along with T3 it plays a crucial role in heart and digestive function, metabolism, brain development, bone health, and muscle control It affects almost all of the body's systems, which means correct thyroxine level is vital for health Both T3 and T4 regulate growth and development, and also influence mental, physical and metabolic activities
Calcitonin	Controls the level of calcium in the blood

The functions of the thyroid gland are to:

- control the metabolic rate by stimulating metabolism
- influence growth and cell division
- influence mental development
- be responsible for the maintenance of healthy skin and hair
- store the mineral iodine, which it needs to manufacture thyroxin
- stimulate the involuntary nervous system and control irritability.

The thyroid gland is controlled by a feedback mechanism. It will increase production to meet the demand for more thyroid hormones at various times, such as during the menstrual cycle, pregnancy and puberty.

> **Study tip**
>
> The endocrine system is complex. Try and think of it as a series of different systems working together to relay messages to keep the body in harmony.
>
> Consider the regulation of body heat. In this case think of:
> - the thyroid gland as a furnace that produces heat
> - the pituitary gland as the thermostat that regulates the furnace
> - the hypothalamus as the controller – the person who sets the thermostat.
>
> In this model, when excess heat (thyroid hormones) from the furnace is detected at the thermostat (the pituitary gland), it turns the thermostat off and production of heat at the furnace stops. As the room cools (the thyroid hormone level drops), the thermostat turns back on (TSH increases) and the furnace (the thyroid) produces more heat (thyroid hormones).

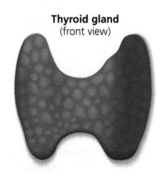
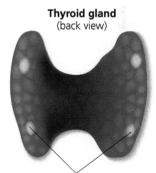

Thyroid gland (front view)   Thyroid gland (back view)

Parathyroid glands

▲ Thyroid and parathyroid glands

Parathyroid glands

These are four small glands situated on the posterior of the thyroid gland. Their principal secretion is the hormone parathormone, which helps to regulate calcium metabolism by controlling the amount of calcium in blood and bones.

Adrenal glands

These are two triangular-shaped glands which lie on top of each kidney. They consist of two parts – an outer cortex and an inner medulla.

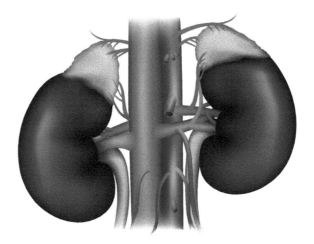

▲ Adrenal glands

Adrenal cortex

The principal hormones secreted by the adrenal cortex are shown in Table 10.5.

Table 10.5 Effects of hormones secreted by adrenal cortex

Hormone	Effects
Glucocorticoids (cortisone and hydrocortisone)	Influence the metabolism of protein and carbohydrates, and utilisation of fats Important in maintaining the level of glucose in the blood so that blood glucose level is increased at times of stress
Mineral corticoids (aldosterone)	Acts on the kidney tubules, retaining salts in the body, excreting excess potassium and maintaining the water and electrolyte balance
Sex corticoids (testosterone, oestrogen and progesterone)	Control the development of the secondary sex characteristics and the function of the reproductive organs

KEY FACT

The production of sex corticoids in the adrenal cortex is important up to puberty. When the ovaries and testes mature, they produce the sex hormones themselves.

Adrenal medulla

The principal hormones secreted by the adrenal medulla are adrenaline and noradrenaline. They are under the control of the sympathetic nervous system and are released at times of stress. Since the release of these hormones is under nervous control, it can happen very quickly.

The effects of adrenaline can be summarised:

1 dilates the arteries, increasing blood circulation and heart rate

2 dilates the bronchial tubes and increases the rate and depth of breathing, thereby increasing oxygen uptake

3 raises the metabolic rate

4 constricts the blood vessels to the skin and intestines, diverting blood from these regions to the muscles and brain to effect action.

The effects of noradrenaline are similar to those of adrenaline and include:

1 vasoconstriction of small blood vessels leading to an increase in blood pressure

2 increase in the rate and depth of breathing

3 relaxation of the smooth muscle of the intestinal wall.

So, the effects of these stress hormones are similar except that:

● **adrenaline** has a primary influence on the *heart*, causing an increase in heart rate

● **noradrenaline** has a greater effect on *blood vessels*, causing peripheral vasoconstriction which raises blood pressure.

KEY FACT

The effects described above can be felt when the body is under stress: a pounding heart, increased ventilation rate, dry mouth and 'butterflies' in the stomach. Stress hormones are broken down slowly, so the effects on the sympathetic nervous system are long-lasting. Over the long term, if levels of these hormones remain elevated, stress-related disorders may result.

Pancreas

The pancreas is known as a dual organ as it has an exocrine and an endocrine function:

● **exocrine function** – the secretion of pancreatic juice to assist with digestion

● **endocrine function** – secretion of hormones by the islets of Langerhans, which are irregular-shaped patches of endocrine tissue located within the pancreas.

Islets of Langerhans

The islets of Langerhans are the groups of endocrine cells in the pancreas which contain alpha, beta and gamma cells that produce glucagon, insulin and somatostatin, respectively.

● Approximately 22% of the cells in the pancreas are **alpha cells** and secrete glucagon, a hormone that plays an active role in regulation of glucose and fat use in the body. The hormone glucagon is released from the alpha cells in response to a low blood glucose level, when the body requires additional glucose, for instance when exercising. Glucagon promotes the conversion of glycogen to glucose.

● **Beta cells** produce insulin, which assists with metabolism in several ways. It helps to regulate the storage of fat and glucose within the body and it stimulates the removal of glucose from the blood. It promotes the conversion of glucose to glycogen.

● The **gamma cells** in the pancreas secrete somatostatin (som-at-o-staa-tin), also known as growth hormone inhibiting hormone (GHIH).

Insulin and blood glucose level

The control of glucose level in the blood is aided by insulin, which initiates the absorption of glucose from the blood by the liver, fat and muscles cells when blood glucose level is high. The glucose is converted to glycogen and later used for energy. Approximately 5% of the liver's overall mass can be used to store glycogen.

Glucagon and blood sugar level

Glucagon has the opposite effects to insulin; in order to maintain the correct balances within the body, glucagon and insulin have to work together. When the body needs additional glucose to meet its energy demands, for example when exercising, glucagon is released.

Glucagon serves to keep the blood glucose level high enough for the body to function well. When blood glucose level is low, glucagon is released and signals the liver to release glucose into the blood.

Somatostatin

Somatostatin affects several areas of the body:

- In the **hypothalamus**, it regulates the secretion of hormones coming from the pituitary gland, including growth hormone and TSH.

- In the **pancreas**, it restricts the secretion of pancreatic hormones, such as glucagon and insulin.

- It is also secreted by the pancreas in response to many factors related to food intake, such as a high level of blood glucose and amino acids.

The sex glands

Testes

The testes are situated in the groin in a sac called the scrotum. They have two functions:

1. the secretion of the hormone testosterone, which controls the development of the secondary sex characteristics in the male at puberty (influenced by LH)

2. the production of sperm (influenced by FSH from the anterior pituitary).

Ovaries

The ovaries are situated in the lower abdomen below the kidneys, and each ovary is attached to the upper part of the uterus by broad ligaments. The two ovaries are the sex glands in the female and have two distinct functions:

1. production of ova at ovulation

2. production of the female sex hormones, oestrogen and progesterone.

Oestrogen is concerned with the development and maintenance of the female reproductive system and the development of the secondary sex characteristics. Progesterone is produced by the ovaries after ovulation. It helps to prepare the uterus for the implantation of the fertilised ovum, develops the placenta if implantation occurs and prepares the breasts for milk secretion.

The ovaries also secrete the following hormones in addition to oestrogen and progesterone:

- **Inhibin** – this hormone inhibits the secretion of FSH towards the end of the menstrual cycle.

- **Relaxin** – this hormone dilates the cervix and promotes widening of the pelvis during childbirth.

Activity

Write the following hormones on pieces of card or paper and place face down in a pile.

- antidiuretic hormone (ADH)
- oxytocin
- FSH (gonadotrophic hormone)
- luteinising hormone (gonadotrophic hormone)
- testosterone
- oestrogen
- progesterone
- insulin
- calcitonin
- melatonin
- glucagon
- parathormone
- adrenaline
- noradrenaline
- glucocorticoids
- mineral corticoids
- sex corticoids
- triiodothyronine (T3)
- thyroxine (T4)
- growth hormone
- thyroid-stimulating hormone (TSH)
- melanocyte-stimulating hormone (MSH)
- adrenocorticotrophic hormone
- prolactin

Now write down the names of the endocrine glands on similar pieces of card or paper and also place face down in a separate pile.

- anterior pituitary gland
- posterior pituitary gland
- thyroid gland
- parathyroid glands
- pineal gland
- islets of Langerhans
- ovaries
- testes
- adrenal cortex (adrenal gland)
- adrenal medulla (adrenal gland)

Working either in groups or individually, match the each hormone to its correct point of secretion.

Natural glandular changes

Puberty

This is the time at which the onset of sexual maturity occurs and the reproductive organs become functional. Changes in both sexes occur, with the appearance of the secondary sexual characteristics such as the deepening of the voice in a boy and growth of breasts in girls. These changes are brought about by an increase in sex hormone activity, due to stimulation of the ovaries and testes by the pituitary gonadotrophic hormones.

The average age for girls to reach puberty is between 10 and 14, although it can occur as early as eight or nine years of age. In boys, the average age is 13 to 16.

In girls, the ovaries are stimulated by the gonadotrophic hormones, FSH and LH. The effects of puberty in girls include:

- the onset of ovulation and the menstrual cycle
- the female reproductive organs becoming functional
- the growth of pubic and axillary hair
- development of breast tissue
- increase in the amount of subcutaneous fat.

In boys, the same gonadotrophic hormones (FSH and LH) stimulate the testes to produce testosterone. The effects of puberty in boys include:

- voice breaking and larynx enlarging
- the growth of muscle and bone
- noticeable height increase
- the development of sexual organs
- the growth of pubic, facial, axillary, abdominal and chest hair
- the onset of sperm production.

The menstrual cycle

Starting at puberty, the female reproductive system undergoes a regular sequence of monthly events, known as the menstrual cycle. The ovaries undergo

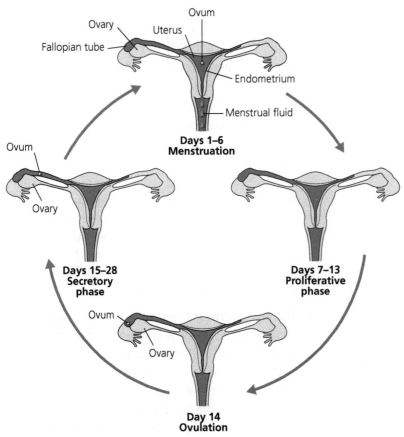

▲ The menstrual cycle

cyclical changes in which a certain number of ovarian follicles develop. When one ovum completes the development process, it is released into one of the fallopian tubes. If fertilisation does not occur, the developed ovum disintegrates and a new cycle begins.

The menstrual cycle lasts approximately 28 days, although it can be longer or shorter than this. There are three stages of the menstrual cycle:

1 **proliferative** (first) phase – days 7 to 14 of the cycle

2 **secretory** (second) phase – days 14 to 28 of the cycle

3 **menstrual** (third) phase – days 1 to 7 of the cycle.

Proliferative phase

At the beginning of the cycle an ovum develops within an ovarian follicle in the ovary. This is in response to a hormone that is released by the anterior lobe of the pituitary gland called FSH, which stimulates the follicles of the ovaries to produce the hormone oestrogen.

Oestrogen stimulates the endometrium (lining of the uterus) to promote the growth of new blood vessels and mucus-producing cells.

When mature, the ovum bursts from the follicle and travels along the fallopian tube to the uterus. This occurs about 14 days after the start of the cycle and is known as **ovulation**.

Secretory phase

A temporary endocrine gland, the corpus luteum, develops in the ruptured follicle in response to stimulation from LH secreted by the anterior lobe of the pituitary gland. The corpus luteum secretes the hormone progesterone, which together with oestrogen causes the lining of the uterus to become thicker and more richly supplied with blood in preparation for pregnancy.

After ovulation, the ovum can only be fertilised during the next eight to 24 hours. If fertilisation does occur, the fertilised ovum becomes attached to the endometrium and the corpus luteum continues to secrete progesterone. Pregnancy then begins. The corpus luteum continues to secrete progesterone

until the fourth month of pregnancy, by which time the placenta has taken over this function.

Menstrual phase

If the ovum is not fertilised, the cycle continues and the corpus luteum shrinks and the endometrium is shed. This is called menstruation. Over a period of about five days, the muscles of the wall of the uterus contract to expel the unfertilised egg, pieces of endometrial tissue and some tissue fluid.

As soon as the level of progesterone drops, due to the breakdown of the endometrium and the corpus luteum, the pituitary gland starts producing progesterone again and, hence, stimulates the ovaries to produce another follicle and a new ovum. The cycle then begins again.

> **In practice**
>
> Due to increased sensitivity, some treatments may be best avoided during menstruation (for instance, waxing, epilation or laser/IPL).

Pregnancy

Pregnancy takes approximately nine calendar months and is divided into three trimesters:

1 **The first trimester** – this is a time of radical hormonal change. During this phase, all of the baby's body systems develop.

2 **The second trimester** – characterised by rapid fetal growth and the completion of systemic development. Blood volume in the mother increases as additional workload is placed on all physiological functions. Cardiac output, breathing rate and urine production increase in response to fetal demands. The uterus enlarges greatly during pregnancy, along with the size of the breasts. Appetite increases in response to the fetal need for increasing amounts of nutrients.

3 **The third trimester** – mostly a weight-gaining and maturing phase, preparing the baby for life outside of the uterus. Posture changes are evident at this stage as the mother gains more weight and internal organs are compressed. The body's connective tissue structure alters by softening, to allow for the expansion needed for the birth.

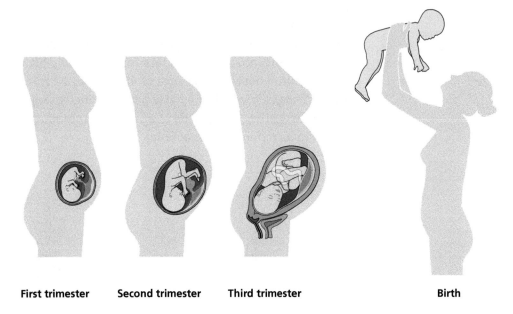

| First trimester | Second trimester | Third trimester | Birth |

▲ Pregnancy

Hormonal changes that occur during pregnancy

During a typical menstrual cycle, the corpus luteum degenerates about two weeks after ovulation. Consequently, the levels of oestrogen and progesterone decline rapidly and the lining of the uterus is not maintained but is cast off as menstrual flow.

If this occurs after implantation, the embryo is spontaneously aborted (miscarried). The mechanism that usually prevents this occurring involves a hormone called human chorionic gonadotrophic hormone (HCG), which is secreted by a layer of embryonic cells that surround the developing embryo. HCG causes the corpus luteum to be maintained in order to establish the pregnancy.

The maintenance of the corpus luteum is important for the first three months, after which the placenta is usually well developed and is able to secrete sufficient oestrogen and progesterone to support the pregnancy.

The secretion of the hormones oestrogen and progesterone is important during pregnancy as they:

- maintain the uterine wall
- inhibit the secretion of the gonadotrophic hormones FSH and LH
- stimulate development of the mammary glands
- inhibit uterine contractions until birth
- cause enlargement of the reproductive organs.

The ovaries and placenta produce inhibin, which inhibits the secretion of the FSH from the anterior lobe of the pituitary, thus preventing the development of ova during pregnancy.

At the end of the gestation period, the level of progesterone falls. Labour cannot begin until the level of progesterone falls, as it inhibits uterine contractions. Oxytocin, secreted by the posterior lobe of the pituitary, stimulates uterine contractions and the ovaries and placenta secrete relaxin, which helps to dilate the cervix and relaxes the ligaments and joints to assist labour.

KEY FACT

Mood, sleep and energy levels are all affected during pregnancy due to the hormonal changes that occur. Some women report extreme tiredness at the start of their pregnancy and experience a surge of energy towards the end. Sleep patterns may be affected due to the activity of the growing foetus and hormone levels may cause emotional disturbances.

Menopause

After puberty, the menstrual cycle normally continues to occur at regular intervals into a woman's late forties or early fifties (most commonly until between the ages of 45 and 55). At this time, there are marked changes in which the cycle becomes increasingly irregular until it ceases altogether. This period in a woman's reproductive life is called the menopause (female climacteric).

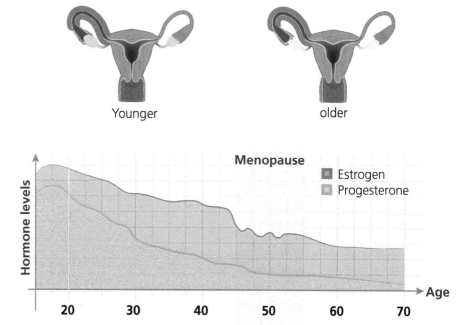

▲ The menopause

During the menopause, there is a change in the balance of the sex hormones. The ovaries cease responding to FSH and this decline in function results in lower levels of oestrogen and progesterone secretion. As a result of reduced oestrogen concentration and lack of progesterone, the female's secondary sexual characteristics undergo varying degrees of change, which may include a decrease in the size of the vagina, uterus and uterine tubes, as well as atrophy of the breasts.

Other changes that occur commonly in response to low oestrogen concentration include a loss of bone matrix leading to an increased risk of osteoporosis, thinning of the skin and dryness of the mucous membrane lining the vagina.

Some women of menopausal age experience unpleasant vasomotor symptoms including sensations of heat in the face, neck and upper body, known as 'hot flushes'. Menopausal women may also experience varying degrees of headache, backache and fatigue, as well as emotional disturbances.

KEY FACT

Many menopausal women take HRT in order to alleviate some of the unpleasant effects of menopause. HRT usually involves administering oestrogens, along with progesterone.

Stress and hormones

Whatever glandular stage of life we are experiencing, one factor is clear: stress has a direct effect on the synchronicity of our hormones.

When the stress level is constantly high, the production of sex hormones is affected, thyroid function slows down and blood sugar level becomes unbalanced. Stress also makes it hard for the body to create 'feel-good' hormones such as serotonin.

Examples of possible symptoms of excessive stress on the endocrine system include amenorrhea (absence of periods), loss of libido and infertility.

KEY FACT
The first step in rebalancing hormones is to minimise stress.

Common pathologies of the endocrine system

Pathologies of the pituitary gland

Acromegaly

During adulthood, if the pituitary gland produces an excess of growth hormone, the hormonal disorder acromegaly develops. This causes an increase in the size of the bones. As time passes, the hands and feet become abnormally large, facial features become coarse and a range of other symptoms develop. Diagnosis of the condition is usually between the ages of 30 and 50, although it can affect people at any age. This disorder is referred to as 'gigantism' if it develops prior to puberty.

Diabetes insipidus

Hyposecretion of ADH by the posterior lobe of the pituitary leads to the disease diabetes insipidus. Symptoms include dehydration, increased thirst and increased output of urine.

Dwarfism

Hyposecretion of the growth hormone during childhood leads to stunted growth, a condition known as dwarfism.

Fröhlich's syndrome (adiposogenital dystrophy)

A rare childhood metabolic disorder characterised by obesity, and abnormal growth and development of the genital organs. It is commonly associated with tumours of the hypothalamus, causing an increase in appetite and reducing the secretion of the gonadotropin hormone.

Gigantism

Hypersecretion of the growth hormone secreted by the anterior pituitary leads to gigantism in children, a disease marked by the rapid growth of the body to extremely large proportions (seven to eight feet in height). If the overproduction occurs in adulthood, this condition is known as acromegaly.

Lorain-Levi syndrome (pituitary dwarfism)

Under-secretion of growth hormone from the pituitary gland in childhood results in smaller-than-average stature.

Simmonds' disease

This is a chronic deficiency of the pituitary gland, a result of a progressive destruction of the anterior lobe of the pituitary gland. It leads to atrophy of the gonads, the thyroid and the adrenal cortex. It can also result in loss of hair on the body.

Pathologies of the thyroid gland

Congenital iodine deficiency syndrome

Hyposecretion of thyroxine leads to this condition (previously known as cretinism) in children, which is a congenital deficiency causing learning difficulties, small stature, coarsening of the skin and hair, and deposition of fat on the body.

Myxoedema

Hyposecretion of thyroxine in an adult leads to myxoedema, which is characterised by the slowing down of physical and mental activity resulting in lethargy, brittle hair, coarse and dry skin and a slow metabolism.

Thyrotoxicosis (hyperthyroidism)

Hypersecretion of thyroxine leads to a condition known as thyrotoxicosis or Graves' disease. Thyrotoxicosis results in an increased metabolic rate, weight loss, sweating, restlessness, increased appetite, sensitivity to heat, raised temperature, frequent bowel action, anxiety and nervousness. When the thyroid gland produces and secretes an excessive amount of thyroxine, it may produce a goitre (an enlargement of the thyroid gland).

In practice

It is helpful to remember that clients with hyperthyroidism are intolerant of heat and those with hypothyroidism are intolerant of cold. Treatments such as laser hair removal or waxing, which involve increased temperature of the skin, may need to be adapted.

Pathologies of the parathyroid gland

Addison's disease

Under-secretion of corticosteroid hormones is responsible for the condition known as Addison's disease. Symptoms include loss of appetite, weight loss, brown pigmentation around joints, low blood sugar, low blood pressure, tiredness and muscular weakness. This disease is treatable by replacement hormone therapy.

Adrenal hyperplasia

Congenital adrenal hyperplasia (CAH) is a genetic disorder characterised by a deficiency in the hormones cortisol and aldosterone and an overproduction of the hormone androgen. CAH is present at birth and affects the sexual development of the child.

Cushing's syndrome

Hypersecretion of the glucocorticoids can lead to a condition known as Cushing's syndrome. This condition results from an excess amount of corticosteroid hormones in the body. Symptoms include weight gain, reddening of the face and neck, excess growth of facial and body hair, raised blood pressure, loss of mineral from bone and sometimes mental disturbances.

Pathologies of the islets of Langerhans

Diabetes mellitus

Hyposecretion can lead to a condition called diabetes mellitus. This condition is due to a deficiency or absence of insulin. The symptoms associated with diabetes include an increased thirst, increased output of urine, weight loss, thin skin with impaired healing capacity, increased tendency to develop minor skin infections and decreased pain threshold when insulin level is low. There are two types of diabetes mellitus:

- **Insulin-dependent diabetes** (early onset) – this occurs mainly in children and young adults and the onset is usually sudden. The deficiency or absence of insulin is due to the destruction of the islet cells in the pancreas. The causes are unknown but there is a familial tendency, suggesting genetic involvement.

- **Non-insulin-dependent diabetes** (late onset) – this type of diabetes occurs later in life and its causes are also unknown. Insulin secretion may be below or above normal, and body cells may become resistant to the effects of insulin. Deficiency of glucose inside the body cells may occur where there is hyperglycaemia and a high insulin level. This may be due to changes in cell walls which block the insulin-assisted movement of glucose into cells. This type of diabetes can be controlled by diet alone, or diet and oral drugs.

In practice

Always obtain a detailed history during the consultation and liaise with the client's GP regarding their type of diabetes. Always ensure that the client brings their glucose and other medications when coming for treatment.

It is important for the therapist to be aware that feedback may be inadequate in those with decreased sensation due to diabetes. Therefore, pressure should be used in treatments only with great care.

Diabetic clients may have acute complications such as hypoglycaemia, resulting in dizziness, weakness, pallor, rapid heartbeat and excessive sweating.

Pathologies of the ovaries

Polycystic ovary syndrome (Stein-Leventhal syndrome)

Hyposecretion of the hormones oestrogen and progesterone in the female can lead to polycystic ovary syndrome, which is characterised by cysts on the ovaries, cessation of periods, obesity, atrophy of the breasts, hirsutism and sterility.

Pathologies of the sex hormones

Gynaecomastia

Hypersecretion of oestrogen and progesterone in the male can lead to muscle atrophy and breast development.

Hirsutism

This is hair growth in the male sexual pattern due to hypersecretion of the hormone testosterone in women and an overproduction of androgens.

Virilism

Hypersecretion of the hormone testosterone in women can lead to virilism (masculinisation), causing an overproduction of androgens.

Other endocrine pathologies

Seasonal affective disorder (SAD)

Hyposecretion of the hormone melatonin is thought to be associated with the condition SAD. Symptoms include depression (typically with the onset of winter), a general slowing down of mind and body, excessive sleeping and overeating.

Stress

Stress can be defined as any factor which affects physical or emotional health. Effects of short-term physical stress are associated with the hormone adrenaline and include an increased heartbeat, rapid breathing, increased sweating, tense muscles, dry mouth, increased frequency of urination and a feeling of nausea. Stress can become negative stress when excess adrenaline is left in the bloodstream following a short-term stress signal. Examples of possible symptoms of excessive stress on the endocrine system include amenorrhea (absence of periods), loss of libido and infertility.

Interrelationships with other systems

The endocrine system

The endocrine system links to the following body systems.

Cells and tissues

Meiosis is the form of cell division involving the formation of sperm in the male and ova in the female.

Skin

MSH produced by the central lobe of the pituitary gland stimulates the production of melanin in the basal cell layer (stratum germinativum) of the skin.

Skeletal

The hormones calcitonin from the thyroid gland and parathormone from the parathyroid glands help to maintain the calcium level in the blood for bone strength and flexibility.

Muscular

Muscles receive additional blood flow in response to the secretion of the hormone adrenaline at times of stress.

Circulatory

Hormones are secreted and carried in the bloodstream to their target organs.

Respiratory

The adrenal glands increase the breathing rate in times of stress to provide more oxygen as fuel for the muscles.

Nervous

The endocrine system works closely with the nervous system in order to maintain homeostasis in the body. The endocrine system is linked to the nervous system by the hypothalamus and the pituitary gland.

Digestive

The production of insulin and glucagon in the pancreas helps to regulate blood sugar level.

Renal

ADH helps to regulate fluid balance in the body

Key words

Adrenal cortex: the outer part of the adrenal gland

Adrenal glands: two triangular-shaped glands which lie on top of each kidney

Adrenal medulla: the inner part of the adrenal gland

Adrenaline: a hormone secreted by the adrenal medulla that increases rates of blood circulation, breathing and carbohydrate metabolism and prepares muscles for exertion

Adrenocorticotrophic hormone (ACTH): a hormone secreted by the pituitary gland that stimulates the adrenal cortex

Alpha cells: cells in the pancreas that secrete glucagon

Antidiuretic hormone (ADH, vasopressin): a hormone secreted by the posterior pituitary that plays an important role in water retention, thirst and blood pressure

Beta cells: cells in the pancreas that secrete the hormone insulin

Calcitonin: a hormone secreted by the thyroid that has the effect of lowering blood calcium

Follicle-stimulating hormone (FSH): a hormone secreted by the anterior pituitary gland which promotes the formation of ova or sperm

Gamma cells: a type of cell in the pancreas which secretes the hormone somatostatin

Glucagon: a hormone formed in the pancreas which promotes the breakdown of glycogen to glucose in the liver

Glucocorticoids: hormones secreted by the adrenal cortex, involved in the metabolism of carbohydrates, proteins and fats

Gonadotrophic hormones: the sex hormones that control the development and growth of the ovaries and testes

Growth hormone: a hormone secreted by the anterior lobe of the pituitary that controls the growth of long bones and muscles

Hormone: a chemical messenger or regulator, secreted by an endocrine gland and which travels via the bloodstream to influence the activity of other organs

Inhibin: a hormone secreted by the gonads that inhibits the production of FSH

Insulin: a hormone produced in the pancreas by the islets of Langerhans that regulates the amount of glucose in the blood

Luteinising hormone (LH): a hormone secreted by the anterior pituitary gland that stimulates ovulation in females and the synthesis of androgens in males

Melanocyte-stimulating hormone (MSH): a hormone secreted by the pituitary gland that is involved in pigmentation changes

Melatonin: a hormone secreted by the pineal gland that inhibits melanin formation and is thought to be concerned with regulating the reproductive cycle

Menopause: permanent cessation of menstruation resulting from the loss of ovarian follicular activity

Menstrual cycle: the cycle of physiological changes affecting the reproductive organs that takes place typically over a month and includes ovulation, thickening of the lining of the uterus and menstruation if fertilisation of the egg has not occurred

Menstrual phase: the phase of the menstrual cycle during which the lining of the uterus is shed (the first day of menstrual flow is considered day 1 of the menstrual cycle)

Mineral corticoids: corticosteroids involved with maintaining the salt balance in the body, such as aldosterone

Noradrenaline: a hormone which is released by the adrenal medulla and by the sympathetic nerves, and which functions as a neurotransmitter

Ovaries: sex glands in the female

Ovulation: discharge of ova from the ovary

Oxytocin: a hormone released by the pituitary gland that causes increased contraction of the uterus during labour and stimulates the ejection of milk into the ducts of the breasts

Parathormone: a hormone that is made by the parathyroid glands and is crucial to maintaining calcium and phosphorus balance

Pineal gland: a pea-sized mass of nerve tissue that is attached by a stalk to the central part of the brain

Pituitary gland: a lobed structure attached by a stalk to the hypothalamus of the brain

Pregnancy: the fetal development period from the time of conception until birth

Prolactin: a hormone secreted by the anterior lobe of pituitary that stimulates the secretion of milk from the breasts following birth

Proliferative phase: the second phase of the menstrual cycle, when oestrogen causes the lining of the uterus to grow

Puberty: the time at which the onset of sexual maturity occurs and the reproductive organs become functional

Relaxin: a hormone secreted by the placenta that causes the cervix to dilate and prepares the uterus for the action of oxytocin during labour

Secretory phase: the second half of the menstrual cycle after ovulation; the corpus luteum secretes progesterone, which prepares the endometrium for the implantation of an embryo; if fertilisation does not occur then menstrual flow begins

Sex corticoids: hormones secreted by the adrenal cortex that control the development of the secondary sex characteristics and the function of the reproductive organs

Somatostatin (growth hormone inhibiting hormone, GHIH): a hormone secreted in the pancreas and pituitary gland which inhibits gastric secretion and somatotropin release

Testes: sex glands in the male

Thymosin: one of the polypeptide hormones secreted by the thymus that control the maturation of T-cells

Thymus: a ductless, butterfly-shaped gland lying at the base of the neck, formed mostly of lymphatic tissue and aiding in the production of T-cells of the immune system

Thyroid: a large ductless gland in the neck which secretes hormones that regulate growth and development through the rate of metabolism

Thyroid-stimulating hormone (TSH): a hormone controlling the growth and activity of the thyroid gland

Thyroxine (T4): the main hormone produced by the thyroid gland, acting to increase metabolic rate and so regulating growth and development

Triiodothyronine (T3): a thyroid hormone similar to thyroxine but having greater potency

Revision summary

The endocrine system

- The endocrine system consists of ductless glands that secrete hormones into the bloodstream.
- Endocrine glands are concerned with the regulation of metabolic processes.
- A **hormone** is a chemical regulator secreted by an endocrine gland into the bloodstream and has the power to influence the activity of other organs.
- The main endocrine glands are the **pituitary** (attached to base of brain), **thyroid** (neck), **parathyroids** (posterior to the thyroid glands), **adrenals** (top of kidneys), **islets of Langerhans** (in the pancreas), **ovaries** (in the female) and **testes** (in the male).
- The principal hormones secreted by the **anterior lobe of the pituitary** include **growth hormone**, thyroid-stimulating hormone **(TSH)**, **adrenocorticotrophic hormone (ACTH)**, **gonadotrophic hormones (FSH and LH)**, **prolactin** and melanocyte-stimulating hormone **(MSH)**.
  - **Growth hormone** controls the growth of long bone and muscle.
  - **TSH** controls the growth and activity of the **thyroid gland**.
  - **ACTH** controls the growth and hormonal output of the **adrenal cortex**.
  - **FSH** and **LH** control the development and growth of the **ovaries** and **testes**.
  - **Prolactin** stimulates the secretion of milk from the breasts following birth.
  - **MSH** stimulates the production of melanin in the basal cell layer of the skin.
- **Hypersecretion** of the growth hormone from the pituitary gland can lead to **gigantism** in childhood and **acromegaly** in adulthood.
- **Hyposecretion** of the growth hormone from the pituitary gland during childhood leads to dwarfism.
- The posterior lobe of the pituitary secretes **ADH** and **oxytocin**.
- **Hyposecretion** of ADH by the posterior lobe of the pituitary can lead to **diabetes insipidus**.

- The **pineal gland** is attached by a stalk in the central part of the brain and secretes a hormone called **melatonin**, which is thought to regulate circadian rhythms and influence mood.
- The **thyroid gland's** principal secretions are **triiodothyronine (T3)** and **thyroxine (T4)**, which regulate metabolism and influence growth and development.
- The thyroid gland also secretes **calcitonin**, which controls the level of calcium in the blood.
- **Hypersecretion** of the thyroid hormones leads to a condition called **thyrotoxicosis**, or Graves' disease.
- **Hyposecretion** of the thyroid hormones leads to congenital iodine deficiency syndrome in childhood and **myxoedema** in adulthood.
- The **parathyroid glands** help regulate calcium metabolism.
- **Hypersecretion** of **parathormone** can lead to renal stones, kidney failure, softening of the bones, and tumours.
- **Hyposecretion** of **parathormone** can lead to a condition called **tetany**.
- The **thymus** gland, located behind the sternum and between the lungs, is only active until puberty.
- The **thymus** gland produces the hormone **thymosin** which stimulates the development of T-cells, who aid the immune system in fighting disease.
- The **adrenal glands** have two parts – an outer **cortex** and an inner **medulla**.
- The principal hormones secreted by the **adrenal cortex** include **glucocorticoids**, **mineral corticoids** and **sex corticoids**.
  - **Glucocorticoids** influence the metabolism of protein, carbohydrates and utilisation of fats.
  - **Mineral corticoids** are concerned with maintaining water and electrolyte balance.
  - **Sex corticoids** control the development of the secondary sex characteristics and the function of the reproductive organs.
- **Hypersecretion** of the **mineral corticoids** can lead to kidney failure, high blood pressure and an excess of potassium in the blood.

- **Hypersecretion** of the **glucocorticoids** can lead to a condition called **Cushing's syndrome**.
- **Hypersecretion** of the **sex corticoids** can lead to **hirsutism** and **amenorrhea** in the female and muscle atrophy and development of breasts in the male.
- **Hyposecretion** of the corticosteroid hormones can lead to a condition called **Addison's disease**.
- The principal hormones secreted by the adrenal medulla include adrenaline and noradrenaline.
- **Adrenaline** and **noradrenaline** are under the control of the **sympathetic nervous system** and are released at times of stress.
- The **pancreas** is an organ with dual functions – exocrine and endocrine.
  - The exocrine function is the secretion of pancreatic juice to assist with digestion.
  - The endocrine function is the secretion of **insulin** from the **islets of Langerhans** cells, which helps regulate blood sugar level.
    - Hypersecretion can lead to hypoglycaemia.
    - Hyposecretion can lead to a condition called diabetes mellitus.
- The **testes** (in the male) have two functions – the secretion of **testosterone** and the production of sperm.
- The **ovaries** (in the female) have two functions – the production of ova and production of the hormones **oestrogen** and **progesterone**.
  - Hypersecretion of the hormone testosterone in women can lead to virilism, hirsutism and amenorrhea.
- Hypersecretion of oestrogen and progesterone in the male can lead to gynaecomastia.
- Hyposecretion of oestrogen and progesterone in the female can lead to polycystic ovary syndrome.
- **Puberty** is a natural glandular change due to stimulation of the ovaries and testes by the pituitary gonadotrophic hormones.
- Starting at **puberty**, the female reproductive system undergoes a regular sequence of monthly events, known as the **menstrual cycle**.
- The **ovaries** undergo cyclical changes, in which a certain number of **ovarian follicles** develop. When one **ovum** completes the development process, it is released into one of the **fallopian tubes**. If **fertilisation** does not occur, the developed ovum disintegrates and a new cycle begins.
- The **menstrual cycle** lasts approximately 28 days, although it can be longer or shorter than this.
- **Pregnancy** takes approximately nine calendar months and is divided into three trimesters.
  - During the **first trimester**, all of the body systems develop.
  - The **second trimester** consists of rapid fetal growth and the completion of systemic development.
  - The **third trimester** is mostly a weight-gaining and maturing process, preparing the baby for life outside of the uterus.
- In the **menopause**, the **ovaries** cease responding to FSH, resulting in lower levels of oestrogen and progesterone secretion.

Test your knowledge questions

Multiple choice questions

1 What is the purpose of the endocrine system?
   a to contribute to the reproductive process
   b to produce and secrete the hormones that regulate body activities
   c to maintain the body during times of stress
   d all of the above

2 Which of the following statements is false?
   a A hormone is a chemical messenger that reaches its destination via the bloodstream.
   b Endocrine glands are ductless glands.
   c All hormones have a quick action.
   d Hormones regulate and co-ordinate various functions in the body.

3 Which of the following secretes the adrenocorticotrophic (ACTH) hormone?
   a posterior lobe of pituitary
   b anterior lobe of pituitary
   c adrenal medulla
   d adrenal cortex

4 Which endocrine gland is responsible for secreting thyroid-stimulating hormone (TSH)?
   a thyroid gland
   b anterior lobe of pituitary
   c posterior lobe of pituitary
   d parathyroid glands

5 Which of the following hormones stimulates the uterus during labour?
   a prolactin
   b oestrogen
   c oxytocin
   d progesterone

6 The hormone parathormone regulates the metabolism of
   a protein
   b calcium
   c carbohydrates
   d fats.

7 Which of the following hormones increases heart and breathing rates during times of stress?
   a insulin
   b noradrenaline
   c testosterone
   d adrenaline

8 Which hormone is responsible for increasing water reabsorption in the kidney tubules?
   a luteinising hormone (LH)
   b antidiuretic hormone (ADH)
   c follicle-stimulating hormone (FSH)
   d oxytocin

9 Where are the islets of Langerhans situated?
   a in the liver
   b in the pancreas
   c in the ovaries
   d in the kidneys

10 Which hormone is concerned with the development of the placenta?
   a prolactin
   b progesterone
   c follicle-stimulating hormone (FSH)
   d oestrogen

Exam-style questions

11 State two functions of the endocrine system.   2 marks

12 List the seven major endocrine glands in the body.   7 marks

13 State two characteristics of a hormone.   2 marks

14 Describe the role of the hypothalamus in relation to the endocrine system.   2 marks

15 Describe the position and function of the thymus gland.   2 marks

11 The reproductive system

Introduction

The reproductive systems are the only systems that are very different, both in terms of structure and function, for men and women. The sex organs (testes in men and ovaries in women) are also endocrine glands. Therefore, there are also sexual differences in the functioning of the endocrine system.

The reproductive system is also the only system that undergoes particular changes at certain times in an individual's life, maturing at puberty, and for women, ceasing to function in the same way after the menopause.

These systems are unique in that they are not vital to the survival of an individual, but they are essential to the continuation of the human species.

OBJECTIVES

By the end of this chapter you will understand:

- functions of the reproductive systems
- structure and functions of the parts of the female reproductive system
- structure and functions of the parts of the male reproductive system
- common pathologies of the reproductive system
- the interrelationships between the reproductive and other body systems.

In practice

It is important for therapists to have a comprehensive knowledge of the reproductive systems in order to be able to understand the effects of the natural glandular changes in the body, which can impact the suitability of certain treatments.

Functions of the reproductive systems

The male and female reproductive systems are specialised to produce the sex hormones responsible for the male and female characteristics and for producing the cells required for reproduction.

The female reproductive system

The function of the female reproductive system is the production of sex hormones and ova (egg cells) which, if fertilised, are supported and protected until birth. The female reproductive system consists of the following internal organs lying in the pelvic cavity:

- the ovaries
- the fallopian tubes
- the uterus
- the vagina.

The external genitalia

The external genitalia is known collectively as the vulva and consists of:

- the **mons pubis** – the exterior of the vulva, a rounded mass of fatty tissue found over the pubic symphysis of the pubic bones
- the **labia majora** and **minora** – lip-like folds at the entrance of the vagina
- the **clitoris** – attached to the symphysis pubis by a suspensory ligament and containing erectile tissue, which can be divided into three major regions:

- the **crura** – two legs of erectile tissue that fan out to support the exterior structures of the clitoris and attach to the underlying tissues; these fill with blood allowing the clitoris to grow in size and harden during sexual stimulation
- the **body** – the main cylindrical region of the clitoris, which extends from the crura and which contains two columns of the erectile tissue
- the **glans** – forms the pointed tip of the clitoris extending outward from the body and beyond the prepuce that covers the rest of the clitoris.

- the **hymen** – a thin layer of mucous membrane at the entrance to the vagina

- the **greater vestibular glands** – the vulva is lubricated by mucus which is secreted from these glands, located in the labia majora, one on either side near the opening of the vagina

- the **vulval vestibule** – this is the central area of the labia minora where the urinary meatus and vaginal opening lie

- the **urinary meatus** – this is where the urethra opens and urine exits the body in females; in males, both urine and semen exit through this opening. In females it is located in the vulval vestibule approximately 25 mm behind the clitoris.

The breasts are accessory glands to the female reproductive system.

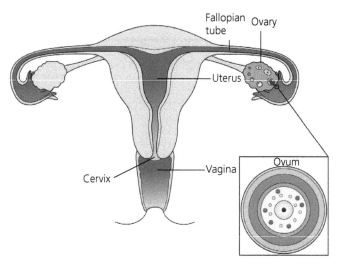

▲ Female reproductive organs

Overview of the female reproductive organs

Table 11.1 The position and functions of the female reproductive organs

Female reproductive parts	Position	Function(s)
Ovaries	Either side of uterus on lateral walls of the pelvis	Production of ova Secretion of oestrogen and progesterone
Fallopian tubes	Extend from the sides of the uterus, passing upwards and outwards to end near each ovary	Convey the ovum from the ovary to the uterus
Uterus	Situated behind the bladder and in front of the rectum	Area in which an embryo grows
Cervix	A tube of tissue that forms a passageway to connect the vagina to the uterus	Functions as a barrier for protection but softens and opens when required allowing sperm to pass through into the uterus During childbirth, opens allowing passage of the baby During ovulation, helps maintain the uterus in a healthy condition to aid egg fertilisation
Vagina	Leads from the cervix to the vulva (connects internal sex organs with external genitalia)	Provides a passageway for menstruation and for childbirth

The ovaries

These are the female sex glands and they lie on the lateral walls of the pelvis. They are almond-shaped organs which are held in place, one on each side of the uterus, by several ligaments. The largest of the ligaments is the broad ligament, which holds the ovaries in close proximity to the fallopian tubes.

The ovary contains numerous small masses of cells called ovarian follicles, within which the ova (egg cells) develop. At the time of birth, there are about two million immature ova in a female's ovaries. Many of the ova degenerate, and at the time of puberty there are only about 400 000 left.

The immature ova (oocytes) lie dormant in the ovary until they are stimulated by a sudden surge in the hormone FSH at the time of puberty. Normally one egg (ovum) ripens and is released each month.

Functions of the ovaries

The ovaries have two distinct functions:

1 the production of ova
2 the secretion of the female hormones oestrogen and progesterone.

Oestrogen and progesterone regulate the changes in the uterus throughout the menstrual cycle and pregnancy. Oestrogen is responsible for the development of the female sexual characteristics, while progesterone, produced in the second phase of the menstrual cycle, supplements the action of oestrogen by thickening the lining of the uterus, ready for the possible implantation of a fertilised egg.

The fallopian tubes

The two fallopian tubes are each about 5 cm long, and extend from the sides of the uterus, passing upwards and outwards to end near each ovary. At the end of each fallopian tube are finger-like projections called fimbriae which encircle the ovaries.

Function of the fallopian tubes

The function of the fallopian tubes is to convey the ovum from the ovary to the uterus. It is swept down the tube by peristaltic muscular contraction, assisted by the lining of ciliated epithelium.

Fertilisation of the ovum takes place within the fallopian tubes and the fertilised egg then passes to the uterus.

The uterus

The uterus is a small, hollow, pear-shaped organ situated behind the bladder and in front of the rectum. It has thick muscular walls and is composed of three layers of tissue:

1 The **perimetrium** – an outer covering which is part of the peritoneum (a serous membrane in the abdominal cavity). It covers the superior (top) part of the uterus.

2 The **myometrium** – a middle layer of smooth muscle fibres. This layer forms 90% of the uterine wall and is responsible for the powerful contractions that occur at the time of labour.

3 The **endometrium** – a soft, spongy mucous membrane lining, the surface of which is shed each month during menstruation.

The uterus can be divided into three parts:

1 The **fundus** is the dome-shaped part of the uterus above the openings of the fallopian tubes.

2 The **body** is the largest and main part of the uterus and leads to the cervix.

3 The **cervix** of the uterus is a thick fibrous muscular structure at the neck of the uterus which opens into the vagina.

Functions of the uterus

The uterus is part of the female reproductive tract which is specialised to receive an ovum, and serves as the area in which an embryo grows and develops into a foetus. After puberty, the uterus goes through a regular cycle of changes which prepares it to receive, nourish and protect a fertilised ovum.

During pregnancy, the walls of the uterus relax to accommodate the growing foetus. If the ovum is not fertilised, the menstrual cycle ends with a short period of bleeding as the endometrium degenerates.

The vagina

The vagina is a 10–15 cm muscular and elastic tube, lined with moist epithelium, which connects the internal organs of the female reproductive system with the external genitalia.

It is made up of vascular and erectile tissue and extends internally from the cervix of the uterus to the vulva on the outside of the body.

During sexual stimulation, the erectile tissues become engorged with blood.

Functions of the vagina

The function of the vagina is for the reception of the male sperm, and to provide a passageway for menstruation and for childbirth.

The wall of the vagina is sufficiently elastic to allow for expansion during childbirth. Between the phases of puberty and the menopause, the vagina also provides an acid environment, due to acid-secreting bacteria, in order to help prevent the growth of microbes that may infect the internal organs.

KEY FACT

The cervix of the uterus dilates during childbirth and the measurement of dilation is used to decide how soon the baby will be born.

Study tip

When learning parts of the female reproductive system, it may be helpful to remember the following mnemonic.

Olivia	**O**varies
Feels	**F**allopian tubes
Unlucky	**U**terus
Contracting	**C**ervix
Vaginosis	**V**agina

Anatomy of the female breast

The female breasts are accessory organs to the female reproductive system, and their function is to produce and secrete milk after pregnancy.

Position

The breasts lie on the pectoral region of the front of the chest. They are situated between the sternum and the axilla, extending from approximately the second to the sixth rib. The breasts lie over the pectoralis major and serratus anterior muscles, and are attached to them by a layer of connective tissue.

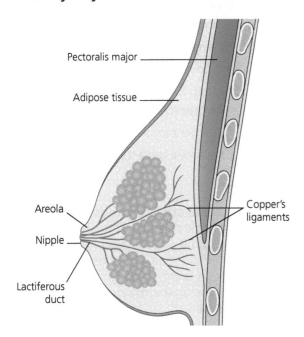

▲ The structure of the female breast

Structure

The breasts consist of glandular tissue arranged in lobules, supported by connective, fibrous and adipose tissue. The lobes are divided into lobules which open up into lactiferous or milk ducts.

The milk ducts open into the surface of the breast at a projection called the nipple. Around each nipple, the skin is pigmented and forms the areola; this varies in colour from a deep pink to a light or dark brown colour.

Glands of Montgomery are sebaceous glands in the areola area, which make oily secretions to keep the areola and the nipple lubricated and protected.

The number of glands can vary greatly, usually from about four to 28 per breast.

A considerable amount of fat or adipose tissue covers the surface of the gland and is found between the lobes. The skin on the breast is thinner and more translucent than the body skin.

Support

The breasts are supported by powerful suspensory Cooper's ligaments, which go around the breast, both ends being attached to the chest wall. The pectoralis major and serratus anterior muscles help to support the ligaments.

If the breast grows large in adolescence or pregnancy, the Cooper's ligaments may become irreparably stretched and the breast will then sag. With age, the supporting ligaments, along with the skin and the breast tissue, become thin and inelastic, and the breasts lose their support.

Physiology of the breast

Lymphatic drainage

The breasts contain many lymphatic vessels, and the lymph drainage, mainly into the axillary nodes under the arms, is extensive.

- About 75% of lymph of the breast drains into the axillary nodes.
- About 20% drains into the internal mammary nodes.
- The remaining 5% drains into the intercostal nodes.

Blood supply

The blood vessels supplying blood to the breast include the internal thoracic artery, a branch of the subclavian artery and the axillary arteries.

The veins of the breast correspond with the arteries, draining into the **axillary** and **internal thoracic veins**.

Nerve supply

There are numerous sensory nerve endings in the breast, especially around the nipple. In lactation, when these touch receptors are stimulated, the impulses pass to the hypothalamus, which stimulates the release of the hormone oxytocin from the posterior lobe of the pituitary. This promotes the flow of milk when required.

Hormones

The hormones responsible for the developing breast are:

- **oestrogen** – responsible for the growth and development of the secondary sex characteristics
- **progesterone** – causes the mammary glands to increase in size if ovum fertilisation and subsequent pregnancy occurs.

Other hormones involved in breast physiology include:

- **prolactin** – a hormone that is released from the anterior pituitary gland and which stimulates milk production after childbirth
- **oxytocin** (ox-ee-toe-sin) – a hormone that is released from the posterior lobe of the pituitary gland.

The two main actions of oxytocin are contraction of the uterus during childbirth and lactation.

KEY FACT

The first secretion from the mammary glands after giving birth is called colostrum. It is rich in antibodies.

Development of the breasts

Puberty

The breast starts out as a nipple which projects from the surrounding ring of pigmented skin called the areola. Approximately two or three years before the onset of menstruation, the fat cells enlarge in response to the sex hormones (oestrogen and progesterone) that are released during adolescence.

KEY FACT

The breasts change monthly in response to the menstrual cycle. The action of the female hormone progesterone increases blood flow to the breast, which increases fluid retention, and the breast may increase in size, causing it to feel swollen and uncomfortable.

Pregnancy

During pregnancy, the increased production of oestrogen and progesterone causes an increase in blood flow to the breast. This causes an enlargement of the ducts and lobules of the breast in preparation for lactation, and there is an increase in fluid retention.

The areola and the nipple enlarge and become more pigmented.

Menopause

The reduction in female hormones during the menopause causes the glandular tissue in the breast to shrink and the supporting ligaments, along with the skin, to become thinner and lose their elasticity. Therefore, during the menopause the breasts begin to lose their support, although the degree of loss is dependent on the original strength of the suspensory ligaments.

Factors determining size and shape

The size of the breast is largely determined by genetic factors. Other factors include:

- the amount of adipose tissue present
- the degree of fluid retention
- the levels of ovarian hormones in the blood and the sensitivity of the breasts to these hormones
- the degree of ligamentary suspension
- the amount of exercise undertaken.

KEY FACT

Exercise may help to strengthen the pectoral muscles, which helps to support the ligaments and increase the uplift of the breast. However, if the wrong type of exercise is undertaken and/or insufficient support is provided for the breasts during exercise, the ligaments may become irreparably stretched.

Reproductive cycles and hormones

Hormones are body chemicals that bring about many changes in the reproductive system throughout life, including at puberty, during the menstrual cycle and ovulation, throughout pregnancy and at the menopause.

Hormones during puberty

The onset of puberty, which is triggered by the brain, produces a number of changes in the reproductive system of both boys and girls.

- Gonadotropin-releasing hormone (GnRH) is released by the hypothalamus, which initiates the first stage of hormonal reactions.

- The pituitary gland receives a signal from this hormone and, in turn, releases FSH along with LH. The process of sexual development is initiated by these hormones, which work in different ways for males and females.

The male sex hormone testosterone is produced in the testes. FSH and LH are the two hormones that are responsible for producing testosterone. During puberty, the development of male secondary sexual characteristics, such as pubic hair, facial hair and muscular development, are due to a surge in these hormone levels.

In females, the ovaries produce oestrogen and progesterone when the brain releases FSH and LH. The development of breasts and a curvier body shape are the result of the female hormones.

The growth spurt in puberty for both males and females is caused by an increase in the secretion of growth hormone.

Hormones during menstruation

The female hormones are responsible for the menstrual cycle. The hormones that are most involved in the process include oestrogen, FSH, LH and progesterone.

Oestrogen

The endometrium builds up in response to the rising level of oestrogen within the female body. This enables the uterus to prepare to receive an egg that has been fertilised. The level of oestrogen decreases if there are no fertilised eggs and no subsequent pregnancy. The decreasing level of oestrogen results in the loss of the built-up uterine lining, leading to menstruation.

Follicle-stimulating hormone (FSH)

The pituitary gland produces this hormone, which is responsible for readying follicles for ovulation. Between 3 and 30 follicles are matured for ovulation each month, but usually only one follicle completes the process and is released as an egg.

Luteinising hormone (LH)

This is the hormone that makes the egg release from the follicle in the ovary. This is called ovulation – the time in the menstrual cycle when the follicle ruptures and releases the egg from the ovary.

During ovulation, the follicle releases the egg due to an increase in LH produced by the pituitary gland, which itself is triggered by the increase in the oestrogen level. Progesterone and oestrogen are then secreted from the burst follicle as the uterus continues to get ready for pregnancy.

Progesterone

This hormone is released by the ruptured follicle (the one that has released an egg). After the egg is released from the follicle, the follicle closes and becomes a corpus luteum. The corpus luteum secretes increasing amounts of progesterone. This rise in the level of progesterone typically causes a rise in body temperature. If no pregnancy occurs, the level of progesterone falls and this, along with the decreasing amount of oestrogen, helps the built-up lining of the uterus to separate and menstruation begins.

Stages of pregnancy

Pregnancy starts with fertilisation and ends with childbirth.

Fertilisation

This is the fusion of a spermatozoon (sperm) with an ovum (egg).

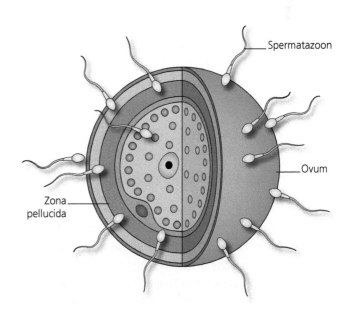

▲ Fertilisation

The spermatozoon penetrates the inner membrane, referred to as the zona pellucida, of the ovum.

This initiates the ovum's final meiotic division and makes the zona pellucida impenetrable to other spermatozoa. After the spermatozoon penetrates the ovum, its nucleus is released into the ovum, the tail degenerates, and its head enlarges and fuses with the ovum's nucleus.

This fusion provides the fertilised ovum, now called a **zygote**, with 46 chromosomes, 23 from the egg and 23 from the sperm.

Pre-embryonic development

The pre-embryonic phase starts with ovum fertilisation and last for two weeks.

As the zygote passes through the fallopian tube, it undergoes a series of mitotic divisions, forming daughter cells, called **blastomeres**, that each contain the same number of chromosomes. The first cell division ends about 30 hours after fertilisation; subsequent divisions occur rapidly.

The zygote then develops into a small mass of cells called a **morula**, which reaches the uterus around the third day after fertilisation. Fluid then masses in the centre of the morula and forms a central cavity. The structure is then called a **blastocyst**.

During the next phase, the blastocyst stays within the zona pellucida, unattached to the uterus. Next, the zona pellucida degenerates and by the end of the first week of fertilisation, the blastocyst attaches to the endometrium.

Formation of the embryo

By day 24, the blastocyst has formed an amniotic cavity containing an embryo. The developing zygote starts to take on a human shape. Each of the three germ layers (ectoderm, mesoderm and endoderm) forms specific tissues and organs in the developing embryo.

The endometrium and part of the blastocyst mesh and development into the placenta, which allows for the passages of nutrients, oxygen and waste to and from baby and mother.

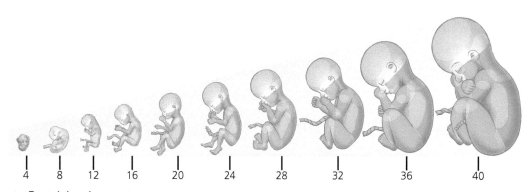

4	8	12	16	20	24	28	32	36	40

▲ Foetal development

Foetal development

Significant growth and development takes place within the first three months following conception.

Month 1

At the end of the first month, the embryo has a definite form. The head, trunk, and the tiny buds that will become the arms and legs are visible. The cardiovascular system has begun to function and the umbilical cord is visible in its most primitive form.

Month 2

During the second month the embryo grows to 2.5 cm in length. The head and facial features develop

as the eyes, ears, nose, lips, tongue and tooth buds form. The arms and legs also take shape. Although the gender of the foetus is not yet visible, all external genitalia are present.

Cardiovascular function is complete and the umbilical cord has a definite form.

From the eighth week, the embryo is called a foetus.

Month 3

During the third month, the foetus grows to 7.5 cm in length. Teeth and bones begin to appear, and the kidneys start to function. The foetus opens its mouth to swallow, grasps with its fully developed hands, and prepares for breathing by inhaling and exhaling amniotic fluid (although its lungs are not functioning properly). At the end of the third month, or first trimester, the foetus's gender is distinguishable.

Months 4 to 9

Over the remaining 6 months, the foetal growth continues, as internal and external structures develop at a rapid rate. In the third trimester, the foetus stores the fats and minerals it will need to live outside of the uterus.

At birth, the average full-term foetus measures 51 cm and weights 7 to 7 ½ lbs (roughly 3 to 3.5 kg).

Birth

Childbirth is divided into three stages; the duration of each stage varies according to the size of the uterus, the woman's age and the number of previous pregnancies.

1   The first stage of labour is when the foetus begins its descent and the cervix starts to dilate, preparing to allow the foetus to pass from the uterus into the vagina.

    During this stage, the amniotic sac ruptures as the uterine contractions increase in frequency and intensity (the amniotic sac can also rupture before the onset of labour).

3   The second stage of labour begins with full cervical dilation and ends with delivery of the foetus.

4   The third stage of labour starts immediately after childbirth and ends with expulsion of the placenta. After the neonate is delivered, the uterus continues to contract intermittently and grows smaller.

Hormones during pregnancy

The ovaries, and then later the placenta, are the main producers of the pregnancy-related hormones that are essential in creating and maintaining the conditions for a successful pregnancy.

Many hormone levels change in the body during pregnancy, with several hormones playing major roles during this time. These include:

- **Human chorionic (cor-ri-on-ik) gonadotrophin (HCG)** – this is only produced during pregnancy and almost exclusively in the placenta. This hormone enters the maternal circulation, preventing the mother's immune system from rejecting the embryo and beginning to change her body to support a pregnancy. HCG hormone levels found in maternal blood and urine increase dramatically during the first trimester.
- **Human placental lactogen (HPL)** – produced by the placenta, this helps in the process of providing nutrition for the foetus and plays a role in stimulating milk glands in the breasts in anticipation of breastfeeding.
- **Oestrogen** – responsible for the development of the female sexual characteristics. Normally formed in the ovaries, oestrogen is also produced by the placenta during pregnancy to help maintain a healthy pregnancy.
- **Progesterone** – produced by the ovaries and by the placenta during pregnancy, progesterone stimulates the thickening of the uterine lining in anticipation of implantation of a fertilised egg.

> **KEY FACT**
>
> Human chorionic gonadotrophin can be detected in the urine as early as 7–9 days after fertilisation and is used as an indicator of pregnancy in most over-the-counter pregnancy tests.

Hormones during labour

The hormone oxytocin has a key role during labour. It is associated with feelings of motherhood and bonding. Prolactin is similarly linked to these feelings.

As labour commences, the regular contractions of the uterus and abdominal muscles are initiated by an increase in the level of oxytocin. These contractions continue to increase in frequency as well as strength, without the assistance of oestrogen and progesterone, which inhibit labour at elevated levels.

For a baby to pass through, the cervix must be dilated to about 10 cm. Oxytocin is one of the hormones that assists with the preparation of the cervix to enable successful dilation during the birth.

During labour, the level of relaxin rises very quickly, which assists the birth process. The increased level of relaxin helps the cervix to soften and become longer, as well as expanding and softening the lower pelvic region.

At the point where the arrival of the baby becomes imminent, increased quantities of the hormones adrenaline and noradrenaline are released. This results in a flood of energy that initiates very strong contractions to help complete the birth process.

Hormones after labour

To reduce bleeding after labour, oxytocin contracts the uterus, effectively restricting the blood flow. This also aids with detachment and expulsion of the placenta. Prolactin, along with oxytocin, aids mother-and-baby bonding by remaining elevated at this stage.

Progesterone and oestrogen fall once the placenta is expelled. This enables the mother to produce a high-density milk that is more suitable for the newborn baby, as opposed to on-going mature milk. This first milk produced by the mother is known as colostrum and contains a much richer content of minerals, protein and vitamins for the baby's first feed.

Oxytocin and prolactin from the pituitary are released during breastfeeding and pass through the blood to the breast. The prolactin promotes the production of the milk and the oxytocin stimulates the delivery of the milk to the mothers' nipples.

Mature milk that continues to nourish baby and aids sleep starts to be produced usually around four days after the birth.

The menopause

The menopause is the period at the end of a woman's reproductive years when the menstrual cycle stops. This is due to the loss of the egg-containing follicles in her ovaries. Once the follicles are gone, progesterone and oestrogen, the hormones that regulate the menstrual cycle, are no longer secreted by the ovaries and menstruation ceases.

A change in the regular period cycle often indicates the start of the menopause. This stage usually lasts for up to four years, but for some women it can last even longer. Approximately 80% of women can experience several different symptoms during the menopause; most common are sweating during the night and hot flushes at varying intervals during the day. These symptoms usually occur within the first year of the menopause and can sometimes be severe enough to affect the quality of sleep, resulting in low energy and a loss of strength.

The role of hormones in menopause

Oestrogen is made up of the three reproductive hormones:

1 oestradiol

2 oestrone

3 oestriol.

During her fertile years, a woman's ability to produce eggs each month is linked with the release of these hormones. The main producer of oestrogen are the ovaries. However, the adrenal glands and, during pregnancy, the placenta also produce small amounts of oestrogen.

Oestrogen works at puberty to stimulate the female characteristics. It governs the reproductive cycle including ovulation and thickening of the uterus in preparation for implantation of a fertilised egg. If there is no pregnancy, the monthly period occurs, and the lining of the uterus is expelled through the vaginal opening.

With age, a woman's ability to become pregnant lessens as the eggs, stored in the ovaries, decrease in number. The amount of oestrogen produced gradually decreases over a number of years, causing changes in the body and various symptoms to arise. This period is known as the **peri-menopause**. The menopause itself usually occurs around the age of 50 to 55. This is when ovulation and the monthly period cycle stop, and a woman is no longer able to become pregnant.

Ovarian follicles reduce during the menopause and the ovaries respond decreasingly to LH and FSH, which are both involved in the process of reproduction. With the release of fewer hormones from the ovaries as age increases, LH and FSH lose their ability to function correctly in order to regulate oestrogen, progesterone and testosterone.

Effects of hormone changes during menopause

The changes in these hormone levels during the menopause can result in decreased health for a number of years. A woman can experience a range of different conditions such as fatigue, mood swings, memory loss, increased sweating during the night and day, hot flushes, anxiety and depression. These can be brought on by changes in the nervous system due to hormonal fluctuations caused by the reduction of oestrogen in the body.

The reduction in oestrogen can cause poor muscle tone and dryness in the vagina, often making sexual intercourse painful or uncomfortable. This is coupled with other changes in mood or sleep patterns, and can result in a reduced libido (little or no interest in sex).

It is not clear whether depression, anxiety or panic attacks are brought on by the menopause, but the onset of these conditions can occur at this stage in a woman's life.

Diagnosis of the menopause can only be made accurately once a woman has stopped having periods for a year or more. It is, however, often the case that a woman knows she has started the menopause due to symptoms she is experiencing.

KEY FACT

To identify the menopause, a test can be performed to check for an elevated level of FSH. If a woman has not had her menstrual period for a year or more and the test results show a consistently elevated FSH blood level of 30 mIU/ml or higher, it is generally considered that she has commenced the menopause.

Hormone replacement therapy (HRT)

A slight change in lifestyle, such as healthier eating and better sleeping habits coupled with regular exercise, is all that may be required for a significant improvement in a woman's comfort during the menopause. However, for some women who experience more unpleasant menopausal symptoms, HRT can be used as treatment. HRT can be taken by pill, administered through skin gels and patches or even implants. HRT medications all contain oestrogen, and some contain progesterone as well.

The oestrogen-only medications are for woman who have had a hysterectomy, whereas for woman who still have their uterus, both oestrogen and progestogen are part of the medication.

Female reproductive changes with ageing

Ovulation usually stops 1 to 2 years before the menopause. As the ovaries reach the end of their reproductive cycle, they become unresponsive to gonadotrophic stimulation. With ageing, the ovaries atrophy and become thicker and smaller.

The vulva also atrophies with age and the tissue shrinks. Atrophy causes the vagina to shorten and the mucous lining to become thin, dry and less elastic.

After the menopause, the uterus shrinks rapidly to half its premenstrual weight. The cervix atrophies and no longer produces mucus for lubrication, and the endometrium and myometrium become thinner.

In the breasts, the glandular, supporting and fatty tissues atrophy and as the Cooper's ligaments lose their elasticity, the breasts become pendulous.

 Activity

Hormones play a major role in regulating body processes. When the body is in balance it is like a finely tuned orchestra with all instruments playing in synchrony.

Discuss with your colleagues what effects stress may have on the following glandular changes in life: puberty, pregnancy and menopause.

Discuss the hormones involved and what effects they may have on the body.

The male reproductive system

The male reproductive system consists of the:

- testes
- epididymis
- vas deferens
- urethra
- penis.

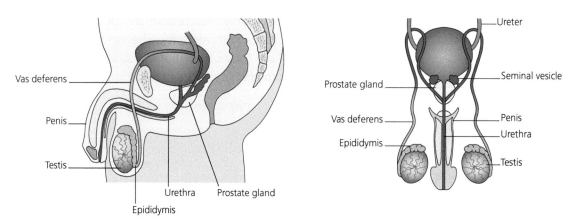

▲ Male reproductive system

Table 11.2 The position and functions of the male reproductive organs

Male reproductive parts	Position	Function(s)
Testes	In the scrotum	Production of sperm
Epididymis (epp-pee-did-ee-mus)	Lies along the posterior border of each testis	Stores sperm until maturation
Vas deferens (vass-def-fer-rens)	Very long tubes leading from the seminiferous tubules of the testes to the urethra	Tubes through which sperm is released
Urethra (you-reeth-ra)	Extends from neck of the bladder through the penis to outside of the body	Provides a common pathway for urine and semen
Penis	Male external sex organ	Excretes urine and ejaculates semen

Study tip

When learning parts of the male reproductive system, it may be helpful to remember the following mnemonic.

Tom	**T**estes
Evades	**E**pididymis
Very	**V**as deferens
Unfortunate	**U**rethra
Pathologies	**P**enis

Testes

The testes are the reproductive glands of the male, and lie in the scrotal sac. Each testis consists of approximately 200 to 300 lobules; these are separated by connective tissue and filled with seminiferous tubules, in which sperm cells are formed. Between the tubules are a group of secretory cells known as the interstitial cells, which produce male sex hormones.

The testes are specialised to produce and maintain sperm cells, and to produce male sex hormones known collectively as **androgens**. **Testosterone** is the most important androgen as it stimulates the development of the male reproductive organs. It is also responsible for the development and maintenance of the male secondary sexual characteristics.

Epididymis

The epididymis (plural: epididimydes) is a coiled tube leading from the seminiferous tubule of the testis to the vas deferens. They store and nourish immature sperm cells and promote their maturation until ejaculation.

Vas deferens

The vas deferens (plural: vas deferentia) is a tube leading from the epididymis to the urethra, through which the sperm are released.

Seminal vesicles

The seminal vesicles are pouches lying on the posterior aspect of the bladder attached to the vas deferens. They secrete an alkaline fluid which contains nutrients and is added to sperm cells during ejaculation.

Ejaculatory ducts

The two ejaculatory ducts are short tubes which join the seminal vesicles to the urethra.

Urethra

The urethra provides a common pathway for the flow of urine and the secretion of semen. A sphincter muscle prevents both functions occurring at the same time.

Urinary meatus

The urinary meatus is located at the tip of the glans penis and is a sensitive part of the male reproductive system. It resembles a vertical slit, which facilitates the flow of urine. Sometimes, the opening may be naturally more rounded, or occur as a result of excessive skin removal during circumcision.

Penis

The penis is the main external sex organ of the male. It is composed of erectile tissue and is richly supplied with blood vessels. When stimulated by sexual activity the blood vessels become engorged with blood and the penis becomes erect. Its function is to convey urine and semen.

Accessory sex glands in the male

Cowper's glands

The Cowper's glands are a pair of small glands that open into the urethra at the base of the penis. These glands produce further secretions to contribute to the seminal fluid, but less than that of the prostate gland or seminal vesicles.

Prostate gland

The prostate gland is a male accessory gland about the size of a walnut. It lies in the pelvic cavity in front of the rectum and behind the symphysis pubis. During ejaculation, it secretes a thin, milky fluid that enhances the mobility of sperm and neutralises semen and vaginal secretions.

Male reproductive changes with ageing

Physiological changes in older men include reduced testosterone production, with in turn may cause decreased libido. A reduced testosterone level also causes the testes to atrophy and soften, and sperm production decreases by around 48–69% between the ages of 60–80. Normally, the prostate glands enlarge with age and its secretions diminish. Seminal fluid also decreases in volume and becomes less viscous.

Common pathologies of the reproductive system

Female pathologies

Amenorrhoea

Amenorrhea is the absence of menstrual periods. Causes may include deficiency of ovarian, pituitary or thyroid hormones, mental disturbances, depression, radical weight loss, stress, excessive exercise or a major change in surroundings or circumstances.

Cancer of the breast

Most breast cancers are detected when the patient notices a breast or axillary lump; mammography screening can confirm whether a lump is potentially cancerous.

Breast cancer can present as redness and pain, puckering of the breast skin or change in breast shape, or discharge from or retraction of the nipple.

Cancer can spread locally, or to the axilla and neck lymph nodes, causing oedema of the arm, or by blood to the lung, bone and liver.

The type of breast cancer can determine whether the spread is rapid or very slow.

In practice

Note that the spread of cancer is determined by the type of breast cancer (some spread rapidly while others are slow growing).

Consult the client's GP or consultant regarding the extent and type of the cancer, and the treatment regime.

Avoid areas that have been exposed to radiation, if a client is having radiotherapy, as these may be sensitive and tender.

Both radio and chemotherapy can reduce a client's immunity and, therefore, therapists should avoid contact if they have an infection.

Clients who have had surgery which involved removal of the axillary nodes are likely to have oedema of the arm. Provided permission for treatment has been granted by the client's GP or consultant, elevating the oedematous arm above heart level throughout a massage can be beneficial. Gently massage the arm with strokes that are directed towards the axilla. Advise the client to open and close their hand tightly six to eight times every few hours (the contraction of the muscles will help venous and lymphatic flow).

Cancer of the cervix

Cervical cancer is asymptomatic in the early stages. Later there may be foul-smelling, blood-stained discharge from the vagina. Lower back pain, loss of weight, unexplained anaemia and pain during intercourse are other symptoms.

Cancer of the ovaries

Ovarian cancer is asymptomatic. Diagnosis is usually made after the cancer has spread extensively. The symptoms are vague and are usually associated with gastrointestinal problems, such as bloating of the abdomen, mild abdominal pain and excessive passage of gas. There may be fluid in the peritoneal cavity in late stages.

Hormone changes may result in abnormal vaginal bleeding.

Dysmenorrhea

This condition is defined as painful and difficult menstruation. It presents with spasms and congestion of the uterus, resulting in cramping lower abdominal pains which start before or with the menstrual flow, and continue during menstruation. It is often associated with nausea, vomiting, headache and a feeling of faintness.

Ectopic pregnancy

This term is used to describe the development of a foetus at a site other than in the uterus. An ectopic pregnancy may occur if the fertilised egg remains in the ovary, or in the fallopian tube, or if it lodges in the abdominal cavity.

The most common type of ectopic pregnancy occurs in the fallopian tube. There is a danger of haemorrhage as growth of the foetus may cause the tube to rupture and bleed. Ectopic pregnancy can be life threatening.

Endometriosis

This is inflammation of the endometrium (the inner lining of the uterus). It presents with abnormal menstrual bleeding, lower abdominal pain and a foul-smelling discharge. Fever and malaise may accompany this condition.

Fibroid

A fibroid is an abnormal growth of fibrous and muscular tissue in the muscular wall of the uterus. Fibroids can cause pain and excessive bleeding, and may become extremely large. Although they do not threaten life, they make pregnancy unlikely.

Some fibroids may be removed surgically; in other cases a hysterectomy may be necessary.

Polycystic ovary syndrome (as known as Stein-Leventhal syndrome)

This is a hormonal disorder in which there is inadequate secretion of the female sex hormones. As a result, the ovarian follicles fail to ovulate and remain as multiple cysts, distending the ovary. Other associated symptoms include obesity, hirsutism, acne and infertility.

Premenstrual syndrome

Premenstrual syndrome is a term for the physical and psychological symptoms experienced 3–14 days prior to the onset of menstruation.

The condition presents with varying symptoms: headache, bloatedness, water retention, backache, changes in co-ordination, abdominal pain, swollen and painful breasts, depression, irritability and craving for sweet foods.

Infertility

Infertility is the inability in a woman to conceive or in a man to induce conception. Female infertility may be due to a failure to ovulate, to obstruction of the fallopian tubes, or endometriosis.

Male pathologies

Cancer of the testis

Slight enlargement of the testis is the first symptom of testicular cancer. It may be accompanied by pain, discomfort and heaviness of the scrotum. Soon there is a rapid enlargement of the testis, which can become hot and red.

Cancer of the prostate

Usually there are no initial symptoms of prostate cancer. If the cancer is located close to the urethra, there may be a frequency of micturition, urgency, difficulty in voiding, blood in urine or blood in the ejaculate. Cancer of the prostate is often diagnosed by rectal examination – the diseased prostate feels nodular and hard. Prostate cancer may spread to the bones, where it produces pain, or causes fractures after trivial injury.

In the advanced stage, as in all cancers, the person loses weight and is anaemic.

Prostatitis

This is inflammation of the prostate gland, which is usually caused by bacteria. This condition presents with a frequent need to urinate and urgency on passing urine (urine may be cloudy). High fever with chills, muscle and joint pain are common. A dull ache may be present in the lower back and pelvic area.

Infertility

Causes of male infertility can include decreased numbers or motility of sperm, or may be due to the total absence of sperm. In both male and female infertility, the cause may also be associated with stress.

Interrelationships with other systems

The reproductive system

The reproductive system links to the following body systems.

Cells and tissues

Ova are the reproductive cells in the female and sperm cells are the reproductive cells in the male.

Skeletal

The pelvis offers protection for the uterus.

Muscular

Smooth muscle is responsible for the passage of ova from the ovaries to the vagina, and sperm from the testes to the urethra.

During orgasm in the female, the muscles of the perineum, uterine wall and the uterine tubes contract rhythmically. During orgasm in the male, motor impulses are transmitted to skeletal muscles at the base of the erectile penis causing them to contract rhythmically.

Circulatory

During erection of the penis, the vascular spaces within the erectile tissue become engorged with blood as arteries dilate and veins are compressed.

During periods of sexual stimulation, the erectile tissues of the clitoris become engorged with blood.

Nervous

Orgasm is the culmination of sexual stimulation; the movement of semen occurs because of sympathetic reflexes.

Endocrine

The ovaries in women and the testes in the male are responsible for the development of the secondary sexual characteristics.

Key words

Clitoris: a female sex organ; its visible button-like portion is near the front junction of the labia minora, above the opening of the urethra

Cooper's ligament: connective tissue in the breast that helps to maintain structural integrity

Corpus luteum: a hormone-secreting structure that develops in an ovary after an ovum has been discharged but degenerates after a few days unless pregnancy has begun

Cowper's gland: either of a pair of small glands which open into the urethra at the base of the penis and secrete a constituent of seminal fluid

Ejaculation: the discharge of semen from the male reproductive organs

Ejaculatory duct: a duct through which semen is ejaculated

Endometrium: the mucous membrane that lines the inside of the uterus

Epididymis: a highly convoluted duct behind the testis, along which sperm passes to the vas deferens

Fallopian tube: one of a pair of tubes along which ova travel from the ovaries to the uterus

Fertilisation: the fusion of a spermatozoon with an ovum

Genitalia: the collective term for male or female reproductive organs

Greater vestibular glands (known as Bartholin's glands): glands that lie in the labia majora, one on each side near the vaginal opening, and secrete mucus which lubricates the vulva

Human chorionic gonadotropin hormone (HCG): a hormone produced by the placenta after implantation

Human placental lactogen (HPL): a hormone, produced by the placenta, which plays a role in stimulating milk glands in the breasts in anticipation of breastfeeding

Hymen: a membrane that surrounds or partially covers the external vaginal opening

Labia majora and **labia minora**: lip-like folds at the entrance of the vagina

Mammary gland: the milk-producing gland in females

Menopause: the ceasing of menstruation and the end of a woman's reproductive life

Mons pubis: a rounded mass of fatty tissue found over the pubic symphysis of the pubic bones at the exterior of the vulva

Myometrium: the middle layer of the uterine wall

Oestrogen: a female hormone that causes development and change in the reproductive organs

Oocyte: an immature ovum

Ovaries: female sex glands that lie on the lateral walls of the pelvis

Ovum/ova: an egg, or eggs within the ovary of the female

Penis: the organ of the male reproductive system through which semen passes out of the body during sexual intercourse; also an organ of urination

Perimetrium: the outer serous layer of the uterus, equivalent to the peritoneum

Pregnancy: the period from conception to birth

Progesterone: a hormone released by the corpus luteum that stimulates the uterus to prepare for pregnancy

Prostate: a gland surrounding the neck of the bladder in males; releases a fluid component of semen

Scrotum: a pouch of skin containing the testicles

Semen: the male reproductive fluid, containing spermatozoa in suspension

Seminal vesicle: one of a pair of glands which opens into the vas deferens near to its junction with the urethra and secretes many of the components of semen

Seminiferous tubule: a coiled tubule of the testis in which spermatozoa are produced

Spermatozoon/spermatozoa: sperm cell(s) that combines with an ovum to form a zygote

Testes: paired male reproductive glands that produce sperm and secrete testosterone

Testosterone: a hormone that stimulates development of male secondary sexual characteristics, produced mainly in the testes

Urethra: the duct by which urine is conveyed out of the body from the bladder, and which in men also conveys semen

Urinary meatus: the opening of the urethra; the point where urine exits the urethra in males and in females, and also where semen exits the urethra in males

Vagina: the muscular tube leading from the external genitals to the cervix of the uterus in women

Vas deferens: the duct that conveys sperm from the testicle to the urethra

Vulva: part of the female reproductive system that contains the external female sex organs

Vulval vestibule: the area between the labia minora where the vaginal opening and the urinary meatus are located

Zygote: an egg that has been fertilised by sperm, and which could develop into an embryo

Revision summary

The reproductive system

- The male and female reproductive systems function to produce:
  - the sex hormones responsible for the male and female characteristics
  - the cells required for reproduction.
- The structures of the female reproductive system include: **ovaries**, **fallopian tubes**, **uterus**, **vagina** and **vulva**.
- The **breasts**, or mammary glands, are also part of the female reproductive system.
  - The **ovaries** lie on the lateral walls of the pelvis and have two distinct functions: the production of **ova** and the secretion of the female hormones **oestrogen** and **progesterone**.
  - The **fallopian tubes** transport **ova** from the **ovaries** to the **uterus**.
  - The **uterus** is situated behind the bladder and in front of the rectum and is designed to receive, nourish and protect a fertilised **ovum**.
  - The **vagina** is a muscular and elastic tube designed for the reception of sperm and to provide a passageway for menstruation and childbirth.
  - The **vulva** is a collective term for the female **genitalia**.
- In older females, levels of **oestrogen** and **progesterone** decrease, causing the menopause. Ovaries atrophy becoming thicker and smaller. The vulva atrophies and tissue shrinks. Atrophy causes the vagina to shorten and the mucous lining to become thin, dry and less elastic.
- After the **menopause**, the **uterus** shrinks rapidly to half its premenstrual weight. The **breasts** atrophy and lose their elasticity and support.
- **Pregnancy** starts with fertilisation and ends with childbirth, and consists of the following stages: fertilisation, pre-embryonic development, formation of embryo, fetal development and birth.
- The structures of the male reproductive system include: **testis**, **epididymis**, **vas deferens**, **ejaculatory ducts**, **urethra**, **seminal vesicles**, **prostate**, **Cowper's gland** and **penis**.
  - The **testes** lie in a scrotal sac; they produce and maintain **sperm** cells, and release the male sex hormone **testosterone**.
  - Each **testis** is filled with a **seminiferous tubule** in which **sperm** cells are formed.
  - The **epididymis** is a coiled tube that leads from the **seminiferous tubule** of the **testis** to the **vas deferens**. It stores and nourishes immature **sperm** cells and promotes their maturation until **ejaculation**.
  - The **vas deferens** leads from the **epididymis** to the **urethra** and is a tube through which the **sperm** are released.
  - The **seminal vesicles** are pouches lying on the posterior aspect of the bladder, attached to the vas deferens. They secrete an alkaline fluid which contains nutrients and is added to **sperm** cells during **ejaculation**.
  - The two **ejaculatory ducts** are short tubes which join the **seminal vesicles** to the **urethra**.
  - The **Cowper's glands** are a pair of small glands that open into the **urethra** at the base of the **penis**. These glands produce further secretions to contribute to the **seminal fluid**.
  - The **prostate** gland lies in the pelvic cavity in front of the rectum and behind the symphysis pubis. During ejaculation, it secretes a thin, milky fluid that enhances the mobility of sperm and neutralises **semen** and vaginal secretions.
  - The **urethra** provides a common pathway for the flow of urine and the secretion of **semen**.
  - The **penis** is composed of erectile tissue and is richly supplied with blood vessels. Its function is to convey urine and **semen**.
- In the male, a decreased level of **testosterone** decreases sexual desire and viable **sperm**; testes also atrophy as muscle strength decreases.

Test your knowledge

Multiple choice questions

1 What is the main function of the ovaries?
   a to accommodate a growing foetus during pregnancy
   b to serve as a site for fertilisation
   c to produce mature ova
   d to receive male sperm

2 What is a fertilised ovum known as?
   a a blastocyst
   b a zygote
   c an embryo
   d a foetus

3 What is the function of the fallopian tubes?
   a to convey ova from the ovary to the uterus
   b to convey ova from the ovary to the vulva
   c to prepare for the implantation of a fertilised ovum
   d to secrete mucus

4 Where is the uterus situated?
   a in front of the bladder and behind the rectum
   b behind the bladder and in front of the rectum
   c on the lateral walls of the pelvis
   d at the entrance of the vulva

5 What is the cervix?
   a a thick muscular structure that opens into the vagina
   b an outer covering of the uterus
   c the largest and main part of the uterus
   d the dome-shaped part of the uterus

6 What is the inner mucous membrane lining of the uterus called?
   a perimetrium
   b perineum
   c myometrium
   d endometrium

7 What is the primary function of the testes?
   a to store seminal fluid
   b to produce and maintain sperm cells
   c development of the male secondary characteristics
   d to nourish immature sperm cells

8 Where in the male reproductive system are sperm cells stored to maturation?
   a vas deferens
   b penis
   c epididymis
   d Cowper's glands

9 Which part of the vulva protects the entrance to the vagina?
   a labia majora
   b labia minora
   c clitoris
   d mon pubis

Exam-style questions

10 Which of the following statements is **true**?
   a The seminal vesicles secrete an alkaline fluid which contains bacteria.
   b Vas deferentia are the tubes through which sperm is released.
   c The prostate gland lies in front of the symphysis pubis and behind the rectum.
   d The male urethra can only serve as a pathway for semen.

11 State two functions of the reproductive system. 2 marks

12 State two functions of each of the following structures:
   a ovaries
   b uterus
   c fallopian tubes. 6 marks

13 Name the hormones responsible for developing the breast tissue. 2 marks

14 Name the hormone that is only produced during pregnancy. 1 mark

15 List the parts of the male reproductive system. 5 marks

12 The digestive system

Introduction

In the digestive system, food is broken down and made soluble before it can be absorbed by the body for nutrition. Food is taken in through the mouth, mechanically broken into smaller particles and chemically broken into subunits. These are absorbed into the bloodstream, from where they can be utilised by the body.

Waste materials that are not required by the body are passed through to be eliminated. Once food has been absorbed by the body, it is converted into energy to fuel the body's activities and into components for growth and renewal. This process is known as metabolism.

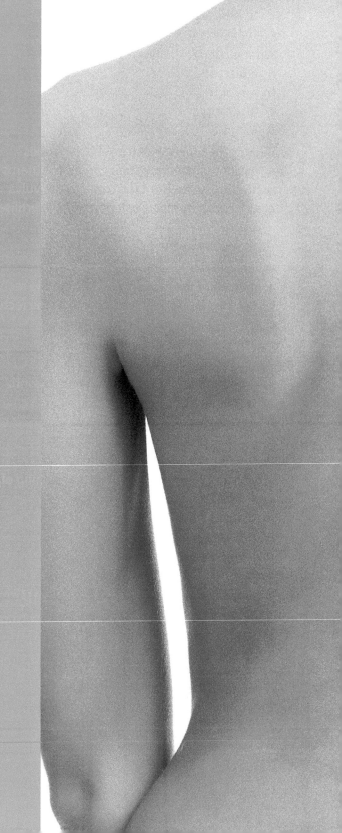

OBJECTIVES

By the end of this chapter you will understand:

- the functions of the digestive system
- the process of digestion; from the ingestion of food to the elimination of waste
- the structure and functions of the organs associated with digestion
- the absorption and utilisation of nutrients in the body
- the sources and functions of the main food groups required for good nutrition and health
- common pathologies of the digestive system
- the interrelationships between the digestive and other body systems.

In practice

It is essential for therapists to have a good knowledge of the process of digestion to understand how the body utilises nutrients for efficient and healthy body function.

Understanding the structure of the digestive system and its links with the parasympathetic nervous system can also help therapists to understand the link between digestive disorders and stress.

Functions of the digestive system

The digestive system serves two major functions:

1   It breaks down food and fluid into simple chemicals that can be absorbed into the bloodstream and transported throughout the body.

2   It eliminates waste products through excretion of faeces via the anal canal.

The structure and function of digestive organs

Digestion occurs in the alimentary tract, which is a long, continuous muscular tube extending from the mouth to the anus. The process of breaking down food is called digestion. Digestion involves the following processes: ingestion, absorption, assimilation and elimination (defaecation).

Ingestion

This is the act of taking food into the alimentary canal through the mouth.

Absorption

This is the movement of soluble materials out through the walls of the small intestine. Nutrients are absorbed through the villi and pass out into the network of blood and lymph vessels to be delivered to various parts of the body.

Assimilation

This is the process by which digested food is used by the tissues after absorption.

Elimination (defaecation)

This is the expulsion of the semi-solid waste called faeces through the anal canal.

The process of digestion

The digestive system serves two major functions:

● **Mechanical digestion** – the breakdown of solid food into smaller pieces by the chewing action of the teeth, known as mastication, and the churning action of the stomach, assisted by peristalsis.

- **Chemical digestion** – the breakdown of large molecules of carbohydrates, proteins and fats into smaller ones by the action of digestive enzymes.

The structure of the digestive system

The digestive system consists of the following parts:

- mouth
- pharynx
- oesophagus
- stomach
- small intestine (consisting of the duodenum, jejunum and the ileum)
- large intestine (consisting of the caecum, appendix, colon and rectum)
- anus.

The **pancreas**, **gall bladder** and the **liver** are accessory organs to digestion.

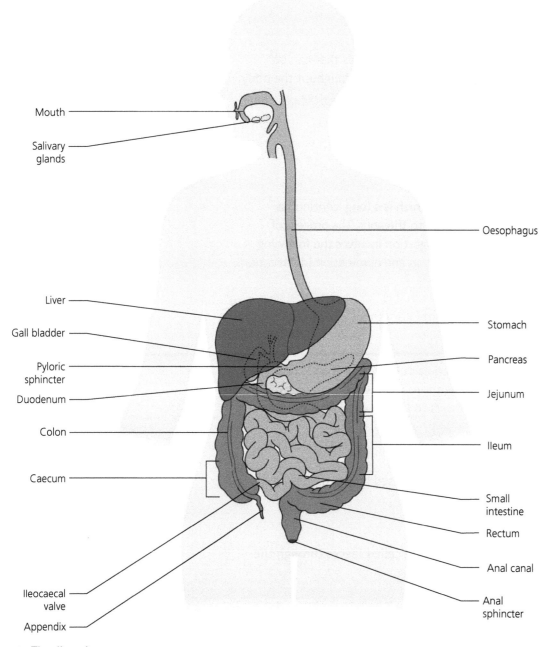

Mouth
Salivary glands
Liver
Gall bladder
Pyloric sphincter
Duodenum
Colon
Caecum
Ileocaecal valve
Appendix

Oesophagus
Stomach
Pancreas
Jejunum
Ileum
Small intestine
Rectum
Anal canal
Anal sphincter

▲ The digestive organs

Overview of the main digestive organs

Table 12.1 The function of the main digestive organs

Digestive organ	Function
Mouth	Commencement of digestion Food is chewed and mixed with saliva
Pharynx	Swallowing projects food down the oesophagus
Oesophagus	Pushes the food onwards to the stomach
Stomach	Mechanical breakdown of food Commences digestion of protein
Small intestine	Chemical breakdown of food Absorption of digested food
Large intestine	Formation and storage of faeces before defaecation
Anus	Defaecation (expulsion of faeces)

The following are accessory organs to digestion:

- The **pancreas** has two main functions – an exocrine function that helps in digestion (production of pancreatic juice) and an endocrine function that regulates blood sugar.
- The **gall bladder** stores bile, which is produced in the liver, until it is needed for digesting fatty foods in the duodenum of the small intestine.
- The **liver** has many functions; the main ones regarding digestion include bile production and excretion, and the metabolism of fats, proteins, and carbohydrates.

Activity

Take a blank body template and draw on the digestive organs, add labels, and then number them in the correct order to reflect the progression of digestion.

Mouth

The digestive system commences in the mouth. Food is broken up into smaller pieces by the action of the jaws and the teeth, and shaped into a ball, or bolus, by the tongue.

The tongue contains many small ridges, known as papillae, which help it to grip and move food around the mouth. Taste buds are hidden in and around some of the papillae and produce the sense of taste by detecting chemicals in food.

Mastication renders the food into small enough pieces to be swallowed and also allows saliva to be thoroughly mixed in.

The smell and sight of food triggers the reflex action of saliva secretion in the mouth. Saliva enters the mouth from three pairs of salivary glands. These are the:

- **sublingual glands** – located in the lower part of the mouth on either side of the tongue
- **submandibular glands** – located inside the arch of the mandible
- **parotid glands** – located superficial to the masseter muscle.

Saliva, containing the enzyme **salivary amylase**, or ptyalin, commences the digestion of starch, or carbohydrates, in the mouth.

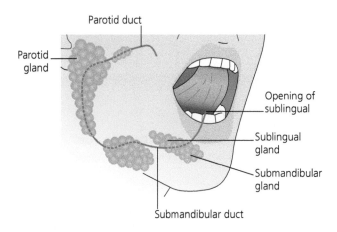

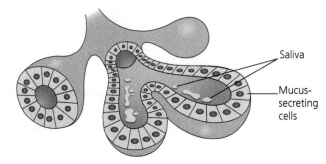

▲ The salivary glands

Pharynx and oesophagus

The ball of food is projected to the back of the mouth. The muscles of the pharynx force the food down the oesophagus, which is a long, narrow tube linking the pharynx to the stomach. A lubricative substance called mucus, secreted from the lining of the oesophagus, makes the food easier to swallow. The food is then conveyed by **peristalsis** down the oesophagus to the stomach.

Stomach

The stomach is a curved J-shaped muscular organ, positioned in the left-hand side of the abdominal cavity below the diaphragm.

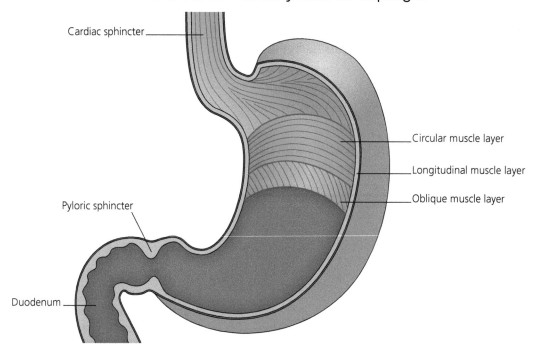

Cardiac sphincter

Circular muscle layer

Longitudinal muscle layer

Oblique muscle layer

Pyloric sphincter

Duodenum

▲ The stomach

Food enters the stomach via the cardiac sphincter, which is a strong circular muscle at the junction of the stomach and the oesophagus. Its function is to control the entry of food into the stomach.

The peritoneum serous membrane lines the abdominal cavity, supporting the alimentary canal and secreting a serous fluid, which prevents friction between different organs.

The stomach consists of four layers as shown in Table 12.2.

Table 12.2 The layers of the stomach

Stomach layer	Description
Muscular coat	Consists of longitudinal, circular and oblique muscle fibres which assist the mechanical breakdown of food
Sub-mucous coat	Made up of areolar tissue containing blood vessels and lymphatics
Mucous coat	Secretes mucus to protect the stomach lining from the damaging effects of the acidic gastric juice
Surface epithelium	Infolded into numerous tubular gastric glands which secrete gastric juice

Rugae

Rugae are folds in the mucous lining of the stomach that allow the stomach to expand when a bolus (ball) of food enters it.

Peritoneum

The peritoneum is a serous membrane that lines the abdominal cavity, supporting the alimentary canal and secreting a serous fluid which prevents friction.

Functions of the stomach

The functions of the stomach are to:

- churn and mechanically break up large particles of food
- mix food with gastric juice to begin its chemical breakdown
- commence the digestion of protein
- absorb alcohol.

Chemical digestion in the stomach

In the presence of food, the endocrine cells in the stomach walls secrete the hormone gastrin, which stimulates the production of gastric juice.

Gastric juice

The main constituents of gastric juice, which is produced and secreted by cells in the stomach wall, are:

- **water**, which helps to liquefy food
- **hydrochloric acid**, which provides the acidic conditions needed for pepsin to become active, kills any germs present in food and prepares it for intestinal digestion

- **mucus**, which is secreted by the neck cells in the stomach wall; it protects the stomach lining from the damaging effects of the acidic gastric juice
- **pepsinogen**, an enzyme precursor that is converted into pepsin in the acidic environment created by gastric juice.

Enzymes in the stomach include:

- **pepsin** – the main gastric enzyme which starts the digestion of proteins, breaking them up into polypeptides
- **rennin** – an enzyme found in the gastric juices of infants that curdles milk protein.

Food stays in the stomach for approximately five hours until it has been churned to a liquid state called chyme. Chyme is then released at intervals into the first part of the small intestine. The exit from the stomach is controlled by the pyloric sphincter, which sits at the junction of the stomach and the duodenum.

The small intestine

The small intestine is approximately three metres long and consists of three parts:

1. the **duodenum**, the first and shortest part of the small intestine
2. the **jejunum**, which lies between the duodenum and the ileum
3. the **ileum**, the longest segment of the small intestine where the main absorption of food takes place.

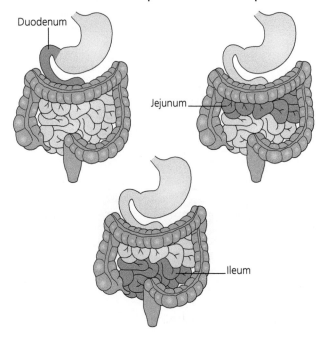

▲ The small intestine

The small intestine consists of the following four layers:

1 peritoneum

2 muscular coat, excluding the oblique fibres

3 sub-mucous layer containing numerous blood and lymph vessels, and nerves

4 circular folds of mucosa, which protect the intestine from bacteria.

The special features of the small intestine are the thousands of minute projections called **villi**, each containing a lymph vessel called a **lacteal**. The villi have a network of capillaries into which the nutrients pass to be absorbed into the bloodstream.

The epithelium of the villi is made up of tall columnar intestinal absorptive cells called **enterocytes** and **goblet cells**. Goblet cells secrete mucin for lubrication of the intestinal contents and protection of the epithelium.

KEY FACT

There are over 400 million villi on the lining of the small intestine, creating a surface area of approximately 250 m². This huge area is necessary for the absorption of water and nutrients.

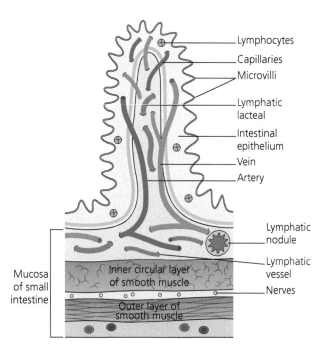

▲ The villi of the small intestine

Chemical breakdown of food in the small intestine

The muscles in the wall of the small intestine continue the mechanical breakdown of food by peristaltic movements, while chemical digestion is brought about by the following juices, which prepare the food to be absorbed into the bloodstream.

- **Bile** is a green alkaline liquid consisting of water, mucus, bile pigments, bile salts and cholesterol. It is produced in the liver and stored in the gall bladder. Its function is to neutralise the chyme and break up any fat droplets in a process called **emulsification**.

- **Pancreatic juice** is produced by the pancreas and the enzymes contained within it continue the digestion of protein, carbohydrates and fats.

- **Intestinal juice** is released by the glands of the small intestine. It neutralises hydrochloric acid coming from the stomach and contains digestive enzymes that facilitate the digestion and absorption of food. It also stimulates the release of gastrointestinal hormones into the bloodstream.

Hormones involved in digestion

Cholecystokinin (CCK)

Cholecystokinin (coal-lee-cyst-ta-ky-nin) is a hormone that is secreted by cells in the lining of the duodenum of the small intestine in response to chyme that has high fat or protein content. CCK stimulates:

- the gall bladder to contract and release bile into the intestine

- the secretion of enzymes by the pancreas.

Secretin

This is a hormone that is released by cells in the duodenum and its role is mainly in improving digestion, while protecting the stomach and intestines.

The main role of secretin is to stimulate the pancreas to secrete digestive juices that are rich in bicarbonates in order to neutralise the acid from the stomach as it passes into the small intestine. Secretin also stimulates the stomach to produce the enzyme pepsin, which helps to break down proteins.

Enzymes in the small intestine

The enzymes made by the pancreas include:

- pancreatic proteases (pro-tea-ay-sis), which help to digest proteins
- pancreatic amylase, which helps to digest sugars (carbohydrates)
- pancreatic lipase, which helps to digest fat.

Pancreatic proteases

- **Trypsin and chymotrypsin** are both digestive enzymes that are produced and secreted by the pancreas. The function of these enzymes is to help break down the large protein molecules that are taken in as food. They are known as protease enzymes because their function is to break down proteins. Without the action of these enzymes, our bodies would be unable to access the amino acids that are essential for tissue building and repair.
- **Enteropeptidase** (also called **enterokinase**, enter-roak-kin-nayys) is an enzyme that is produced by the glands of Brunner in the membrane lining of the duodenum and is involved in digestion. It converts trypsinogen into its active form trypsin, resulting in the subsequent activation of pancreatic digestive enzymes.

Pancreatic amylase

Pancreatic amylase acts on more complex sugars (carbohydrates), hydrolysing (breaking down) dietary starch into disaccharides and trisaccharides, which are then converted by other enzymes to glucose to supply the body with energy.

Pancreatic lipase

Pancreatic lipase is the principal enzyme that breaks down fat molecules in the digestive system. It converts triglycerides (with three fatty acid chains) to monoglycerides (with one fatty acid chain).

- **Monoglycerides** consist of one fatty acid chain.
- **Diglycerides** consist of two fatty acid chains.
- **Triglycerides** consist of three fatty acid chains.

Carbohydrate digestion

Carbohydrate digestion is completed by the following enzymes, which split disaccharides into monosaccharides:

- **maltase** – splits maltose into glucose
- **sucrase** – splits sucrose into glucose and fructose
- **lactase** – splits lactose into glucose and galactose.

Mono, di and polysaccharides

- **Monosaccharides** are the simplest form of carbohydrate (simple sugar). Monosaccharides are used by the body's cells to produce energy. Examples of monosaccharides are glucose, fructose and galactose.
- **Disaccharides** are formed by two monosaccharides joined together (double sugar). Disaccharides are usually used to store energy. Examples include maltose, sucrose and lactose.
- **Polysaccharides** are large molecules – long-chain carbohydrates made up of multiple monosaccharides. Polysaccharides are used for energy storage (examples include starch and glycogen) or structural support (an example is cellulose).

Protein digestion

Protein digestion is completed by peptidases (pep-tied-days-es), which split short chain **polypeptides** into **amino acids**.

Peptidase, also known as protease, is an enzyme with a very important role in the hydrolysis of proteins. Hydrolysis is the process of breaking down bigger molecules into smaller parts, in this case breaking proteins into peptides and then into even smaller units called amino acids.

Amino acids are joined together by peptide bonds to form proteins:

- two amino acids join together to form a **dipeptide**
- three amino acids join together to form a **tripeptide**
- more than three amino acids join together to form a **polypeptide**.

Pancreatic enzymes split polysaccharides into di- and tri- peptides:

- pH rises to about neutral, enabling the next enzymes in the process to accomplish the final breakdown of the polypeptide strands
- protein-digesting enzymes from the pancreas and small intestine continue working until almost all pieces of protein are broken into strands of two or three amino acids, dipeptides and tripeptides, or into single amino acids.

Absorption of digested food

The absorption of the digested food takes place in the jejunum, but mainly in the ileum. It occurs by diffusion from the intestinal tract, through the walls of the villi, into the rich network of blood capillaries. Each villus also contains a lymph vessel called a lacteal, into which **fatty acids** and **glycerol** can pass.

Simple sugars from carbohydrate digestion and amino acids from protein digestion pass into the bloodstream via the villi and are then carried to the liver via the hepatic portal vein to be processed. Vitamins and minerals also travel across to the blood capillaries of the villi and are absorbed into the bloodstream to assist in normal body functioning and cell metabolism. The products of fat digestion pass into the intestinal lymphatics and are carried through the lymphatic system before they eventually reach the blood circulation.

How the body's nutrients are assimilated

Once all the nutrients have been absorbed into the bloodstream they are transported to the body's cells for metabolism as shown in Table 12.3.

Table 12.3 The uses of nutrients in metabolism

Nutrient	Origin	Use in metabolism
Glucose	End product of carbohydrate digestion	Used to provide energy for the cells to function
Amino acids	End products of protein digestion	Used to produce new tissues, repair damaged cell parts and to formulate enzymes, plasma proteins and hormones
Fatty acids and glycerol	End products of fat digestion	Used to provide energy, in addition to glucose. Those fats which are not required immediately by the body are used to build cell membranes and some are stored under the skin (where they provide insulation) or around vital organs such as the kidneys and the heart

When the nutrients in food have been absorbed and assimilated by the body, any undigested material that remains passes into the large intestine, where it is eventually eliminated from the body.

Large intestine

The large intestine is formed of the **caecum**, **appendix**, **colon** and **rectum**. It coils around the small intestine and is characterised by:

- three bands of longitudinal muscle
- deep, longitudinal folds of mucosa, which increase in the rectum
- numerous tubular glands, which secrete mucus from their goblet cells.

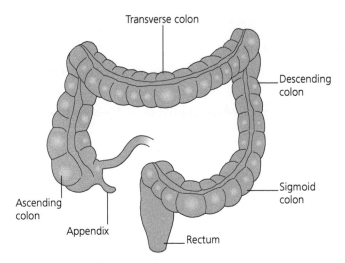

Transverse colon

Descending colon

Sigmoid colon

Ascending colon

Appendix

Rectum

▲ The large intestine

The parts of the large intestine are:

- the **caecum**, a small pouch to which the appendix is attached and into which the ileum opens through the ileocecal valve
- the **appendix**, which has no known function in humans
- the **colon**, the main part of the large intestine, which is divided into four sections:
  - the **ascending colon**, which is the part that passes upwards on the right side of the abdomen from the caecum to the lower edge of the liver
  - the **transverse colon**, which is the longest and most mobile part; it extends across the abdomen from right to left below the stomach
  - the **descending colon**, which is the part that passes downwards along the left side of the abdominal cavity to the brim of the pelvis
  - the **sigmoid colon**, which is the S-shaped part of the large intestine between the descending colon and the rectum.
- the **rectum**, which is the last part of the large intestine, extending from the sigmoid colon to the anal canal. It is firmly attached to the sacrum and ends about 5 cm below the tip of the coccyx, where it becomes the anal canal. Faeces are stored in the rectum before defaecation
- the **anus** is an opening at the lower end of the alimentary canal. Faeces are discharged through the anal canal. The anus is guarded by two sphincter muscles:
  - the **internal sphincter**, which is composed of smooth muscle under involuntary control
  - the **external sphincter**, which this is composed of skeletal muscle under voluntary control.

The anus remains closed through the action of these sphincters, except during defecation.

The functions of the large intestine

The functions of the large intestine are:

- absorption of most of the water from the faeces in order to conserve moisture in the body
- formation and storage of faeces, which consist of undigested food, dead cells and bacteria
- production of mucus to lubricate the passage of faeces
- the expulsion of faeces out of the body through the anus.

Overview of the accessory organs to digestion

- The **liver** has many important functions in metabolism. One of these is to regulate the nutrients that are absorbed from the small intestine to make them suitable for use in the body's tissues.
- The **gall bladder** stores bile and releases it when needed.
- The **pancreas** secretes pancreatic juice, which contains enzymes to continue the digestion of protein, carbohydrates and fats. It also secretes the hormone insulin, which is important in carbohydrate metabolism.

The liver

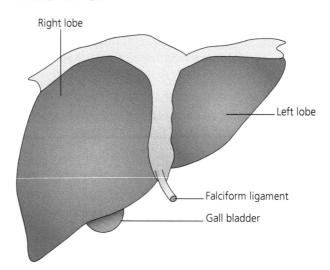

Right lobe

Left lobe

Falciform ligament

Gall bladder

▲ The liver

The liver is the largest gland in the body and is situated in the upper right-hand side of the abdominal cavity under the diaphragm. It has a soft reddish-brown colour and four lobes. It is made up of cells called hepatocytes. The liver receives oxygenated blood from the hepatic artery *and* deoxygenated blood from the hepatic portal vein. Blood from the digestive tract, which is carried in the hepatic portal vein, brings newly absorbed nutrients into the sinusoids and nourishes the liver cells. The liver is a vital organ with many important functions in the metabolism of food. One of these is to regulate the nutrients absorbed from the small intestine to make them suitable for use in the body's tissues. Other functions are described in Table 12.4.

Functions of the liver

Table 12.4 Significance of the liver functions

Function	Significance
Secretion of bile	Bile is manufactured by the liver but is stored and released by the gall bladder to assist the body in the breakdown of fats.
Regulation of blood sugar levels	When the blood sugar level rises after a meal, the liver cells store excess glucose as glycogen. Some glucose may also be stored in the muscle cells as muscle glycogen. When both these stores are full surplus glucose is converted into fat by the liver cells.
Regulation of amino acid levels	As our bodies cannot store excess protein and amino acids, they are processed by the liver. Some are removed by the liver cells and are used to make plasma proteins. Some are diverted for use by the cells in the body's tissues, while the rest are deaminated and excreted as urea in the kidneys.
Regulation of the fat content of blood	The liver is involved in the processing and transporting of fats. Those absorbed in the diet are used for energy and excess fats are stored in the tissues.
Regulation of plasma proteins	The liver is active in the breakdown of worn-out red blood cells.
Detoxification	The liver detoxifies harmful wastes (for example, from the diet and drugs) and excretes them in bile or through the kidneys.
Storage	The liver stores vitamins A, D, E, K and B12 and the minerals iron, potassium and copper. The liver can also hold up to a litre of blood. During exercise, the liver supplies extra blood and so increases oxygen transport to the muscles.
The production of heat	Due to its many functions, the liver generates heat. This keeps the body warm.

The gall bladder

The gall bladder is a pear-shaped organ attached to the posterior and inferior surface of the liver by the cystic and bile ducts.

The **Sphincter of Oddi** is the smooth muscle ring that surrounds the common bile duct at the point at which it enters the duodenum.

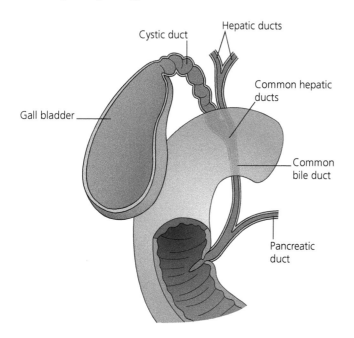

▲ The gall bladder

Functions of the gall bladder

The gall bladder concentrates and stores bile that is produced by the liver until it is needed. It releases bile into the common bile duct for delivery to the duodenum.

Bile

Bile is a thick alkaline liquid that is produced in the liver as a result of the breakdown of red blood cells. Bile is partially an excretory product and partially a digestive secretion. Bile salts (sodium and potassium) play a role in emulsification and breakdown of large fat globules.

Bile is a slightly alkaline liquid, with a pH of 7–8. It consists of water, bile acids, bile salts, cholesterol, phospholipids, electrolyte chemicals and bile pigments.

The most important bile pigments are bilirubin and biliverdin. Bilirubin is orange–yellow and is oxidised from biliverdin, which is green.

When worn-out red blood cells are broken down, substances such as iron and globin are recycled but some haemoglobin is broken down to form bilirubin, which is excreted into the bile ducts. Bilirubin is eventually broken down in the intestines and one of its breakdown products gives the faeces their normal brown colour.

The pancreas

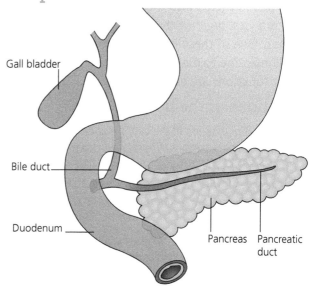

▲ The pancreas

The pancreas is situated behind the stomach, between the duodenum and the spleen. It is divided into a head, body and tail. The head is the expanded portion that fits into the c-shaped curve of the duodenum. The pancreas is composed of numerous lobules, each containing secretory alveoli (small sac-like cavities) which contain cells that produce pancreatic juice. The islets of Langerhans, which produce insulin, are located between the alveoli.

Functions of the pancreas

The pancreas has two functions:

1 **Exocrine function** – the pancreas secretes pancreatic juice, which contains water, alkaline salts, the enzymes lipase, pancreatic amylase, trypsinogen and chymotrypsinogen. The alkalinity of pancreatic juice helps to neutralise the acidity of chyme from the stomach and allows the pancreatic and intestinal enzymes to work.

2 **Endocrine function** – the islets of Langerhans are endocrine glands that secrete the hormone **insulin** into the bloodstream. The insulin circulates around the body in the blood and is important in carbohydrate metabolism.

KEY FACT

Without insulin, glycogen cannot be stored in the liver and muscles, and glucose cannot be oxidised to produce energy.

Study tip

When studying the digestion, it is helpful to break down the process into smaller sections that are easier to digest!

1 **The mouth** starts the process of mechanical breakdown of food.
2 The **pharynx** swallows the partially broken down food.
3 The **oesophagus** conveys food from pharynx to the stomach.
4 The **stomach** is concerned with the mechanical breakdown of food.
5 The **small intestine** is concerned with chemical breakdown of food. (Bile is added from the gall bladder and pancreatic juice is added from the pancreas.)
6 The **large intestine** is concerned with absorption of digested nutrients.
7 The **anus** is concerned with the absorption of water from the remaining indigestible food matter and the formation, storage and elimination of waste as faecal matter.

Nutrition

Nutrition is the utilisation of food to facilitate growth and to maintain the normal working of the body. Poor nutrition can have a dramatic effect on our general health, energy level, sleep pattern and stress response.

Table 12.5 Sources and functions of nutrients

Food group	Dietary sources	Main functions
Carbohydrates (starches and sugars)	Bread, cereals, potatoes, fruit and sugars	Body's main source of energy, required for the metabolism of other nutrients such as proteins and fats
Proteins	First-class proteins are fish, milk, egg and meat Second-class proteins include pulses, beans and peas	Necessary for the growth and repair of the body tissues and are used in the production of hormones and enzymes
Fats (classified as saturated or unsaturated, depending on whether they are solid (saturated) or liquid (unsaturated) at room temperature)	Meat, milk, cheese, butter and eggs	Sources of stored energy Also offer support and protection for the body and are used to build cell structures
Water (although not usually considered as food, it is an essential nutrient needed by every part of the body)	Fresh water, fruit and vegetables	Aids digestion and elimination Essential in maintaining the body's fluid balance and in the transport of substances around the body
Fibre (although fibre is not broken down into nutrients, it is a component for effective digestion)	Pulses, peas, beans, brown rice, wholemeal bread, jacket potatoes and green leafy vegetables	Aids digestion and bowel functioning Provides bulk in food to satisfy the appetite

Food group	Dietary sources	Main functions
Vitamins (divided into two groups according to whether they are soluble in water or fat)		Essential for normal physiological and metabolic functioning of the body Regulate the body's processes and contribute to its resistance to disease
Vitamin A (fat soluble)	Carotene in carrots, liver, kidney, eggs, dairy products, fish and liver oils	Essential for healthy vision, healthy skin and mucous membrane
Vitamin D (fat soluble)	Fish liver oils, fatty fish, margarine and eggs Also synthesised in the skin using ultraviolet light	Essential for healthy teeth and bones, it maintains the blood calcium level by increasing calcium absorption from food
Vitamin E (fat soluble)	Peanuts, wheatgerm, milk, butter and eggs	Inhibits the oxidation of fatty acids that help form cell membranes
Vitamin K (fat soluble)	Green leafy vegetables, cereals, liver and fruit	Essential for blood clotting
Vitamin B1 (water soluble)	Egg yolk, liver, milk, wholegrain cereals, vegetables and fruit	Necessary for the steady release of energy from glucose
Vitamin B2 (water soluble)	Milk, liver, eggs and yeast	Essential for using energy released from food
Vitamin B5 (water soluble)	Wholegrain cereals, yeast extract, liver, beans, nuts and meat	Involved in the breakdown of glucose to release energy
Vitamin B6 (water soluble)	Wholegrain cereals, yeast extract, liver, meat, nuts, bananas, salmon and tomatoes	Necessary for the metabolism of protein and fat
Vitamin B12 (water soluble)	Liver, kidney, milk, eggs and cheese	Necessary for the formation of red blood cells in bone marrow Also involved in protein metabolism
Folic acid (water soluble)	Liver, kidney, fresh leafy vegetables, oranges and bananas	Essential for the normal production of red and white blood cells
Vitamin C (water soluble)	Citrus fruits and blackcurrants	Assists in the formation of connective tissue and collagen Helps prevent bleeding and aids healing
Minerals		Provide the body with materials for growth and repair, and for the regulation of body processes Needed in trace amounts and used to build bone, work muscles, support various organs and transport oxygen and carbon dioxide
Calcium	Milk, egg yolk, cheese and green leafy vegetables	Essential for the formation of healthy bones and teeth, blood coagulation and the normal function of muscles and nerves

Food group	Dietary sources	Main functions
Iron	Liver, kidney, red meats, egg yolk, nuts and green vegetables	Essential for the production of haemoglobin in red blood cells
Phosphorus	Cheese, eggs, white fish, wholemeal bread, peanuts and yeast extract	Important in the formation of bones and teeth, muscle contraction and the transmission of nerve impulses
Sulfur	Egg yolk, fish, red meat and liver	An important component of structural proteins (those in the skin and hair)
Sodium and chlorine	Table salt, bacon, kippers and is found in all body fluids	Maintains fluid balance in the body Necessary for the transmission of nerve impulses and contraction of muscle
Magnesium	Green vegetables and salad	Important for the formation of bone and is required for the normal functioning of muscles and nerves

Common pathologies of the digestive system

Anorexia nervosa

This is a psychological illness in which clients starve themselves, or use other techniques such as vomiting or laxatives, to induce weight loss. They are motivated by a false perception of their body image and a phobia of becoming fat. The result is a severe loss of weight with amenorrhea. Starvation can be life threatening.

Appendicitis

This is an acute inflammation of the appendix. The main symptom is abdominal pain located centrally and in the right lower abdomen over the appendix. It is usually treated by surgical removal (appendectomy).

Bulimia

This is a psychological illness which is characterised by overeating (bingeing), followed by self-induced vomiting.

Cancer of the colon

In the early stages, the signs and symptoms are vague and related to the location of the cancer. A dull abdominal pain may or may not be present. General symptoms include loss of weight, fatigue, anaemia and weakness. If the tumour is on the right side of the abdomen (caecum or ascending colon) symptoms of obstruction appear slowly, as tumours in this region generally tend to spread along the walls of the gut without narrowing the lumen. If on the left side (descending colon, sigmoid colon or rectum) the signs of obstruction appear earlier in the disease. There is constipation or diarrhoea with passage of pencil-shaped or ribbon-like stools. The blood in the stools may be red or dark in colour.

Cancer of the gall bladder

Indigestion and colicky pain may be present, especially after a fatty meal. The pain is located in the upper right quadrant of the abdomen and may be referred to the back, right shoulder, right scapula or between the scapula.

Cancer of the liver

The most common type of liver cancer is that which has spread from other areas of the body – a metastatic carcinoma. Spread usually occurs from those areas which supply the liver with blood. Commonly, liver cancer is due to secondary spread from the stomach, intestine or pancreas.

Cancer can also arise from the liver tissue – primary cancer. Liver cancer may be present as a swelling in the upper right quadrant, associated with jaundice or fluid in the abdomen. Other general symptoms may include weight loss, weakness and loss of appetite. This type of cancer is often well advanced when diagnosed, whether arising from the liver or secondary to cancer elsewhere in the body.

Cancer – oral

This may be caused by chronic irritation of the mucosa of the oral cavity, as in tobacco chewing. A recurrence of chronic ulcers of the mouth can lead to this type of cancer. Oral cancer may appear as a non-healing, slow-growing red ulcer or as a growth. Usually it is painful and firm to the touch.

Cancer of the pancreas

The person presents with severe weight loss and pain in the lower back. The pain increases a few hours after taking food and is worse when lying down. If the tumour is growing around the bile duct, obstruction may result in jaundice and diarrhoea. The accumulation of bilirubin under the skin causes severe itching. The jaundice may be so severe that the skin may turn from yellow to green or black as the bilirubin changes in structure. The reduction in bile slows down the absorption and digestion of fat, causing clay-coloured, foul-smelling stools and diarrhoea. The cancer spreads directly and rapidly to the surrounding tissues, including the lymph nodes and liver. The kidneys, spleen and blood vessels may also be involved. The symptoms may vary according to the tissues affected.

Cancer of the stomach

In the early stages, the person has chronic pain or discomfort in the upper part of the abdomen. Since the symptoms are vague, this cancer is often not diagnosed until it has spread considerably. There is weight loss, anaemia, loss of appetite and the person will feel easily fatigued. Vomiting is common and often the vomit contains blood. A mass may be felt in the upper abdomen. Indigestion and acidity is not relieved by medication.

Cirrhosis of the liver

Cirrhosis refers to a distorted or scarred liver as a result of chronic inflammation. The functional liver cells are replaced by fibrous or adipose connective tissue. The symptoms of cirrhosis include jaundice, oedema in the legs, uncontrolled bleeding and sensitivity to drugs. Cirrhosis may be caused by hepatitis, alcoholism, certain chemicals that destroy the liver cells or parasites that infect the liver.

Colitis

This is an inflammation of the colon. The usual symptoms are diarrhoea, sometimes with blood and mucus, and lower abdominal pain.

Constipation

This condition presents as a difficulty in passing stools or infrequent evacuation of the bowels. The causes may be dietary, due to reduced fibre and fluid intake, or due to certain medications or intestinal obstruction.

Diabetes mellitus

This is a carbohydrate metabolism disorder in which sugars are not oxidised to produce enough energy due to lack of the pancreatic hormone insulin. The accumulation of sugar leads to its appearance in the blood, then in the urine. Symptoms of diabetes mellitus include thirst, loss of weight and excessive production of urine.

Diabetes insipidus

This is a rare metabolic disorder in which a person produces large quantities of dilute urine and is constantly thirsty. It is due to a deficiency of the hormone ADH which regulates reabsorption of water in the kidneys. It is treated by administration of the hormone.

Diarrhoea

This condition presents with frequent bowel evacuation or the passage of abnormally soft or liquid faeces. It may be caused by intestinal infections or other forms of intestinal inflammation, such as colitis or irritable bowel syndrome.

Gall stone

This is a hard pebble-like mass which is formed within the gall bladder. The condition may be asymptomatic, or indigestion and colicky pain may be present. Changes in the composition of bile cause cholesterol and/or the bile pigment bilirubin to form stones. Stagnation of bile and inflammation of the gall bladder increase the concentration of bile and promote stone formation.

> ### In practice
> In the case of a client with gall stones, avoid massage to the upper right quadrant of the abdomen.

Haemorrhoids

This condition presents with abnormal dilatation of veins in the rectum. It is caused by increased pressure in the venous network of the rectum. If the haemorrhoids are chronic, they may be seen or felt as soft swellings in the anus.

Heartburn

This is a burning sensation felt behind the sternum and often appearing to rise from the abdomen up the oesophagus towards or into the throat. It is caused by regurgitation of the acidic stomach contents.

Hepatitis

This is an inflammation of the liver caused by viruses, toxic substances or immunological abnormalities.

- **Hepatitis A:** this infection is highly contagious and is transmitted by the faecal–oral route, via ingestion of contaminated food, water or milk. The incubation period is 15 to 45 days.
- **Hepatitis B:** this is also known as serum hepatitis and is more serious than hepatitis A. It lasts longer and can lead to cirrhosis, cancer of the liver and a carrier state. It has a long incubation period of 1.5–2 months. The symptoms may last from weeks to months. The virus is usually transmitted through infected blood, serum or plasma. However, it can spread by oral or sexual contact as it is present in most body secretions.
- **Hepatitis C:** this form can cause acute or chronic hepatitis and can also lead to a carrier state and liver cancer. It is transmitted through blood transfusions or exposure to blood products. Most clients with hepatitis are jaundiced but they can appear to be entirely healthy.

Hepatitis as a side-effect of drugs and alcohol intake is not infective.

Hernia

This is an abnormal protrusion of an organ or part of an organ through the wall of the body cavity in which it normally lies.

Hiatus hernia

This is the most common type of hernia and occurs when part of the stomach is protruding into the chest. This may cause no symptoms at all, but it can cause acid reflux when acid from the stomach passes to the oesophagus, causing pain and heartburn.

Jaundice

This is a yellowing of the skin or whites of the eyes caused by excessive bilirubin (bile pigment) in the blood. It is caused by a malfunctioning gall bladder or obstructed bile duct.

Irritable bowel syndrome (IBS)

This is a common condition in which there is recurrent abdominal pain and bloating, with constipation and/or diarrhoea. Clients with stressful lifestyles are more vulnerable to this illness. They defaecate infrequently, usually in the morning, but may feel that their bowel is not empty or they may pass pellet-like stools.

> **In practice**
>
> Remember that, in IBS, the lower abdomen can be painful and tender.
>
> Clients with this condition may need quick access to the toilet.
>
> Advise clients to avoid wind-producing foods, e.g. onion, beans.
>
> Relaxation by any form of therapy is often helpful.

Stress

Stress can be defined as any factor that affects physical or emotional wellbeing. Signs of stress in the digestive system include the development of ulcers, irritable bowel syndrome and indigestion.

Ulcer

This is a break in the skin or a break in the lining of the alimentary tract which fails to heal and is accompanied by inflammation. Peptic, duodenal and gastric ulcers can present with increased acidity, epigastric pain (in the upper central region of the abdomen) and heartburn. This may be worse when hungry or after consumption of irritating foods, such as spicy or fatty foods, and alcohol. It can present with similar symptoms to a hiatus hernia and reflux.

Interrelationships with other systems

The digestive system

The digestive system links to the following body systems.

Cells and tissues

In areas of the digestive system, such as the small intestine where absorption of nutrients is required, there is a thin lining of simple epithelium to allow for speedy absorption.

Skin

One of the skin's functions is to produce vitamin D, which helps in the absorption of calcium in the small intestine.

Skeletal

The maxilla and mandible, the larger bones in the face, support the jaw and teeth when food in ingested in the mouth.

Muscular

The action of peristalsis, which propels the food through the digestive tract, is due to the involuntary contraction of the smooth muscle in the alimentary canal. Skeletal facial muscles, such as masseter and buccinator, assist in chewing.

Circulatory

Nutrients are carried in the body to nourish the cells and tissues, and waste products are carried away by the blood to be eliminated.

Lymphatic

Lymphatic vessels called lacteals (in the villi of the small intestine) assist digestion by absorbing the products of fat digestion.

Respiratory

Oxygen absorbed from the lungs activates glucose from the digestive system to produce energy for cell metabolism.

Nervous

All the organs of the digestive system are stimulated by nerve impulses.

Endocrine

The pancreas secretes insulin from cells called the islets of Langerhans to help control blood sugar level.

Key words

Absorption: the movement of soluble materials out through the walls of the small intestine and into the bloodstream

Alimentary tract: a tube of the digestive system that extends from the mouth to the anus

Amino acids: the building blocks of polypeptides and proteins

Anus: the opening at the end of the alimentary canal through which solid waste matter leaves the body

Appendix: a tube-shaped sac attached to and opening into the lower end of the large intestine

Ascending colon: the first main part of the large intestine, which passes upwards from the caecum on the right side of the abdomen

Assimilation: the process by which digested nutrients are used by the tissues after absorption

Bile: a greenish-brown alkaline fluid that is secreted by the liver and stored in the gall bladder

Caecum: a pouch connected to the junction of the small and large intestines

Cholecystokinin (CCK): a hormone that is secreted by cells in the duodenum and stimulates the release of bile into the intestine and the secretion of enzymes by the pancreas

Chyme: acidic fluid that passes from the stomach to the small intestine, consisting of gastric juices and partly digested food churned together

Chymotrypsin: a digestive enzyme that helps to digest proteins in food

Chymotrypsinogen: a precursor of the digestive enzyme chymotrypsin

Colon: part of the large intestine and the final part of the digestive system; its function is to reabsorb fluids and process waste products from the body to prepare them for elimination

Defaecation: the discharge of faeces from the body

Digestion: the process of digesting food

Diglyceride: lipids with two fatty acid chains

Dipeptide: a short protein consisting of only two amino acids linked together by one peptide bond

Disaccharide: (also called a double sugar) a sugar formed when two monosaccharides (simple sugars) are joined together

Duodenum: the first part of the small intestine immediately beyond the stomach, leading to the jejunum

Elimination: the expulsion of the semi-solid waste called faeces through the anal canal

Emulsification: the process by which fat globules are broken up into smaller droplets by the action of bile salts

Enterocytes: absorptive cells in the small intestine

Enteropeptidase (enterokinase): an enzyme produced by cells of the duodenum that converts trypsinogen into its active form trypsin, resulting in the subsequent activation of pancreatic digestive enzymes

Enzyme: a protein catalyst that speeds up the rate of a chemical reaction in a living organism

Faeces: waste matter remaining after food has been digested

Fatty acids: the building blocks of the fat in our bodies and in the food we eat, usually joined together in groups of three, forming a molecule called a triglyceride

Gall bladder: the small sac-shaped organ beneath the liver, in which bile is stored after secretion by the liver and before release into the intestine

Gastric juice: a thin, watery, acid digestive fluid secreted by glands in the mucous membrane of the stomach

Glycerol: with fatty acids, part of a fat molecule

Hydrochloric acid: primary digestive acid that prevents harmful bacteria from entering the stomach, and helps in breakdown of proteins

Hydrolysis: the process of breaking down bigger molecules into smaller parts

Ileum: third portion of the small intestine, between the jejunum and the caecum

Ingestion: the act of taking food into the alimentary canal through the mouth

Intestinal juice: secretions by glands lining the walls of the intestines secretion

Jejunum: the part of the small intestine between the duodenum and ileum

Lacteals: the lymphatic vessels of the small intestine which absorb digested fats

Liver: a large, reddish-brown, glandular organ located in the upper right side of the abdominal cavity

Monoglyceride: lipid with one fatty acid chain

Monosaccharides: a simple sugar, consisting of one sugar unit that cannot be further broken down into simpler sugars

Mucus: a sticky substance that is used as a digestive lubricant (for food passing down the alimentary canal); also lines the wall of the stomach to protect against the acidic environment

Oesophagus: part of the alimentary canal which connects the throat to the stomach

Pancreas: a large gland behind the stomach that secretes digestive enzymes into the duodenum, and secretes insulin and glucagon to help control blood sugar level

Pancreatic amylase: an enzyme secreted by the pancreas into the small intestine to convert starches into sugars

Pancreatic juice: the clear alkaline digestive fluid secreted by the pancreas

Pancreatic lipase: an enzyme that breaks down fats, produced by the pancreas

Papillae: small ridges in the tongue that help grip and move food around the mouth

Pepsin: the chief digestive enzyme in the stomach, which breaks down proteins into polypeptides

Peptide: a compound consisting of two or more amino acids linked in a chain

Peptidase: an enzyme that breaks down peptides into amino acids

Pharynx: serves as a pathway for the movement of food from the mouth to the oesophagus

Polypeptides: a chain of amino acids

Polysaccharide: long-chain carbohydrate made up of smaller carbohydrates called monosaccharides, typically used by our bodies for energy or to help with cellular structure

Rectum: the final section of the large intestine, terminating at the anus

Rugae: coiled sections of tissue in the mucosal and submucosal layers of the stomach, that allow the stomach to expand

Saliva: watery liquid secreted into the mouth by salivary glands, providing lubrication for chewing and swallowing, and aiding digestion

Salivary amylase: a digestive enzyme produced by the salivary glands that converts starches to sugars

Secretin: a hormone released into the bloodstream by the duodenum (especially in response to acidity) to stimulate secretion by the liver and pancreas

Sigmoid colon: the S-shaped last part of the large intestine, leading into the rectum

Sphincter of Oddi: smooth muscle ring that surrounds the common bile duct of the gall

bladder at the point at which it enters the duodenum

Transverse colon: the middle part of the large intestine, passing across the abdomen from right to left below the stomach

Triglyceride: a lipid with three fatty acid chains

Tripeptide: a peptide consisting of three amino acids joined by peptide bonds

Trypsin: a digestive enzyme that breaks down proteins in the small intestine, secreted by the pancreas as trypsinogen

Trypsinogen: an inactive substance secreted by the pancreas, from which the digestive enzyme trypsin is formed in the duodenum

Villi: the tiny finger-shaped processes of the mucous membrane of the small intestine that serve in the absorption of nutrients

Revision summary

The digestive system

- **Digestion** is the process of breaking down food. It involves **ingestion**, **mechanical** and **chemical digestion**, **absorption**, **assimilation** and **elimination**.

- Digestion occurs in the **alimentary tract**, which extends from the mouth to the anus.

- **Peristalsis** is the co-ordinated, rhythmical contraction of the muscles in the wall of the alimentary tract.

- The digestive system consists of the **mouth**, **pharynx**, **oesophagus**, **stomach**, **small intestine**, **large intestine** and **anus**.

- The accessory organs to digestion are the **pancreas**, **gall bladder** and **liver**.

- The digestive process commences in the **mouth** where food is broken down by mastication and mixed with **saliva**.

- **Saliva** contains the enzyme **salivary amylase** which commences the digestion of **starch** in the mouth.

- The muscles of the **pharynx** force the food down the **oesophagus** to the **stomach**.

- In the **stomach**, food is mixed with **gastric juice** containing the enzyme **pepsin** which starts the

breakdown of proteins; **hydrochloric acid**, to kill germs present in food and to prepare it for intestinal digestion; and **mucus**, which protects the stomach lining from the damaging effects of the acidic gastric juice.

- The functions of the **stomach** are the mechanical breakdown of large particles of food and mixing food with gastric juice by churning to begin the digestion of protein and to absorb alcohol.

- Food stays in the stomach for approximately five hours until it has been churned to a liquid state called **chyme**.

- **Chyme** is then released at intervals into the first part of the **small intestine**.

- The **small intestine** consists of three parts – **duodenum**, **jejunum** and **ileum** (where absorption of food mainly takes place).

- Special features of the small intestine are the thousands of tiny projections called **villi**, a network of capillaries into which the nutrients pass to be absorbed into the bloodstream.

- The **small intestine** continues the mechanical breakdown of food by **peristalsis**, while the chemical digestion is brought about by **bile** (released by the gall bladder), enzymes in **pancreatic juice** (released by the pancreas) and **intestinal juice** (released by the walls of the small intestine), all of which prepare the food to be absorbed into the bloodstream.

- The function of **bile** is to neutralise the **chyme** and break up any fat droplets by **emulsification**.
- The enzymes contained within **pancreatic juice** continue the digestion of protein (**trypsin**), carbohydrates (**pancreatic amylase**) and fats (**pancreatic lipase**).
- **Intestinal juice** is released by the glands of the small intestine and completes the final breakdown of nutrients, including of **complex sugars** to **glucose** and of **protein** to polypeptides and then **amino acids**.
- Protein digestion is completed by **peptidases** which split short chain **polypeptides** into **amino acids**.
- The absorption of digested nutrients takes place in the **jejunum** and mainly in the **ileum**.
- Each villus contains a lymph vessel called a **lacteal** into which fatty acids and glycerol can pass.
- Simple sugars from carbohydrate digestion and amino acids from protein digestion pass into the bloodstream via the **villi** and are then carried to the **liver** via the **hepatic portal vein** to be processed.
- The **lacteals** (intestinal lymphatics) absorb products of fat digestion and carry them through the lymphatic system before they reach the blood circulation.
- Vitamins and minerals travel across to the blood capillaries of the villi and are absorbed into the bloodstream. They are important in cell metabolism and normal body function.
- **Glucose**, the end product of carbohydrate digestion, is used to provide energy for cells to function.
- **Amino acids**, the end products of protein digestion, are used to produce new tissues, repair damaged cell parts and formulate enzymes, plasma proteins and hormones.
- **Fatty acids** and **glycerol** are the end products of fat digestion.

- Fats are used primarily to provide energy, in addition to glucose. Those fats which are not required immediately by the body are used to build cell membranes and some are stored under the skin (where they insulate the body) or around vital organs such as the kidneys and the heart.
- When the nutrients in food have been assimilated by the body, undigested matter is passed into the large intestine and is eventually eliminated from the body.
- The **large intestine** is made up of the **caecum**, **appendix**, **colon** and **rectum**.
- The **colon** is the main part of the large intestine and is divided into **ascending**, **transverse**, **descending** and **sigmoid** colons.
- The **rectum** is the last part of the large intestine where **faeces** are stored before **defecation**.
- The functions of the large intestine are the absorption of most of the water from the faeces, formation and storage of faeces, production of mucus to lubricate the passage of faeces and the expulsion of faeces out of the body.
- The **anus** is an opening at the lower end (anal canal) of the alimentary canal, through which faeces are discharged.
- The **liver** is the largest gland in the body and is an accessory organ to digestion with many metabolic functions.
- The functions of the **liver** include the secretion of bile, regulation of blood sugar level, regulation of amino acids, regulation of the fat content of blood, regulation of plasma proteins, detoxification, glycogen storage and the production of heat.
- The **pancreas** is also an accessory organ to digestion. Its exocrine function is the secretion of pancreatic juice.
- The **gall bladder** is attached to the posterior and inferior surface of the liver and its function is to store bile produced by the liver until it is needed.

Test your knowledge questions

Multiple choice questions

1 The alimentary tract is a long continuous muscular tube extending from the:
   a mouth to anus
   b stomach to anus
   c small intestine to anus
   d large intestine to anus.

2 Which of the following structures completes digestion?
   a stomach
   b gall bladder
   c small intestine
   d large intestine

3 Which of the following statements is **true**?
   a The liver is situated in the upper left-hand side of the abdominal cavity.
   b The liver's internal structure is made up of cells called hepatocytes.
   c Bile is stored in the liver and released by the pancreas.
   d When blood sugar level is low, liver cells stores excess glucose.

4 Which of the following is produced in the stomach?
   a bile
   b pancreatic juice
   c pepsin
   d maltase

5 Where does protein digestion start?
   a mouth
   b small intestine
   c stomach
   d pancreas

6 Food stays in the stomach for approximately how long before it is churned to a liquid state?
   a two hours
   b five hours
   c three hours
   d one hour

7 Which of the following is responsible for the chemical reactions of digestion?
   a homeostasis
   b enzymes
   c absorption
   d peristalsis

8 Salivary amylase commences:
   a carbohydrate digestion
   b protein digestion
   c fat digestion
   d vitamin and mineral digestion.

9 What are the main constituents of gastric juice?
   a gastrin and pepsin
   b gastrin and pepsinogen
   c pepsin, hydrochloric acid and mucus
   d gastric amylase

10 Which structure produces the enzyme trypsin?
   a duodenum
   b liver
   c pancreas
   d gall bladder

Exam-style questions

11 State two functions of the digestive system.
                                    2 marks

12 a What is meant by the term *enzyme*?   1 mark
   b Give an example of a digestive enzyme.
                                    1 mark

13 a Where in the digestive system does the digestion of protein commence?   1 mark
   b Where in the digestive system does the absorption of digested food take place?
                                    1 mark

14 a Define the term *peristalsis*.   1 mark
   b Where in the digestive system does peristalsis occur?   1 mark

15 a What is bile?   1 mark
   b State two constituents of bile.   1 mark
   c Which part of the digestive system produces bile?   1 mark

13 The renal system

Introduction

The kidneys and their associated structures are all part of the excretory system, along with the skin, lungs and the intestines, which also contribute to the job of waste elimination from the body.

The renal system, also referred to as the urinary system, is made up the kidneys, ureters, bladder and urethra, which are involved in the processing and elimination of normal metabolic waste from the body, sometimes likened to the body's 'plumbing'.

OBJECTIVES

By the end of this chapter you will understand:

- the functions of the renal system
- the structure and functions of the individual parts of the renal system (kidneys, ureters, urinary bladder and urethra)
- common pathologies of the renal system
- the interrelationships between the renal and other body systems.

In practice

It is essential for therapists to have a working knowledge of the renal system in order to understand how fluid balance is controlled in the body and to have an appreciation of the role of the kidneys in detoxification.

Functions of the renal system

Maintaining homeostasis

The prime function of the renal system is to help maintain homeostasis by controlling the composition, volume and pressure of blood. It does this by removing and restoring selected amounts of water and dissolved substances.

Regulation of the composition and volume of body fluids

Waste products, such as urea and uric acid, along with excess water and mineral salts must be removed from the body in order to maintain good health. If these waste materials were allowed to accumulate in the body they would cause ill health. The primary function of the renal system is to regulate the composition and the volume of body fluids in order to provide a constant internal environment for the body.

Structures of the renal system

The renal system parts are shown in Table 13.1.

Table 13.1 Location and function of parts of renal system

Part of renal system	Location	Function
Kidneys × 2	Posterior wall of the abdomen on either side of the spine (between twelfth thoracic vertebra and the third lumbar vertebra)	Main functional organs of the renal system Site where blood is filtered and urine is processed
Ureters × 2	Long thin tubes that lead from each kidney down to the bladder	Transport urine from the kidneys to the bladder
Urinary bladder × 1	Lies in the pelvic cavity behind the symphysis pubis	Collects and temporarily stores urine
Urethra × 1 (you-reeth-ra}	Extends from the neck of the bladder to the outside of the body	Tube through which urine is discharged from the bladder and out of the body In men, also acts as conducting channel for semen

Kidneys

The kidneys are bean-shaped organs lying on the posterior wall of the abdomen on either side of the spine between the level of the twelfth thoracic vertebra and the third lumbar vertebra. Due to the position of the liver, the kidney on the right side of the body is slightly higher than the one on the left.

Structure of the kidney

A kidney has an outer fibrous renal capsule and is supported by adipose tissue. It has two main parts:

1 **Outer cortex:** this reddish-brown part is where fluid is filtered from the blood.

2 **Inner medulla:** this part is paler in colour and is made up of conical-shaped sections called renal pyramids. The renal pyramids consist mainly of tubules that transport urine from the cortex of the kidney, where urine is produced, to the **calyces**, or cup-shaped cavities in which urine collects before it passes through the ureter to the bladder. The point of each pyramid, called the papilla, projects into a **calyx**.

The medulla is the area where some materials are selectively reabsorbed back into the bloodstream.

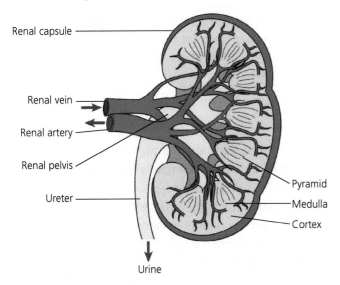

▲ Structure of a kidney

The **renal pelvis** is a large funnel-shaped cavity in the centre of the kidney that collects urine from the renal pyramids in the medulla and drains it into the ureter. The medial border of the kidney is called the hilus and is the area where the renal blood vessels enter and leave the kidney.

Nephrons

The cortex and the medulla contain tiny blood filtration units called nephrons (nef-frons). Nephrons are the functional units of the kidney and they extend from the renal capsule through the cortex and medulla to the cup-shaped renal pelvis. Nephrons are approximately 2–4 cm long and a single kidney has more than a million nephrons.

Urine production

Urine is produced by three processes:

1 filtration

2 selective reabsorption

3 secretion and collection.

1 Filtration

Blood enters the medulla of the kidney from the renal artery. Inside the kidney, the renal artery splits into a network of capillaries called the **glomeruli**, which filter the waste. Almost encasing each **glomerulus** is a sac called the **Bowman's capsule**.

The **afferent arteriole** brings blood to the glomerulus and the **efferent arteriole** takes blood away from the glomerulus.

The blood pressure in the glomerulus is maintained at a high level, assisted by the fact that the afferent arteriole that feeds into the glomerulus has a larger diameter than the efferent arteriole leaving it. This pressure forces fluid out through the walls of the glomerulus, together with some of the substances of small molecular size that are able to pass through the capillary walls, into the Bowman's capsule. This process is called **simple filtration**.

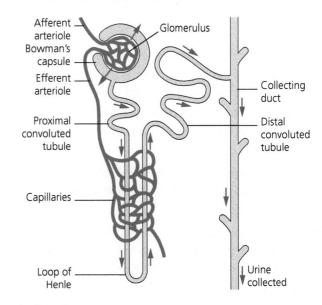

▲ A nephron

343

KEY FACT

Urine is the body's primary waste product and so the release of urine is the final step of all metabolism.

2 Selective reabsorption

The filtered liquid continues through a series of twisted tubes called the **convoluted tubules**, which are surrounded by capillaries. The tubules of the nephron that lead away from the Bowman's capsule are known as the **proximal convoluted tubules** and they straighten out to form a long loop called the **loop of Henle**. There are then another series of twists called the **distal convoluted tubules**, which lead to a straightened collecting duct, which in turn leads to the pelvis of the kidney and then to the ureter.

The composition of the filtered liquid alters as it flows through the convoluted tubules. Some substances contained within the waste, such as glucose, amino acids, mineral salts and vitamins, are reabsorbed back into the bloodstream.

This reabsorption process is selective as the amounts of these substances that pass back in to the blood depend on the level already present in the bloodstream. The reabsorption of salts and water is variable to achieve maintenance of a stable pH and electrolyte (sodium and potassium) balance of body fluids. Excess water, salts and the waste product urea are all filtered and processed through the kidneys. Filtered blood leaves the kidney via the renal vein.

3 Secretion and collection

Some substances are not completely removed from the blood in the glomerular filtrate, the residue of medicinal drugs for example. These substances are passed from the blood into the distal convoluted tubule by secretion in order that they may be excreted in the urine.

Secretion involves the transfer of water, hydrogen ions, creatinine, drugs, and urea from the blood into the collecting duct.

The waste remaining in the distal convoluted tubule (now known as urine) then flows via a collecting tubule to the renal pelvis of the kidney. From here it passes into the ureter to be passed to the bladder

and urethra, from where it is excreted through a process known as **micturition**.

Study tip

Remember, the processing of urine works like any other filtration system. For example, if we imagine the Bowman's capsule to be a teapot and liquid (in this case hot water) is poured into the pot to brew tea.

Inside the teapot, the tea brews. It is filtered as the tea is poured through a tea strainer and into a cup.

Just like tea through a tea strainer, the composition of the urine alters as it flows through the convoluted tubules. Some substances contained within the waste, such as glucose, are reabsorbed back into the bloodstream (think of adding sugar to the tea).

The tea in the cup equates to urine in the bladder. Urine is passed from the kidney via the ureter to the urinary bladder where it is stored before being released via the urethra.

The composition of urine

Urine is the concentrated filtrate from the kidneys. Its composition is approximately 96% water, 2% urea and 2% other substances, such as chloride salts, sodium ions, potassium ions, creatinine (a waste product from the normal breakdown of muscle tissue), other dissolved ions, inorganic and organic compounds (such as proteins, hormones and metabolites).

Urine is a pale watery fluid varying in colour according to its composition and quantity. It is usually acidic and its pH varies between 4.5 and 8, depending on blood pH. A typical average is around 6.0.

Much of the variation in the pH of urine occurs due to diet. For example, a high protein diets result in more acidic urine, but vegetarian diets generally result in more alkaline urine (but both within the typical pH range of 4.5–8).

The salts, chiefly sodium chloride, must be removed from the body (or reabsorbed back into the body) in sufficient quantities to keep the blood at its normal pH and to maintain the usual water and electrolyte balance. As maintenance of pH and salt concentration are both essential to the normal function of blood and cells, the kidneys are of paramount importance.

The amounts of certain substances in urine can be a good indication of health status. Urine tests can be used to diagnose some disorders. If urine contains glucose, this may indicate diabetes. Protein in the urine may indicate kidney failure. Urine tests can be used to confirm pregnancy, since a fertilised ovum releases a hormone which the mother excretes in the urine.

Factors that affect urine production

The kidneys maintain our water balance by producing urine of different concentrations.

- When the level of water in our blood plasma is low, more water is reabsorbed back into the bloodstream and the urine becomes more concentrated as a result.
- When the level of water in our blood plasma is high, less water is reabsorbed back into the bloodstream and the urine becomes more dilute as a result.

Factors affecting urine production include:

- **Body temperature** – if body temperature increases (due to exercise, fever or environmental factors), water is lost from the body in sweat. If you sweat, the amount of water in your body decreases. The kidneys compensate for this by producing a smaller volume of more highly concentrated urine.

- **Physical activity/exercise** – during exercise, blood flow to the kidneys is reduced due to an increase in the activity of the sympathetic nervous system. The reduction in blood flow to the kidneys is important in order to maintain blood pressure, due to the fact blood vessels dilate in the exercising muscles.

  Due to decreased blood flow in the kidneys, the volume of fluid filtered also decreases during exercise, resulting in a fall in urine production. In order to maintain equilibrium, the kidneys respond by conserving sodium and reabsorbing water, which in turn results in decreased urine production.

- **Cold weather** – in cold weather, the body tries to keep the core warm by constricting its blood vessels and reducing the flow of blood to the skin (vasoconstriction). Constriction of blood vessels causes the blood pressure to rise, because the same amount of blood has less space to flow through.

In order to regulate blood pressure, the kidney filters out some of the excess fluid from the blood, to reduce its volume. As the bladder fills up with the excess fluid, we feel the urge to urinate. Relieving a full bladder causes heat loss, so urinating sooner helps preserve your core warmth.

- **Hot weather** – when the body temperature rises during hot weather, the body responds by producing sweat, which removes body heat when it evaporates. The production of urine decreases because sweating causes dehydration.

- **High dietary salt intake** – high salt intake can result in increased water reabsorption in the kidneys, which reduces the volume of urine produced.

- **Water consumption** – the more we drink, the more dilute our blood plasma becomes. The kidneys respond by producing dilute urine to get rid of the excess water.

- **Tea, coffee and alcohol consumption** – tea, coffee and alcohol are diuretics which increase the volume of urine produced. Alcohol, in particular, causes the kidneys to produce a greater volume of more dilute urine. This can lead to dehydration.

- **Stress and anxiety** – these are common causes of frequent urination.

  Part of the body's stress response includes causing the body to eliminate waste as quickly as possible, hence the need to go to the toilet more frequently when anxious or stressed.

- **Medication (diuretics)** – diuretic drugs help rid the body of excess water and sodium by increasing a person's urinary output. Some diuretics act on the kidneys directly and others increase blood flow to the kidneys; either way these drugs cause a person to urinate more frequently to lose water from the body.

- **Blood pressure** – when the blood pressure inside the kidney tubules rises, less water is reabsorbed and the volume of urine increases. When the blood pressure inside the kidney tubules falls, more water is reabsorbed into the blood and the volume of urine decreases.

 **Activity**

The amount of urine produced by the body varies greatly according to various internal and external factors. Split into two groups. Write each of the following factors on a card and divide the cards between the two groups.

- High blood pressure
- Low blood pressure
- Medication (diuretics)
- Stress and anxiety
- Cold weather
- Hot weather
- High salt intake
- Alcohol
- Body temperature
- Exercise

Now work with your group colleagues to decide which factors increase and which decrease urine production. Indicate this by placing an arrow in the top right-hand corner of each card:

- an upwards arrow to denote an increase in urine production
- a downwards arrow to denote a decrease.

Swap cards with the other group and check their assessment of each factor. Did you all agree?

Functions of the kidney

The functions of the kidney are to:

- filter impurities and metabolic waste from blood, preventing poisons from accumulating in the body
- regulate the water and salt balance in the body
- maintain the normal pH balance of blood
- form urine
- regulate blood pressure and blood volume.

Role of the kidneys in fluid balance

The amount of fluid taken into the body must equal the amount of fluid excreted from it, in order for the body to maintain a constant internal environment. The balance between water intake and water output is controlled by the kidneys.

- **Water intake** – water is taken into the body as liquid in food and drink. It is absorbed during digestion.

- Some water is used up in the cells' metabolic activities.
- **Water output** – water is lost from the body in the following ways:
  - through the kidneys as urine
  - through the alimentary tract in the faeces
  - through the skin as sweat
  - through the lungs as saturated exhaled breath.

Hormones involved in water reabsorption

Antidiuretic hormone (ADH)

The kidneys are responsible for regulating the amount of water contained within the blood. The amount of water reabsorbed into the blood is controlled by ADH (also known as **vasopressin**), which is stored and released into the blood by the posterior lobe of the pituitary gland.

The release of ADH is triggered by dehydration. The hypothalamus in the brain detects when the water concentration of blood is low and triggers the release of ADH. An increase in the level of ADH increases the amount of water that is reabsorbed from the nephron back into the blood.

The reabsorption of water from the nephron into the blood decreases the volume of urine produced and increases the hydration level of the blood. This mechanism brings the water in the blood back to an acceptable level.

When the blood hydration level is back to normal, the hypothalamus moderates the secretion of ADH by the pituitary.

> **KEY FACT**
> This important **negative feedback** mechanism between the nervous and endocrine systems maintains the blood water concentration within normal limits and is the means by which fluid balance is controlled in the body.

Aldosterone

Aldosterone is a hormone that is secreted by the adrenal cortex of the adrenal glands. It regulates the reabsorption of sodium and water in the kidneys.

It plays a vital role in regulating blood pressure by acting on the kidney and the colon to increase the amount of sodium that is reabsorbed into the bloodstream and to increase the amount of potassium that is excreted in the urine.

Calcitonin

Calcitonin is a hormone that is secreted by the thyroid gland. The main action of calcitonin is to reduce the calcium level in the blood and it does this in two ways:

1 It inhibits the activity of the cells that are responsible for breaking down bone. (When bone is broken down, the calcium from the bone is released into the bloodstream.) The calcitonin, by its inhibitory action, reduces the amount of calcium being released into the bloodstream.

2 It also has a role in decreasing the reabsorption of calcium in the kidneys, resulting in a lower level of calcium in the bloodstream.

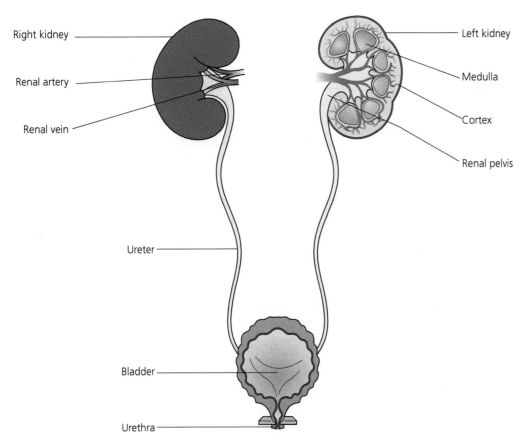

▲ The urinary organs

Ureters

The ureters are two very fine muscular tubes which transport urine from the renal pelvis of the kidney to the urinary bladder. They consist of three layers of tissue:

1 an outer layer of **fibrous tissue**

2 a middle layer of **smooth muscle**

3 an inner layer of **mucous membrane**.

Function of the ureters

Ureters propel urine from the kidneys into the bladder by the peristaltic contraction of their muscular walls.

Urinary bladder

This is a pear-shaped sac which lies in the pelvic cavity behind the symphysis pubis. The size of the bladder varies according to the amount of urine it contains. The bladder is composed of four layers of tissue:

1 a **serous membrane** which covers the outer surface

2 a layer of **smooth muscular** fibres

3 a layer of **adipose tissue**

4 an inner lining of **mucous membrane**.

Functions of the urinary bladder

The urinary bladder stores urine. It expels urine out of the body, assisted by the muscular wall of the bladder, the lowering of the diaphragm and the contraction of the abdominal cavity.

The expulsion of urine from the bladder is called **micturition** and is a reflex over which there is voluntary control. When the volume of urine in the bladder causes it to expand, stretch receptors in the bladder wall are stimulated to trigger urination. The micturition reflex causes the detrusor muscle in the wall of the bladder to contract and the internal urethral sphincter to relax. It is the combination of both the micturition reflex and voluntary relaxation of the urethral sphincter that allows urination to occur.

Urethra

The urethra (you-reeth-ra) is a canal which extends from the neck of the bladder to the outside of the body. The length of the urethra differs in males and females. The female urethra is approximately 4 cm in length, whereas the male urethra is longer at approximately 18–20 cm in length. The exit from the bladder is controlled by a round sphincter of muscles which must relax before urine can be expelled.

The urethra is composed of three layers of tissue:

1 a muscular coat which is continuous with that of the bladder

2 a thin spongy coat which contains a large number of blood vessels

3 a lining of mucous membrane.

Function of the urethra

The urethra serves as a tube through which urine is discharged from the bladder to the exterior. The urethra is longer in a male and it also serves as a conducting channel for semen.

Common pathologies of the renal system

Cancer of the bladder

This usually presents with blood in the urine, and urgency and pain on passing urine. Secondary symptoms may arise if the cancer has spread to the lungs, liver, lymph nodes and neighbouring tissues.

Cystitis

This is an inflammation of the urinary bladder, usually caused by infection of the bladder lining. Common symptoms are pain just above the pubic bone, lower back or inner thigh, blood in the urine and frequent, urgent urination with a burning or painful sensation. This condition is very common in women due to the shorter length of the female urethra.

> **In practice**
>
> Encourage a client with cystitis to increase their intake of fluids (water and cranberry juice). If symptoms of persist, they may need GP assessment and advice.
>
> Massage over the lower abdomen should be avoided, as this may be painful and risks inducing spasm.

Incontinence

This is a condition in which the individual is unable to control urination voluntarily. Loss of muscle tone and problems with innervation are associated with this condition.

Kidney stones

These are insoluble deposits of substances in the urine, which form solid stones in the renal pelvis of the kidney, ureter or bladder. This condition can be extremely painful. Stones are usually removed by surgery.

Nephritis

This is a general, non-specific term used to describe inflammation of the kidney. Glomerulonephritis (also known as Bright's disease) is an inflammation of the glomeruli in the kidneys. This condition is characterised by blood in the urine, fluid retention and hypertension.

Pyelonephritis

This is a bacterial infection of the kidney. In acute pyelonephritis there is pain in the back, high temperature and shivering fits. Treatment is usually with antibiotics.

Urinary tract infection

This is a bacterial infection of one or more of the structures of the renal system. Symptoms include fever, lower back pain, frequency of urination, a burning sensation on passing urine (urine may be bloodstained and cloudy). If the infection is severe, there may be pus as well as blood in the urine.

> **In practice**
>
> In the case of a client with a urinary tract infection, all forms of therapeutic treatment should be avoided until the infection has cleared.

Interrelationships with other systems: the renal system

The renal system links to the following body systems.

Cells and tissues

Transitional epithelium lines renal system organs, such as the bladder, which changes shape when stretched.

Skin

Like the renal system, the skin is also an excretory organ. When the skin loses excess water through sweating, the kidneys release less water in the urine to help maintain the body's fluid balance.

Skeletal

The kidneys and the bones of the skeleton help to control the amount of calcium in the blood by storing some in the bones and excreting some from the body in urine.

Muscular

Smooth muscle is responsible for the passage of urine through the urinary tract.

Circulatory

The kidneys filter the blood to avoid accumulation of poisons in the body.

Nervous

The relaxation and contraction of the bladder, and closing and opening of the sphincter muscles is under the control of the autonomic nervous system (sympathetic and parasympathetic nervous systems).

Digestive

Water is an essential nutrient which is needed by every part of the body to aid the metabolic processes. It is ingested in the diet and absorbed during the process of digestion. The colon absorbs most of the water from the faeces in order to conserve moisture in the body.

Key words

Aldosterone: a hormone that stimulates absorption of sodium by the kidneys and so regulates water and salt balance

Antidiuretic hormone (ADH): a hormone that increases the amount of water absorbed by the kidney and so increases blood pressure

Bowman's capsule: a cup-shaped structure around the glomerulus of each nephron of the kidney; acts as a filter to remove organic wastes, excess inorganic salts, and water

Calcitonin: a hormone secreted by the thyroid gland to lower levels of calcium (and phosphorous) in the blood; it can also decrease the resorption of calcium in the kidneys, again leading to lower blood calcium level

Calyx (calyces): a cup-shaped cavity inside the medulla of the kidney in which urine collects before it passes through the ureter to the bladder

Creatinine: a chemical waste product from muscle metabolism that is filtered, along with other waste products, from the blood and leaves the body via urine

Detrusor: a muscle in the wall of the bladder

Distal convoluted tubule: a portion of kidney nephron between the loop of Henle and the collecting tubule

Glomerulus: a cluster of capillaries around the end of a kidney tubule

Kidney: one of a pair of organs that lie on the posterior of the abdominal cavity and produce urine

Loop of Henle: the portion of a nephron that leads from the proximal convoluted tubule to the distal convoluted tubule

Medulla: the innermost part of the kidney

Micturition: the act of urinating

Nephron: the functional unit of the kidney that filters blood and forms urine

Proximal convoluted tubule: the convoluted portion of the nephron that lies between Bowman's capsule and the loop of Henle

Renal artery: the branch of the abdominal aorta that supplies oxygenated blood to the kidney

Renal pyramids: cone-shaped masses of tissue that make up the medulla (inner part of a kidney's structure)

Renal vein: a blood vessel that drains blood from the kidneys

Ureter: a muscular tube by which urine passes from the kidney to the bladder

Urethra: a tube through which urine is discharged from the bladder to the exterior

Urinary bladder: a pear-shaped sac which lies in the pelvic cavity behind the symphysis pubis; stores urine

Urine: a watery, typically yellowish fluid stored in the bladder and discharged through the urethra

Revision summary

The renal system

- The organs that contribute to the elimination of wastes in the body are the kidneys, lungs, skin and the digestive system.
- The organs of the renal system are the **kidneys**, **ureters**, **urinary bladder** and **urethra**.
- The **kidneys** are bean-shaped organs lying on the posterior wall of the abdomen.
- The **kidney** has two main parts – the outer **cortex** where fluid is filtered from blood and the inner **medulla**, which is the area where some materials are selectively reabsorbed back into the bloodstream.
- The **cortex** and the **medulla** contain tiny blood filtration units called **nephrons**.
- Urine is produced by three processes – **filtration**, **selective reabsorption** and **collection**.
- Blood to be processed enters the kidneys via the **renal artery**.
- **Filtration** takes place inside a network of capillaries in the **nephron** called the **glomerulus**.
- The sac encasing the **glomerulus** is called the **Bowman's capsule**.
- The filtered liquid then continues through a series of twisted tubes called the **convoluted tubules**, to the **loop of Henle** and the **distal convoluted tubule** before passing to the **collecting duct** and to the **renal pelvis**.
- The composition of the filtered liquid alters as it flows through the **convoluted tubules**.
- Some substances in the filtrate such as glucose, amino acids, mineral salts and vitamins are reabsorbed back into the bloodstream via the **renal vein**.
- From the **distal convoluted tubule** the filtrate then flows into the **collecting duct** (as urine) and passes to the pelvis of the kidney to be passed to the **ureter** and **bladder**.
- The composition of urine is 96% water, 2% urea and 2% other substances (uric acid, creatinine, sodium ions, potassium ions, phosphates, chloride salts, sulfate salts, excess vitamins and drug residues).
- Functions of the kidneys include filtration of impurities and metabolic waste from blood, regulation of water and salt balance, formation of urine and regulation of blood pressure and volume.
- The **ureters** are muscular tubes that transport urine from the pelvis of the kidney to the urinary bladder.
- The **urinary bladder** is a pear-shaped sac which lies in the pelvic cavity behind the symphysis pubis. It functions as a storage organ for urine.
- The **urethra** is a canal which extends from the neck of the bladder to the outside of the body.
- The urethra serves as a tube through which urine is discharged from the bladder to the exterior and as a conducting channel for semen in men.

Test your knowledge
Multiple choice questions

1 What is the function of the kidneys?
   a filtering of impurities from the blood
   b regulation of water and salt balance
   c formation of urine
   d all of the above

2 Which of the following is **not** considered an excretory organ?
   a digestive system
   b skin
   c muscular system
   d respiratory system

3 Blood is filtered inside which section of the kidney?
   a glomerulus
   b Bowman's capsule
   c loop of Henle
   d proximal convoluted tubule

4 What is the blood filtration unit inside a kidney is known as?
   a hilus
   b renal pyramid
   c nephron
   d medulla

5 Which of the following best describes the position of the kidneys?
   a posterior of abdomen between the level of twelfth thoracic and fifth lumbar vertebrae
   b posterior of thorax, between the level of twelfth thoracic and fifth lumbar vertebrae
   c posterior of abdomen between the level of twelfth thoracic and third lumbar vertebrae
   d posterior of thorax, between the level of twelfth thoracic and third lumbar vertebrae

6 Which of the following statements is **true**?
   a Filtered blood leaves the kidney via the renal artery.
   b Excess water, salts and urea are all filtered and processed through the kidneys.
   c Blood to be processed enters the medulla from the renal vein.
   d The renal artery splits into a network of capillaries called the Bowman's capsule.

7 Which hormone is responsible for controlling water reabsorption in the kidneys?
   a insulin
   b antidiuretic hormone (ADH)
   c oxytocin
   d adrenocorticotrophic hormone

8 What is the function of the ureter?
   a to propel urine from the bladder to the exterior
   b to store urine
   c to filter impurities
   d to propel urine from the kidneys to the bladder

9 Where is the bladder situated?
   a in the abdominal cavity behind the intestines
   b in the pelvic cavity behind the symphysis pubis
   c on the posterior of the abdominal cavity
   d behind the urethra

10 Which of these occurs in micturition?
   a contraction of the detrusor muscle and relaxation of the internal urethral muscle
   b relaxation of the detrusor muscle and contraction of the internal urethral muscle
   c contraction of the anal sphincter and relaxation of the bladder
   d relaxation of the anal sphincter and contraction of the bladder

Exam-style questions

11 State two functions of the renal system.   2 marks

12 What is the name given to the blood filtration units inside a kidney?   1 mark

13 a In which part of the kidney does simple filtration take place?   1 mark
   b In which part of the kidney is the composition of the filtered liquid altered?   1 mark

14 a Name the canal that extends from the neck of the bladder to the outside of the body.   1 mark
   b State the difference in this structure between males and females.   2 marks

15 a State three factors that may cause urine production to increase.   3 marks
   b State two factors that may cause urine production to decrease.   2 marks

Index